SPACE TELESCOPE SCIENCE INSTITUTE

SYMPOSIUM SERIES: 12
Series Editor S. Michael Fall, Space Telescope Science Institute

UNSOLVED PROBLEMS IN STELLAR EVOLUTION

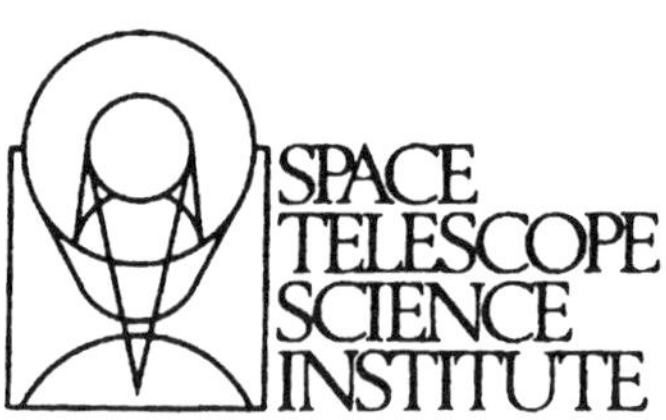

Other titles in the Space Telescope Science Institute Symposium Series.

UNSOLVED PROBLEMS IN STELLAR EVOLUTION

Proceedings of the Space Telescope Science Institute Symposium,
held in Baltimore, Maryland
May 4–7, 1998

Edited by

MARIO LIVIO
Space Telescope Science Institute, Baltimore

Published for the
Space Telescope Science Institute

PUBLISHED BY THE PRESS SYNDICATE OF THE UNIVERSITY OF CAMBRIDGE
The Pitt Building, Trumpington Street, Cambridge, United Kingdom

CAMBRIDGE UNIVERSITY PRESS
The Edinburgh Building, Cambridge CB2 2RU, UK   http: //www.cup.cam.ac.uk
40 West 20th Street, New York, NY 10011-4211, USA   http: //www.cup.org
10 Stamford Road, Oakleigh, Melbourne 3166, Australia

First published 2000

Printed in the United States of America

Typeset in LATEX by the authors

*A catalog record for this book is available from
the British Library*

Library of Congress Cataloging-in-Publication Data is available

ISBN 0 521 78091 8 hardback

Contents

Contents

Participants

An, Jin	Ohio State University
Andersen, Johannes	Copenhagen University
Bahcall, John	Institute for Advanced Study
Balachandran, Suchitra	University of Maryland
Balick, Bruce	University of Washington
Barba, Rodolfo	Space Telescope Science Institute
Barnes, Sydney	Lowell Observatory
Blake, Melvin	York University
Bludman, Sidney	University of Pennsylvania
Bobrowsky, Matt	Orbital Sciences Corporation
Bond, Howard	Space Telescope Science Institute
Bonnell, Ian	Institute for Astronomy
Brocato, Enzo	Osservatorio Astronomico Ternamo
Brown, Timothy	National Center for Atmospheric Research
Carr, John	Naval Research Laboratory
Catelan, Marcio	NASA/Goddard Space Flight Center
Charbonnel, Corinne	Laboratoire d'Astrophysique de Toulouse
Chieffi, Alessandro	Instituto di Astrofisica Spaziale
Churchwell, Edward	University of Wisconsin
Cook, Kem Holland	Lawrence Livermore National Laboratory
Crowther, Paul	University College London
D'Cruz, Noella	University of Virginia
De Marchi, Guido	European Southern Observatory
de Mello, Duilia	Space Telescope Science Institute
Diaz-Miller, Rosa	Space Telescope Science Institute
Dorman, Ben	NASA/Goddard Space Flight Center
Duerbeck, Hilmar	Muenster University
Dwarkadas, Vikram	University of Washington
Eggleton, Peter	Institute of Astronomy
Elmegreen, Bruce	IBM Corporation
Famiano, Michael	Ohio State University
Favata, Fabio	European Space Agency
Feigelson, Eric	Penn State University
Ferguson, Harry	Space Telescope Science Institute
Ferraro, Francesco	Osservatorio Astronomico di Bologna
Figer, Donald	University of California
Franz, Otto	Lowell Observatory
Fullton, Laura	Space Telescope Science Institute
Glassgold, A. E.	New York University
Godon, Patrick	Space Telescope Science Institute
Gonzalez, Guillermo	University of Washington
Grabowski, Udo	Astronomical Institute Basel
Hajian, Arsen	United States Naval Observatory
Hanson, Margaret	University of Arizona, Steward Observatory
Hauser, Mike	Space Telescope Science Institute
Heap, Sara	NASA/Goddard Space Flight Center
Heaton, Harold	The Johns Hopkins University
Heideman, John	

Hindsley, Robert	United States Naval Observatory
Houdashelt, Mark	University of Maryland
Hubrig, Swetlana	University of Potsdam
Hummel, Christian	United States Naval Observatory
Hunter, Donald	United States Naval Observatory
Iben, Icko	University of Illinois
Igea, Javier	New York University
Ivanova, Natalia	Oxford University
Kalogera, Vicky	Harvard-Smithsonian Center for Astrophysics
Kavelaars, J. J.	McMaster University
Krautter, Joachim	Landessternwarte Heidelberg
Krishnamurthi, Anita	JILA, University of Colorado
Kudritzki, Rolf	Universitäts-Sternwarte München
Lame, Nancy	University of Maryland
Lançon, Ariane	Observatoire de Strasbourg
Landsman, Wayne	NASA/Goddard Space Flight Center
Lebreton, Yveline	Paris-Meudon Observatory
Leitherer, Claus	Space Telescope Science Institute
Lennon, Daniel	Universitäts-Sternwarte München
Liebert, James	University of Arizona
Limongi, Marco	Osservatorio Astronomico di Roma
Lindqvist, Michael	Onsala Space Observatory
Livio, Mario	Space Telescope Science Institute
Long, Knox	Space Telescope Science Institute
Lubow, Steve	Space Telescope Science Institute
Lutz, Julie	University of Washington
Maeder, Andre	Geneva Observatory
Massa, Derck	NASA/Goddard Space Flight Center
Mateo, Mario	University of Michigan
Matthews, Grant	University of Notre Dame
McDavid, David	Limber Observatory
Mitalas, Romas	University of Western Ontario
Mowlavi, Nami	Observatoire de Geneve
Mustapha, Mouhcine	Observatoire Astronomique de Strasbourg
Najita, Joan	Space Telescope Science Institute
Narayan, Ramesh	Harvard-Smithsonian Center for Astrophysics
Nelemans, Gijs	University of Amsterdam
Nota, Antonella	Space Telescope Science Institute
Paltrinieri, Barbara	Osservatorio Astronomico di Bologna
Panagia, Nino	Space Telescope Science Institute
Pasquali, Anna	European Southern Observatory
Pilachowski, Catherine	National Optical Astronomy Observatory
Pinsonneault, Marc	Ohio State University
Podsiadlowski, Philipp	Oxford University
Pols, Onno	Instituto de Astrofisica de Canarias
Pompei, Emanuela	Ohio State University
Provencal, Judith	University of Delaware
Ramirez, Solange	Ohio State University
Ratnatunga, Kavan	Carnegie Mellon University
Renzini, Alvio	European Southern Observatory

Romano, Patrizia   Ohio State University
Rood, Robert   University of Virginia
Rucinski, Slavek   Canada-France-Hawaii Telescope Corporation
Sahu, Kailash   Space Telescope Science Institute
Salim, Samir   Ohio State University
Salpeter, Edwin   Cornell University
Sandage, Allan   Carnegie Observatories
Sarajedini, Ata   San Francisco State University
Schaerer, Daniel   Laboratoire d'Astrophysique Toulouse
Schulte-Ladbeck, Regina   University of Pittsburgh
Seitter, Waltraut   Muenster University
Shara, Mike   Space Telescope Science Institute
Shu, Frank   University of California, Berkeley
Siess, Lionel   Space Telescope Science Institute
Sigurdsson, Steinn   Institute of Astronomy
Soderblom, Dave   Space Telescope Science Institute
Stahler, Steven   University of California, Berkeley
Starniero, Oscar   Osservatorio Astronomico di Collurania
Starrfield, Sumner   Arizona State University
Stephens, Andrew   Ohio State University
Sweigart, Allen   NASA/Goddard Space Flight Center
Thorsett, Stephen   Princeton University
Toomre, Juri   University of Colorado
Vesperini, Enrico   University of Massachusetts, Amherst
Wade, Richard   Pennsylvania State University
Walborn, Nolan   Space Telescope Science Institute
Waldman, Roni   Hebrew University Jerusalem
Wallerstein, George   University of Washington
Willson, Lee Anne   Iowa State University
Wing, Robert   Ohio State University
Winget, Donald   University of Texas
Woosley, Stanford   University of California, Santa Cruz
Wyckoff, Eric   Space Telescope Science Institute
Ya'ari, Atara   Racah Institute of Physics
Yi, Sukyoung   NASA/Goddard Space Flight Center
Yuan, Yongquan   Ohio State University
Zoccali, Manuela   Universita di Padova

Preface

The Space Telescope Science Institute Symposium on "Unsolved Problems in Stellar Evolution" took place during 4–7 May 1998. When I asked Icko Iben, Jr. to give the summary talk of the symposium, he told me: "There are many unsolved problems in stellar evolution, but unfortunately no one is working on them!"

Well, apparently this is one of the very rare occasions in which Icko was proven wrong. As the invited reviews presented here show, fundamental problems in stellar evolution still attract considerable interest, and extensive research. Furthermore, with the dramatic improvements in observational techniques, and in the processes of data reduction and analysis, new problems keep surfacing.

These proceedings represent a part of the invited talks that were presented at the symposium, in order of presentation. I thank the contributing authors for preparing their papers.

I thank Sharon Toolan and Ron Meyers of ST ScI for their help in preparing this volume for publication.

Mario Livio
Space Telescope Science Institute
Baltimore, Maryland
May, 1998

Introduction

By E. E. SALPETER

Physics and Astronomy Departments, Cornell University, Ithaca, NY

When Fred Hoyle and I were awarded the Crafoord Prize in October 1997, for our early work on nucleosynthesis and stellar evolution, we had the pleasure of listening to a two-day symposium on related topics. On some topics I had done little listening or reading over the intervening thirty years, so I got a bird's eye view of the long range development of the subject. This exposure was not quite enough for me to make substantial comments or predictions at the present occasion, so I will mainly reminisce (i.e. brag) about the Good Old Days forty to fifty years ago. I will use this as an excuse to give some "Advice To Young Players," but first some musings on topics that have not changed as much over the last forty years as we had expected.

There obviously are some topics where the present day status is excitingly new and was not anticipated forty years ago (at least not by me). One example of this is the importance of hydrodynamic plus nuclear instabilities and the necessity to carry out three dimensional computations: In the 1950s we could make fun of biophysics colleagues for inventing a spherical cow to make calculations easier, whereas we knew that our stars really were spherical. I gather that, for modern stellar evolution computations, a star is no more spherical than a cow! There obviously are other topics where, in spite of further progress, basically "things are still the same": It is not so obvious that this can happen in three different ways—(1) success, (2) failure and (3) neither:

(1) One example of success, in the sense of "already done then," is the purely nuclear aspect of nuclear reactions in stars (in contrast to the stellar structure aspect). The classic BBFH paper is forty years old, but still sounds pretty modern.

(2) One example of failure, in the sense of "still not done today," is the theory of formation of extreme stellar population II stars. When I first met Walter Baade in 1951, he apologized for not knowing how stars could form so quickly in the hostile environment of low matter density and large UV flux in our galactic halo. Baade was confident then that this puzzle would be solved quickly (quicker than all the complex nuclear physics, anyway), but here we are, still not understanding either extreme stellar population II nor population III.

(3) one example of "neither success nor failure" is the Initial Mass Function, which does not seem to vary much from quiescent to disturbed regions, nor from young to old galaxies. This invariance is good for me personally, because authors still refer to my 1955 paper on estimating the IMF, but it is disastrous for the profession as a whole: We had all hoped that strong dependence of the IMF on external conditions would shed light on star formation itself. This has not happened, but I hope that Frank Shu, Steve Stahler and Bruce Elmegreen will enlighten us about star formation nevertheless.

I want to illustrate two pieces of "Advice To Young Players" with my own reminiscences. The first is on the importance of being "at the right place at the right time." My own best example was to travel to Cal. Tech. in the summer of 1951 to work with Willy Fowler on nuclear astrophysics. It was the right place because of Willy himself, plus all the other nuclear experimentalists, as well as astronomers and astrophysicists. It was the right time, because the right nuclear experimental results were pouring out of the Kellogg Radiation Laboratory, including the beryllium near resonance in the collision of two alpha particles. I was able to systematically explore all the possibilities for proceeding beyond hydrogen and helium and to use the beryllium resonance for the triple alpha reaction. This example is more poignant because of Ernest Opik, who was a very brilliant man but was usually at the wrong place at the wrong time (usually ten years too early): He wrote a paper on the triple alpha reaction a year before I did, but (a) was not aware of the beryllium resonance and got too small a reaction rate, (b) published only in "Contributions from the Armagh Observatory," so the rest of us were not aware of his work until much later.

My other piece of advice is to occasionally deviate from a systematic, planned professional program and, instead, to do a piece of research with *chuzpah*. When applied to research, this Yiddish word means having the guts to make shortcuts, to have the gall to make unwarranted assumptions with childlike enthusiasm. My best example was my 1955 paper on the IMF and the processing of heavy elements through star births and deaths: Martin Schwarzschild could have done this work more easily and earlier, but he was too professional to disregard all the uncertainties and did not tackle the problem. An even more clear cut example was Fred Hoyle's prediction of the carbon resonance for the triple alpha reaction—just to make abundance ratios come out right! I had all the technical equipment for this calculation three years earlier when I wrote my triple alpha paper, but I just did not have the guts (nor did I have Fred's originality). I am making propaganda for using some chuzpah occasionally, just because modern National Science Foundation guidelines for writing and judging Grant Proposals insist on the complete opposite—carefully planned research programs with all steps fully justified beforehand. With today's guidelines I could have done my triple alpha work (which was part of a systematic project), but there is no way that I could done my IMF work or that Fred could have done his carbon resonance prediction. So my advice is: write your proposals in the politically correct manner, but sometimes just goof off and use chuzpah.

The "Good New Days" are an improvement over the old regarding cross-fertilization and applying computational techniques to very different problems. Interchanging codes between stellar evolution, chemical combustion and weather prediction already has started, but one can go even further afield: My two most recent papers, published in medical journals, are on the epidemiology of tuberculosis but benefitted from work on heavy element processing through star deaths and births. The processing of tuberculosis bacillus through humans in the U.S. has gone on for only 300 years compared with 10 billion years for the iron nucleus in our Galaxy, but that is not so different when counted in numbers of generation. Even the controversies over unsolved problems are similar—infall into the Galaxy is important but uncertain and so is illegal immigration into the U.S. from countries with larger tuberculosis rates. I hope all the unsolved problems will be solved by the end of this symposium but, in case some are not, you might even read some medical journals!

Pre-Main-Sequence evolution

By STEVEN W. STAHLER

Berkeley Astronomy Department

1. Introduction

Prior to their ignition of hydrogen, stars undergo a phase of bulk contraction. This process drives both the central density and temperature upward, until nuclear fusion can begin. Somewhat paradoxically, the star's luminosity, which stems from the released gravitational binding, is relatively high at this early stage. The luminosity generally decreases as a result of contraction, reaches a minimum just at the point of hydrogen ignition, and then rises again during post-main-sequence evolution. Thus, observed pre-main-sequence stars are relatively bright, a fact which aids in their detection. Several thousand low-mass examples have now been cataloged, despite the fact that the relevant lifetime is only about 0.1 percent that of the main-sequence phase.

A snapshot of the star taken during the contraction epoch would show it to be very nearly in hydrostatic equilibrium. That is, the compressive force of self-gravity closely balances the internal pressure gradient. Evolution is driven by radiative emission from the photosphere, which gradually drains thermal energy. Note that the star's interior temperature rises even as its heat content falls. Thus, the entire object is characterized by negative heat capacity. Note also that the inward velocity of any mass element is always well below the local sound speed. The contraction is therefore termed *quasi-static*, to distinguish it from more rapid, dynamical motion.

This basic physical picture was already well established by the 1950s, when Henyey constructed the first computer models of young stars (Henyey et al. 1955). At the time, the role of thermal convection in stellar interiors was still not fully appreciated, even in mature objects like the Sun (for an historical account, see Stahler 1988). Henyey assumed that pre-main-sequence stars would be radiatively stable, and found that they contracted at nearly constant luminosity. However, the opacity in the subphotospheric layers is enhanced by the presence of H^- ions. Hayashi (1961) first realized that the H^- opacity forces the star to radiate at nearly constant *surface temperature*. The luminosity L_* is therefore high when the radius and surface area are both large. In particular, L_* is more than can be transported by radiation alone, and the interior is convectively unstable.

Within a few years after Hayashi's discovery, the extensive numerical models of Iben (1965) and Ezer & Cameron (1967) added more quantitative details. The theory now assumed the form it was to retain for many years. Each stellar mass has its own pre-main-sequence track in the HR diagram, which it follows down to the ZAMS. Early on, the star indeed contracts almost vertically, as found by Hayashi. The track then veers sharply to the left, and proceeds to the ZAMS along the more horizontal path obtained by Henyey. This change reflects a transition to radiative stability with falling luminosity.

Figure 1 shows the traditional set of tracks, as obtained by Ezer & Cameron. Each track is labeled by the corresponding mass is solar units. Also indicated at a number of points are the times required to evolve along the track to the position in question. It is immediately apparent that more massive stars evolve faster than their low-mass counterparts. The great utility of pre-main-sequence theory is its capability of supplying both a mass and an age for any observed object with known bolometric luminosity and

"

effective temperature. These data not only help us to analyze individual stars, but to discern the star formation history of entire clusters.

Despite the systematic drop in contraction time with mass, all the tracks in Figure 1 are qualitatively the same. Each begins by descending vertically from a very high luminosity, before turning onto its more horizontal, radiatively stable portion. In contrast, Figure 2 displays results from a recent calculation (Palla & Stahler 1998). Although the individual tracks are still recognizable, the landscape has markedly changed. The upper portion of the diagram, corresponding to the highest luminosities, is now inaccessible to pre-main-sequence stars. So too is the area to the left containing objects of the highest effective temperature. In the new picture, the most massive stars have no pre-main-sequence contraction phase at all, but appear directly on the ZAMS.

What has brought about this change? How, in turn, have these revisions impacted our understanding of stellar formation? My purpose in this review is to answer these two questions. It should be recognized at the outset that the theoretical developments, initiated over 20 years ago, still continue, although the key ideas are probably secure. During the same period, our empirical knowledge of pre-main-sequence stars has increased tremendously. Here, I can only sketch the major advances, and indicate the ongoing effort to accommodate these observations with a more refined theoretical treatment.

2. Protostars and the birthline

Most of the alterations to pre-main-sequence theory reflect the impact of new *initial conditions*. Compared to Hayashi and his contemporaries, we now have a better, though still imperfect, understanding of how each star begins its contraction phase. This knowledge in turn translates into the stellar *birthline*, a rather well defined upper boundary for the set of evolutionary tracks in the HR diagram. Referring again to Figure 2, it is the presence of the birthline that now restricts the luminosities to relatively modest values. In addition, the joining of this curve to the ZAMS sets the upper mass limit for all pre-main-sequence objects.

Earlier calculations of quasi-static contraction needed to adopt specific initial conditions, whatever their physical plausibility. In the absence of relevant observational data, researchers set the stellar radii at very large values. The distended young stars had correspondingly high luminosities, as is evident in Figure 1. Hayashi (1966) attempted to justify this choice on theoretical grounds, arguing that the youngest stars gain their mass too quickly to radiate away appreciable energy. In reality, the hope was that all memory of the earliest epoch would quickly be lost during subsequent evolution.

This idea does receive support from the temporal behavior of contracting stars. As we have just noted, any self-gravitating object must lose a large portion of its total energy content in order to reduce its size substantially. Now the virial theorem tells us that this energy is comparable to that from gravitational binding alone. Thus a star of mass M_*, radius R_*, and luminosity L_* contracts over the Kelvin-Helmholtz time scale, given by

$$t_{\mathrm{KH}} = \frac{G M_*^2}{R_* L_*}$$

$$= 3 \times 10^7 \,\mathrm{yr} \left(\frac{M_*}{1 \,\mathrm{M_\odot}} \right)^2 \left(\frac{R_*}{1 \,\mathrm{R_\odot}} \right)^{-1} \left(\frac{L_*}{1 \,\mathrm{L_\odot}} \right)^{-1} . \tag{2.1}$$

The luminosity, in turn, is that due to a radiating blackbody:

$$L_* = 4 \pi R_*^2 \sigma_B T_{\mathrm{eff}}^4 , \tag{2.2}$$

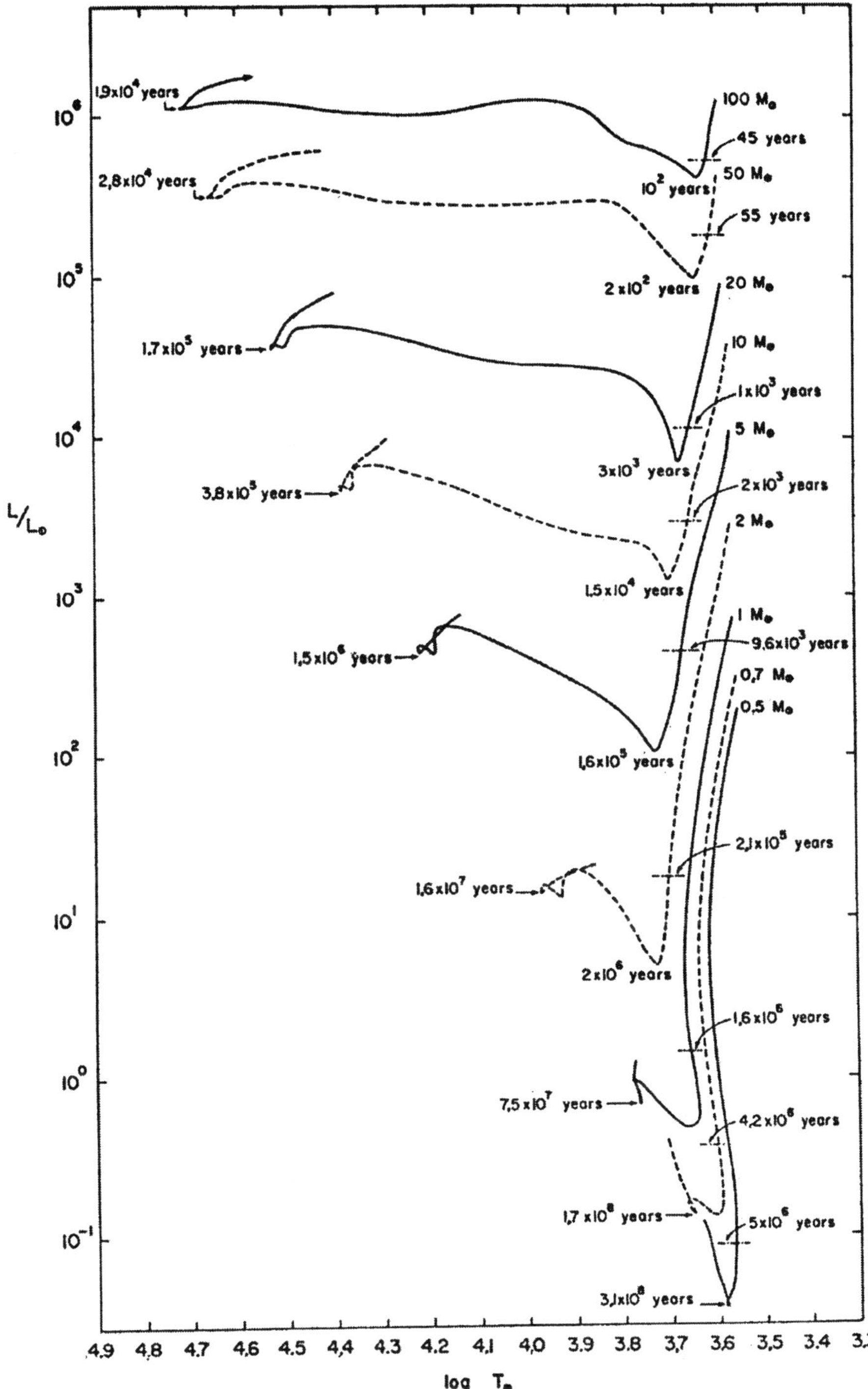

FIGURE 1. Pre-main sequence tracks in the traditional theory (Ezer & Cameron 1967). Each track is labeled by the appropriate mass. Dashed lines and corresponding times denote the point at which the stars cease to be fully convective. The times to reach the ZAMS are also indicated.

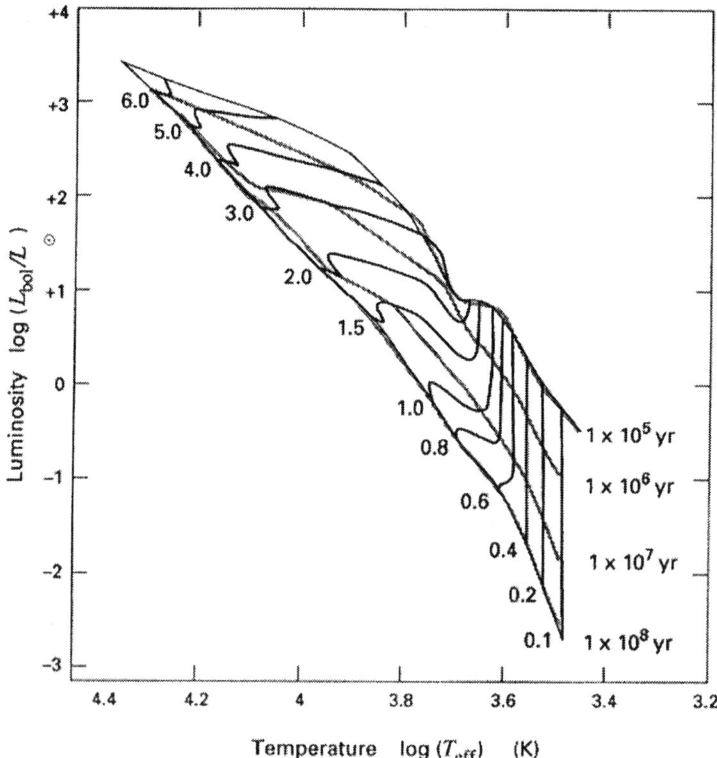

FIGURE 2. Recent calculation of pre-main-sequence tracks (Palla & Stahler 1998). The upper curve is the stellar birthline, and the lower one is the ZAMS. Each track is labeled by the mass in solar units. Also shown are isochrones for various times.

where $T_{\rm eff}$ is the star's effective temperature. As the star contracts with nearly constant $T_{\rm eff}$, we see that $t_{\rm KH}$ *increases* as R_*^{-3}. Thus, the contraction slows down markedly, and any early, distended state indeed occupies a small fraction of the total time to reach the ZAMS.

From an observational perspective, their relatively brief evolutionary lifetimes would make the brightest pre-main-sequence objects relatively uncommon. The difficulty is that stars with the highest L_*-values depicted in Figure 1 are *never* seen. Given the large sample sizes, one must conclude that the adopted initial conditions were mistaken. A star actually begins contracting at a much smaller radius and luminosity than was originally envisioned. Although the contraction indeed slows down, the time spent in the earliest state is not negligible. In other words, initial conditions do matter.

The three decades since the original pre-main-sequence calculations have witnessed a substantial growth in our knowledge of stellar formation. Astronomers of the 1960s had no knowledge of molecular clouds, now established as the breeding grounds for stars of all mass. Nor did they understand well the mechanism of gravitational collapse, i.e. the transformation from diffuse cloud fragments to compact stars. Needless to say, these central developments have impacted the theory of pre-main-sequence evolution.

Radio astronomers have by now supplied detailed maps of molecular clouds over a vast range of size scales. Largest are the *giant complexes*, elongated structures some 50 pc in size, and containing perhaps $10^5 \, M_\odot$ of gas, mostly in the form of molecular hydrogen (Blitz 1993). Complexes are self-gravitating swarms of clumps, containing about $10^3 \, M_\odot$.

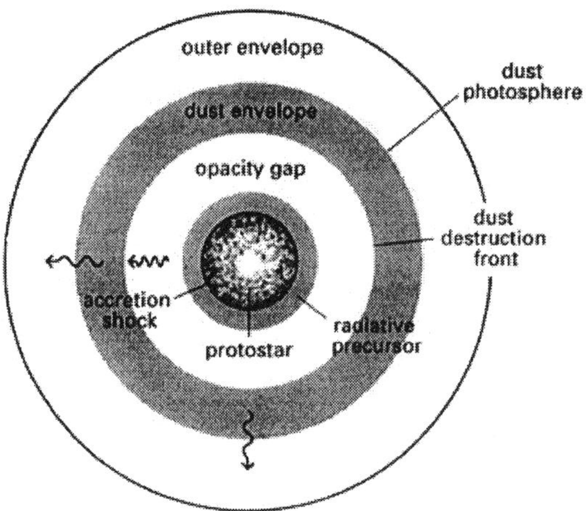

FIGURE 3. Structure of a protostar and its environment. The range of distances is greatly compressed in this schematic rendering. Adapted from Stahler et al. (1980).

These clumps, in turn, are dotted with numerous *dense cores* (Myers & Benson 1983). Each core measures some 0.1 pc in diameter, has a mass of 3 $M_\odot$ or so, and shares the same low temperature (from 10 to 20 K) of its parent cloud. Although the origin of these smallest units remains mysterious, we do know that they are responsible for individual stars.

The first phase of stellar evolution, then, is not the pre-main-sequence, but that earlier epoch when a dense core is undergoing gravitational collapse. The hydrostatic object forming at the center of this converging flow is known as a *protostar*. Theoretical work on protostars began almost 30 years ago, with the first direct simulation of a collapsing, spherical cloud (Larson 1969). It is the subsequent development of this theory that has led to our present, altered conception of pre-main-sequence stars.

Figure 3, adapted from Stahler et al. (1980), depicts schematically a protostar and its surrounding cloud environment. Material in the collapsing dense core rains down onto the surface of the star, forming a strongly radiating accretion shock front. The ultraviolet and X-ray photons produced here work their way back upstream through the infalling gas. As they do so, they elevate the temperature of the envelope to the point where all dust grains thermally sublimate. Since dust is the cloud's chief source of opacity, the protostar is surrounded by a relatively transparent region known as the *opacity gap*. Beyond this region, photons diffusing through the dusty envelope become degraded to infrared wavelengths. The full accretion luminosity then leaves the cloud at the *dust photosphere*, typically located about 10 AU from the central star.

A more complete theoretical description endows the parent dense core with a finite degree of rotation. Some portion of the infalling matter then misses the star itself, and instead forms a circumstellar disk. This material then spirals onto the stellar surface. The structure and growth of the protostar are largely determined by the total mass accretion rate, conventionally designated $\dot{M}$. The geometrical pattern of infall, i.e. the proportion coming through a disk or by direct infall, is less significant than the overall magnitude and temporal behavior of this rate. The former is set by the basic physics of

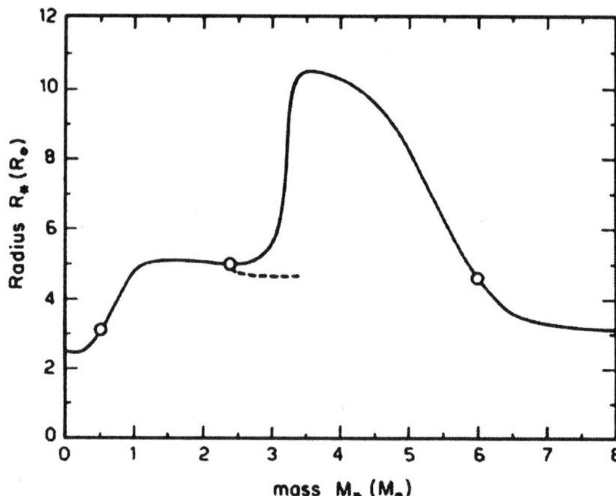

FIGURE 4. Protostar radius as a function of mass (Palla & Stahler 1991). The open circles mark the onset of deuterium-induced convection at the center, appearance of an internal radiatively stable region, and the onset of central convection due to hydrogen burning. The short dashed curve was obtained by artificially ignoring the deuterium shell burning.

gravitational collapse (Shu 1977). If the parent cloud was largely supported by thermal pressure, its subsequent collapse yields a characteristic accretion rate toward the center:

$$\dot{M} \approx \frac{a_T^3}{G} \, . \tag{2.3}$$

Here, a_T is the isothermal sound speed in the dense core, a quantity depending only on the ambient temperature T and molecular weight μ. In a molecular cloud with T between 10 and 20 K, equation (2.3) yields $\dot{M}$-values ranging from 10^{-6} to 10^{-5} $M_\odot$ yr^{-1}.

The most detailed protostar calculations focus on the star itself, and represent the infalling cloud envelope solely through the mass accretion rate. Collapse simulations designed to cover the large-scale dynamics of the cloud find that $\dot{M}$ has a rather slow variation with time, at least beyond an initial, transient phase. Thus, it is worthwhile examining the idealized case of a protostar accumulating mass at a strictly constant rate.

Figure 4 displays the results of such a theoretical calculation (Palla & Stahler 1991). Here we see the protostellar radius R_*, displayed as a function of the mass M_* for $\dot{M} = 1 \times 10^{-5} \, M_\odot$ yr^{-1}. The first thing to notice is that the radii are not much larger than the corresponding main-sequence values. This result contradicts the original assumption of Hayashi and his contemporaries. Protostars have modest sizes because they lose significant amounts of energy while accumulating their mass. Radiation is generated at the accretion shock front bounding the stellar and disk surfaces. As we have seen, this energy readily escapes the cloud by virtue of the opacity gap.

The relatively small sizes of protostars create elevated internal temperatures. Specifically, the stars become hot enough for deuterium to ignite, in the reaction

$$^2\text{H} + \text{p} \rightarrow \, ^3\text{He} + \gamma \, . \tag{2.4}$$

Although the number fraction of this isotope is only of order 10^{-5}, heat from the fusion swells the star's radius. This swelling begins at subsolar mass, and is evident as the first knee in the curve of Figure 4. The second, even more dramatic rise, occurs when deuterium subsequently ignites in a subsurface shell. At still higher mass, $M_* \gtrsim 4 \, M_\odot$, the protostar largely exhausts its deuterium supply and starts to contract under the influence of self-gravity. Its temperature rise concurrently accelerates, and ordinary hydrogen ignites by the time M_* reaches about $8\,M_\odot$. Even more massive protostars are essentially main-sequence objects still gathering material from their surrounding clouds.

Deuterium-generated energy not only swells the protostar, but acts as an effective thermostat, locking the central temperature to a value not far from 10^6 K. One important consequence is that the mass-radius relationship is rather insensitive to the adopted $\dot{M}$-value, at least for changes of about a factor of ten. On the other hand, the protostar's history would be very different indeed if no deuterium at all were present. The short dashed curve in Figure 4 was obtained by artificially ignoring the fusion reaction in the subsurface shell. In this hypothetical situation, the pronounced swelling would not occur at all.

Protostar calculations like the one illustrated assume a continuous flow of mass onto the central object. Naturally, something must eventually truncate this flow to produce stars of finite mass. One possible factor is rotation of the parent cloud. We have remarked, however, that rotation merely shifts the infall trajectories from the stellar surface to a circumstellar disk. It is likely that internal torques within the disk then promote transfer of this material onto the star. Another possibility is that the star itself blows back its infalling envelope through the action of a strong wind. Finally, infall may cease because the remaining dense core is mechanically prevented from collapsing.

We shall later revisit these last two possibilities. Even without a definitive answer to the question of infall termination, however, we can still address our original concern of pre-main-sequence initial conditions. As long as infall ends relatively quickly, the curve in Figure 4 yields the radius of any star of mass M_* at the beginning of its quasi-static contraction phase. The stipulation here is that $\dot{M}$ must decrease substantially over a time that is brief compared to t_{KH} for the mass in question. Provided this condition is met, we may construct a sequence of initial pre-main-sequence models with a range of masses and the appropriate radii. These objects, unlike their protostellar antecedents, are optically visible, and so may be placed in the HR diagram. The resulting curve is the birthline depicted in Figure 2.

The existence of the birthline is thus motivated by basic theoretical considerations (Stahler 1983). From an observational perspective, the curve should represent the *upper envelope* to the positions of young stars in the HR diagram. That is, no stars should be found above the birthline, whether they are observed in isolation or in heavily populated clusters. So far, the voluminous photometric observations of pre-main-sequence stars largely conform to this expectation. For example, Hillenbrand (1997) recently placed over 900 stars of the Orion Nebula Cluster in the HR diagram, after obtaining their bolometric luminosities from I-band observations. The crowded diagram (see Figure 5) shows a reasonably sharp upper boundary, which coincides well with the theoretical birthline.

3. Quasi-static contraction

3.1. *Low-mass objects*

The presence of a birthline in the HR diagram means that pre-main-sequence tracks now fall into two broad categories. Those of lower mass depart from the birthline vertically downward. Observationally, these correspond to T Tauri stars. In the second category are tracks that begin more horizontally. Here, the theoretical models have radiatively stable interiors, and represent the observed class of Herbig Ae and Be stars. Figure 2 shows that the dividing point is about $2\,M_\odot$.

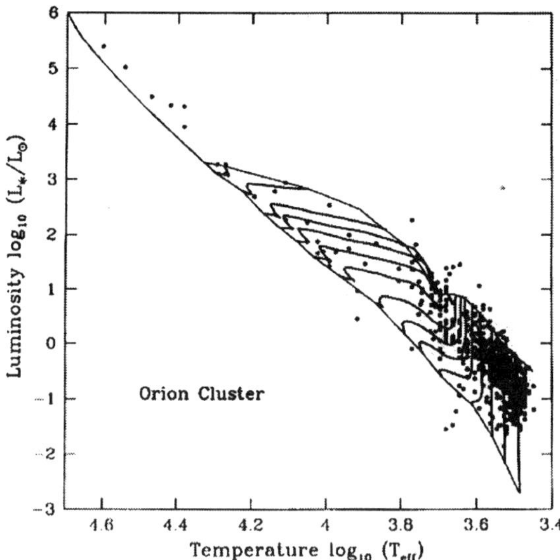

FIGURE 5. HR diagram of the Orion Nebula Cluster (Palla & Stahler 1998), based on the data of Hillenbrand (1997). The theoretical birthline, tracks, and ZAMS are identical to those in Figure 2, but the ZAMS has been extended to include massive stars.

We may understand this difference, i.e. predict the fate of any newly formed pre-main-sequence star, by comparing two luminosities. One is the surface value L_*, as given by the blackbody formula of equation (2.2). The second is L_{rad}, the maximum value that can be transported through the interior by radiation alone. This critical luminosity actually varies spatially within a star, and also depends on the specific stellar model. Dimensional arguments show, however, that it must be approximately

$$L_{\mathrm{rad}} = L_1 \left(\frac{R_*}{\mathrm{R}_\odot} \right)^{-1/2} \left(\frac{M_*}{\mathrm{M}_\odot} \right)^{11/2} , \qquad (3.5)$$

where $L_1 \approx 1\,\mathrm{L}_\odot$ (Cox & Giuli 1968). Note the steep mass dependence, reflecting the temperature sensitivity of the interior opacity.

Consider now a sequence of very young pre-main-sequence stars. For sufficiently low mass, L_{rad} is much less than L_*. These objects are radiating more energy than can be transported by radiation. Their interiors are thus fully convective. On the other hand, stars of greater mass have $L_{\mathrm{rad}} > L_*$, and are radiatively stable. Because L_{rad} climbs so rapidly with M_*, the transition between the two cases occurs at a rather well-determined mass value, even considering plausible uncertainties in the initial stellar radii.

Schwarzschild's criterion dictates that stars become convective once their specific entropy s decreases outward, i.e. when $s(M_r)$ is a declining function of the internal mass coordinate M_r. In practice, however, convection is so efficient at transporting energy that $s(M_r)$ is nearly uniform, apart from a thin super-adiabatic layer near the surface. Now the heat equation governing evolution of the entropy reads

$$\frac{\partial L_{\mathrm{int}}}{\partial M_r} = \epsilon - T \frac{\partial s}{\partial t} . \qquad (3.6)$$

Here, $L_{\mathrm{int}}(M_r)$ is the interior luminosity crossing any mass shell, and ϵ the local nuclear energy generation rate. Prior to deuterium ignition, we may ignore this latter term. The star loses heat during contraction, so that $\partial s/\partial t$ is negative. It follows that L_{int} is a monotonically increasing function of M_r, rising from zero at the center to the surface value L_*.

The kind of quasi-static contraction we have just described may be termed *homologous*. Here, the stellar interior at any one time is simply a rescaled version of its earlier state. This simplification arises from the spatial uniformity of the specific entropy. In mathematical terms, the star may be accurately represented as an $n = 3/2$ polytrope, i.e. the pressure P and density ρ are related at every location through

$$P = K \rho^{5/3} \ . \tag{3.7}$$

The coefficient K is a surrogate for the specific entropy, and also decreases with time.

Even if a pre-main-sequence object begins its life as a convective object, Figure 2 indicates that it eventually becomes radiatively stable, at least for $M_* \gtrsim 0.5\,\mathrm{M_\odot}$. We may again understand this transition in terms of our two luminosities. Contraction reduces R_* and therefore, by equation (2.2), L_* as well. On the other hand, equation (3.5) shows that L_rad weakly *increases* at fixed M_*. Thus, the original inequality $L_* > L_\mathrm{rad}$ may eventually be reversed, and the star become stable against convection. In stars of the lowest mass, L_* is so much greater than L_rad initially that the two cannot reach equality before hydrogen ignites and halts further contraction. That is, the star settles onto the ZAMS while still fully convective.

The transition of a star to radiative stability occurs first at the center, where the temperature is highest and the interior opacity a minimum. The luminosity in the spreading radiative region obeys equation (3.5), at least in order of magnitude. Thus, the average L_rad rises slowly as R_* continues to drop. The surface luminosity must keep pace with L_rad, so L_* also increases. This rise is evident in the tracks of Figure 2.

From equation (2.2), T_eff can no longer be fixed, but is forced to increase. The tracks now veer to the left in the HR diagram.

This picture of low-mass young stars accords well with a variety of observational data (for a review, see Stahler & Walter 1993). We note, however, that the evidence for convection itself is entirely circumstantial. The phenomena actually observed are consistent with *surface activity*, an enhanced version of that present in the Sun and nearby dwarf stars. Thus, many T Tauri stars exhibit periodic fluctuations in light that indicate large, rotating starspots. Prominent optical emission lines, such as Ca H and K, are thought to stem from chromospheric heating. So too is the X-ray emission that has been especially useful in locating pre-main-sequence objects in dense cloud environments.

It is conventional to divide T Tauri stars into two types, *classical* and *weak-lined*. Objects in the first category have been studied for over 50 years (Joy 1945). They display, besides the emission lines and other signs of surface activity, excess broadband flux in both the infrared and ultraviolet regimes. Some of their emission lines exhibit P Cygni profiles, signaling the presence of an outflowing wind. Weak-lined stars derive their name from the suppression or absence of Hα emission, another diagnostic of heated, possibly outflowing gas that appears in the classical objects. Originally dubbed "naked" stars by Walter (1986), they have normal broadband energy distributions for their spectral types.

In the HR diagrams of young clusters and associations, the two types are thoroughly intermingled close to the birthline. The weak-lined variety predominates, however, as one approaches the ZAMS. The generally accepted interpretation of this pattern is that classical stars are surrounded by disks of dust and gas. Stellar luminosity heats the dust, causing it to reradiate at infrared and longer wavelengths. Apparently, the disks vanish as stars contract toward the ZAMS. Note finally that many classical stars also display *inverse* P Cygni profiles. The suggestion is that matter from the disk may be spilling onto the stellar surface, perhaps along magnetic field lines (see, e.g. Kenyon et al. 1996).

In any case, the associated mass transfer rate of roughly $10^{-8}\,M_\odot$ is too low to cause substantial deviation from the constant-mass theoretical tracks of Figure 2.

The variety of T Tauri broadband spectra implies that some pre-main-sequence stars first appear with substantial disks, while others seem to lack them entirely. As we have noted, disks represent material from the parent dense core that had too much angular momentum to impact the star directly during collapse. The production of objects both with and without disks *may* therefore point to a range in cloud rotation rates. However, it will be difficult to quantify this idea before we have a better understanding of how the infall process actually terminates.

There is a smooth morphological progression from T Tauri stars with disks to other stars exhibiting far greater levels of infrared emission. The so-called "flat-spectrum" and even more extreme "Class I" sources have far too much flux at long wavelengths to be attributable to disk reradiation. Indeed, the typical Class I spectral energy distribution shows a broad maximum near 100 μm, far beyond the near-infrared peaks of classical T Tauri stars (Lada 1987). What is the nature of these more embedded objects?

One critical observation is that Class I sources are invariably located near the centers of dense cores of molecular gas. Thus, their infrared and millimeter radiation plausibly stems from heated dust within more massive circumstellar envelopes. A number of researchers have suggested that these envelopes are collapsing onto their central stars, i.e. that the objects are true protostars (e.g. Adams et al. 1987). One should then observe, in addition to the star's photospheric luminosity, the additional energy release from infall. Indeed, current theory predicts that this infall component should actually dominate the total. Observationally, however, the bolometric luminosities of Class I sources are quite similar to those of optically visible T Tauri stars, whether classical or weak-lined (Kenyon & Hartmann 1995). It thus appears that, while some pre-main-sequence stars appear promptly as revealed objects on the birthline, others remain embedded for some time in remnant cloud material. How this gas is mechanically suspended is not well understood, but the ambient magnetic field almost certainly plays a role. In this regard, it will be interesting to see if high-resolution mapping at millimeter wavelengths eventually reveals systematic differences between dense cores with and without internal stars.

3.2. *Stars of intermediate mass*

Consider next the evolution of objects above the transition near $2\,M_\odot$. Here $L_{\rm rad}$, the maximum luminosity carried by radiation, exceeds the surface value L_*. The interior opacity is now low enough that the star can have a fully radiative interior. As we have noted, $L_{\rm rad}$ is not just an upper limit, but also a good estimate for the *actual* luminosity under the condition of radiative stability. The inequality $L_{\rm rad} > L_*$ then implies that the surface cannot emit all the energy generated by the deep interior through gravitational contraction. Very young stars of intermediate mass are out of thermal balance, and must readjust accordingly.

Figure 6 displays a temporal sequence of luminosity profiles from detailed modeling of a $3.5\,M_\odot$ star (Palla & Stahler 1993). Plotted is $L_{\rm int}(M_r)$, starting when the object first appears at the birthline. At this earliest time, we see that $L_{\rm int}$ reaches a maximum value of about $60\,L_\odot$, while the surface luminosity L_* is only about $10\,L_\odot$. What becomes of the excess energy? From the heat equation (3.6), $\partial s/\partial t$ is now positive for all mass shells between the luminosity peak and the surface. The energy being released by contraction thus heats these layers, at least temporarily.

Equation (3.6) also implies that the entropy is still falling inside of the luminosity maximum. This region therefore contracts as in lower-mass objects. Most of the energy expelled is transferred to the outer layers, which actually expand. The structural

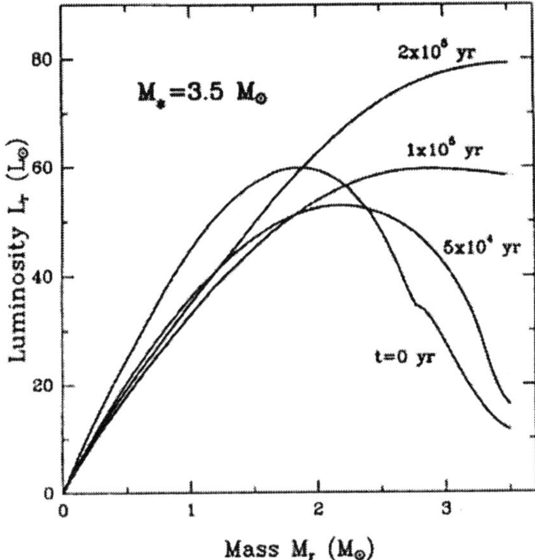

FIGURE 6. Evolution of the internal luminosity in a $3.5\,M_\odot$ pre-main-sequence star (Palla & Stahler 1993). The luminosity crossing any shell is plotted against the mass interior to that shell, for the indicated times.

evolution of the star is now *nonhomologous*, and the earlier description in terms of polytropes breaks down. In physical terms, the star has accumulated its mass at a rate which outpaced its ability to thermally relax.

Heating of the outer layers persists until the surface temperature T_{eff} climbs substantially. The luminosity L_* then rises as well, eventually matching the interior peak value. This development may be followed through the alteration of the profiles in Figure 6. In effect, the star has released a thermal pulse to the surface. From a more observational perspective, the object may be said to undergo a *luminosity jump* (Stahler 1989), a feature evident in the evolutionary tracks of Figure 2. The duration of this phenomenon is the thermal relaxation time, or roughly 10^5 yr at the relevant masses. Note that this period is shorter than t_{KH} in equation (2.1). In the present case, the energy content of the star refers only to its outer layers, and is therefore less than the full GM_*^2/R_*.

The observed objects corresponding to these theoretical models are the Herbig Ae and Be stars, first identified nearly 40 years ago (Herbig 1960). While these stars are rare compared to their low-mass counterparts, they have come under increasing scrutiny in recent years (Thé et al. 1994). There is still no empirical evidence for the thermal transition that should occur at these masses, including the luminosity jump. On the other hand, the very existence of this class is a testament to the essential correctness of the theoretical approach. It should be apparent by now that the location of the birthline in the HR diagram depends solely on the mass-radius relation for accreting protostars. We have seen how those of intermediate mass swell because of deuterium fusion (recall Figure 4). Were it not for this effect, the birthline would be shifted downward toward the ZAMS, excluding the region where Herbig Ae and Be stars are actually found (see, e.g. Larson 1972). Physically, smaller-radius protostars would have higher central temperatures and would ignite ordinary hydrogen earlier, thus joining the ZAMS at a lower mass.

Pre-main-sequence stars of intermediate mass also display, to varying degrees, excess broadband emission at infrared wavelengths. Subtracting off the likely photospheric contribution to the spectral energy distribution, one often finds that the long-wavelength flux comprises a large fraction of the total observed luminosity. Hillenbrand et al. (1992) first argued that this emission must stem from accretion through a circumstellar disk, since a more isotropic distribution of dust would have obscured the visible radiation that is actually seen. However, the required accretion rates can be so high that the disk would still need continual replenishment from an external envelope of dust and gas (Palla & Stahler 1993). In addition, Di Francesco et al. (1994) have spatially resolved the infrared emission in a number of cases, a clear demonstration that the source is much larger than a disk. We are led once more to a picture in which remnant cloud material hovers about the star for a protracted time.

Even more puzzling than the infrared excesses are the signs of surface activity in a large fraction of Herbig Ae and Be stars. Emission in $H\alpha$ and other lines is present in these cases, and frequently displays the P Cygni profiles indicative of winds; other objects have substantial X-ray luminosities (Catala 1989). Now the theoretical models do have transient surface convection, powered by the diminishing supply of deuterium. However, convection always vanishes by the time $T_{\rm eff}$ increases to the range appropriate for A and B stars. Could the X-rays and winds instead originate in lower-mass companions that have not been spatially resolved? Or do they signal residual infall, as is believed for T Tauri stars? Further observations are needed to sort out this issue.

The thermal relaxation characterizing pre-main-sequence stars also occurs in protostars, provided they accumulate sufficient mass. Thus, the steep increase of R_* seen in Figure 4 is caused both by deuterium shell burning and by the shift of internal entropy accompanying the thermal pulse. At these masses, the protostar's luminosity stems more from internal energy release than from external infall. The time scale $t_{\rm KH}$ is now significantly shorter than the evolutionary time $M_*/\dot{M}$, and the object rapidly contracts even as it accretes. The mass at which ordinary hydrogen ignites, about $8\,{\rm M}_\odot$ for $\dot{M} = 1 \times 10^{-5}\,{\rm M}_\odot\,{\rm yr}^{-1}$, also marks the point in the HR diagram where the birthline joins onto the ZAMS. Observations of Herbig Ae and Be stars indeed find an upper mass limit close to this value (e.g. Berilli et al. 1992).

Stars of even higher mass never undergo pre-main-sequence contraction. They exist either as obscured protostars, objects on the ZAMS, or post-main-sequence stars rapidly depleting their nuclear reserves. It is doubtful, moreover, that their protostar phase resembles that at lower masses. Powerful winds from these objects drive back infalling matter, a tendency that is even present to a lesser extent at lower masses (Wilkin & Stahler 1998). The extraordinary luminosities of massive stars also disrupt the envelope, through radiation pressure acting on the grains (Wolfire & Cassinelli 1987). A number of researchers have pointed out that anisotropic infall might lessen the radiative effect (e.g. Nakano 1989). For example, gas in a rotating envelope would not fall directly onto the star, but onto a disk, where it could be shielded by the high optical thickness. However, the disk itself must remain intact for an extended period, despite the star's harsh ionizing flux. Observations of disks in Orion being ablated by the Trapezium stars show that this is no easy matter (O'Dell et al. 1993).

How then do massive stars form? It is significant that the youngest such objects appear close to the density peaks of crowded stellar clusters. The more commonly observed OB associations are the expanding descendants of such groups, while systems such as the Orion Nebula Cluster represent them at an earlier, more compact stage. In this latter example, numerical calculations have demonstrated that the massive stars have not had time to drift to their present, central locations, but must have formed *in situ* (Bonnell

& Davies 1997). One possibility is that these objects originate through the coalescence of previously existing stars (Larson 1990; Bonnell et al. 1998; Stahler et al. 1998). The interaction cross sections of the stars themselves, or even binaries, are probably too small to be relevant, even in the densest observed clusters. However, we have noted that pre-main-sequence stars appear to remain for some time within their parent cloud cores. The merger of these gaseous entities and the subsequent joining of their internal stars is an attractive picture that remains to be developed.

4. Using the tracks

4.1. *Young binaries*

We now shift our attention from the basic theory of pre-main-sequence evolution to more practical application of its results. Our focus will also be on stellar groups rather than individual young objects. Star formation is a collective phenomenon, and pre-main-sequence theory is helping to develop our understanding of the emerging systems. We begin with the simplest groups of all, binary stars.

It has long been appreciated that most stars belong to binary, or less frequently, multiple systems. Within the best studied group of G dwarfs in the Solar neighborhood, a complete survey reveals that nearly 60 percent have companions (Duquennoy & Mayor 1991). A similar figure holds for the more common M dwarfs, although the sampling here is not as thorough (Fischer & Marcy 1992). What is the origin of this multiplicity? Historically, the two main contending ideas have been the mutual capture of originally distant cluster members and the local fragmentation of some parent entity.

Among the exciting advances of the past few decades has been the realization that pre-main-sequence stars have a multiplicity comparable to their mature counterparts. This discovery was made through a variety of techniques, including near-infrared array-detector surveys (e.g. Ghez et al. 1993). In some cases, the companion stars are themselves optically visible, but most are known only through a limited number of longer-wavelength observations. A critical means to decide between the capture and fragmentation hypotheses is the measurement of relative *ages* within binaries. It is well established that most clusters form stars over a period of order 10^7 yr (Iben & Talbot 1966). If the two members of a typical binary have an age difference less than this, we can be assured that they formed *in situ*, rather than through capture in the larger cluster.

Obtaining the age of any young star requires placing it in the HR diagram, so we are restricted to optically visible pairs. The most favorable cases are spectroscopic binaries, ideally those which are both double-lined and eclipsing. Here, analysis of the composite light curve yields reliable measurements of L_* and T_{eff} for both stars. After locating each star in the diagram, we may read off both its age and *mass* from the appropriate pre-main-sequence track. But the two masses also follow from analysis of the individual velocity variations, along with knowledge of the system's inclination angle. There is no reason *a priori* why these *spectroscopic* values for the two masses should match the *photospheric* ones found from the tracks. Eclipsing binaries therefore present unique opportunities both for measuring stellar ages and for gauging the accuracy of the pre-main-sequence tracks themselves.

Unfortunately, these systems are also very rare among young stars. The best studied example to date is TY CrA (Casey et al. 1998). Here the primary is a Herbig Ae star and the secondary is in the T Tauri class. The light-curve analysis gives the two masses as 3.0 $M_\odot$ and 1.6 $M_\odot$. These figures agree well with the more precise spectroscopic determinations. The stellar ages are not so straightforward, since the primary falls very

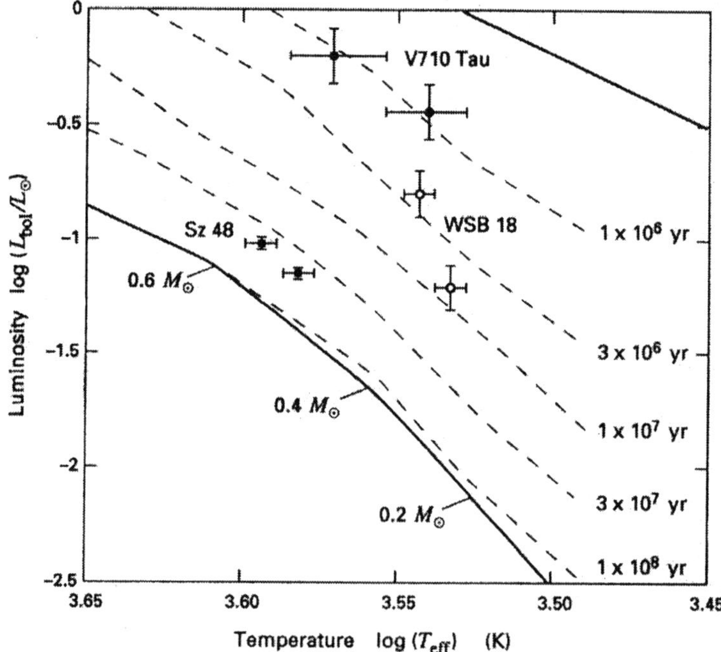

FIGURE 7. Positions in the HR diagram of three wide, pre-main-sequence binaries. Data from Kenyon & Hartmann (1995) and Brandner & Zinnecker (1997).

close to the ZAMS. If we assign it 2×10^6 yr, or the age of a $3\,\mathrm{M}_\odot$ star just arriving at the ZAMS, then the secondary lies on the appropriate isochrone. Thus, the components of TY CrA are plausibly coeval, but the data do not permit a more reliable assessment.

Age assignments are more trustworthy when both stars lie well above the ZAMS. No eclipsing binaries of this kind are known, but there are a number of non-eclipsing, double-lined systems. An example is the weak-lined T Tauri star V773 Tau and its companion (Welty 1995). In the absence of eclipses, one must obtain T_{eff} and L_* for each star by careful analysis of both the composite spectral energy distribution and the shifting, narrow-band spectrum. One then finds that the components of V773 Tau have roughly the same age, which again happens to be about 2×10^6 yr. Now the photometric masses for the primary and secondary in this case are $M_1 = 1.7\,\mathrm{M}_\odot$ and $M_2 = 1.2\,\mathrm{M}_\odot$, respectively. We may compare their ratio, $M_1/M_2 = 1.4$, to the value of 1.32 obtained spectroscopically from the velocity amplitudes. The similarity of these two estimates is gratifying, and bolsters our confidence in using the tracks to obtain both ages and masses.

Discerning the individual luminosities of the binary members becomes easier when we consider wider, spatially resolved pairs. In this case, however, we have no spectroscopic mass ratios for comparison. Figure 7, adapted from Brandner & Zinnecker (1997) and Kenyon & Hartmann (1995), shows three systems taken from Taurus-Auriga and Orion. For two of them, the components lie close to the displayed theoretical isochrones, while the ages within WSB 18 apparently differ by about a factor of three. The sampling here is representative in a statistical sense. That is, roughly a third of visual, pre-main-sequence binaries have component ages that do not match.

Some binaries in this category are proving, under closer scrutiny, to be triples or even quadruples. The nature of the remaining pairs with discrepant ages remains unclear. In any case, we may be confident that most binaries do consist of stars that were born nearly simultaneously. No such coincidence would be expected from the random coupling of objects drawn from a much larger cluster.

Application of pre-main-sequence tracks has thus eliminated one major hypothesis for binary origins. Since binaries are seen everywhere in the HR diagram from the birthline to the ZAMS, they must arise in the protostar phase. Now the surveys of G and M-type binaries find that the mass distributions of the primary and secondary are both consistent with the field-star initial mass function. The implication is that the members form independently, but in spatial proximity. Currently, the most attractive picture is that the parent body is a single cloud core which developed more than one protostellar condensation. Quantitative exploration of this idea has begun (e.g. Bonnell et al. 1991), and will eventually require a generalization of our notions of cloud stability and collapse.

4.2. *Cluster evolution*

While they are occasionally observed in relative isolation, the vast majority of pre-main-sequence stars exist in large aggregates. Objects of solar mass or less form within T associations. Here, several hundred stars are scattered throughout a molecular cloud complex. The stellar population is not strongly peaked in density, and constitutes a minor fraction of the total cloud mass of roughly 10^4 $M_\odot$. This morphology is quite distinct from the next type of group, the OB associations. Best known for the massive stars at their centers, these also contain a much larger population of lower-mass members. Proper motions of the stars show them to be dispersing into space. Finally, about 10 percent of stars form within gravitationally bound clusters such as the Pleiades. Here, the projected stellar density is centrally concentrated, but few if any massive stars are present. The internal motions in this case are consistent with long-term stability of the system.

Why do stars form within such groups? What factors determine whether a given complex or clump spawns an OB association or a bound cluster? Our understanding of these matters is woefully inadequate, in part reflecting our ignorance of the clouds' mechanical structure. Before we can begin to tackle the root issue of how these entities form stars *en masse*, we must quantify the most significant patterns in the empirical data. Here, pre-main-sequence theory has proved invaluable by allowing us to reconstruct the star formation histories of observed systems.

Let us begin with the very youngest groups. The advent of near-infrared array detectors in the 1980s gave observers access to deeply obscured stars that are invisible optically by virtue of the surrounding dust. Surveys have now revealed entire clusters that were previously unknown or at best sparsely sampled (Zinnecker et al. 1993). It has become evident that these embedded populations represent precursors to the traditionally recognized stellar groups. Since most of the cluster members cannot be placed in the HR diagram, our knowledge is generally confined to luminosity functions. Most are built up from data obtained at a single near-infrared waveband, usually K (2.22 μm). More useful from a theoretical perspective are *bolometric* luminosity functions. Requiring much broader wavelength coverage, these are still only available in a few cases, and for a limited subset of the relevant populations. Nevertheless, it is instructive to see what theory predicts for their character and temporal evolution (Fletcher & Stahler 1994a).

Suppose we know the rate at which all the dense cores within a parent cloud initiate gravitational collapse. Let us denote this quantity as $C(t)$. Suppose further that each core builds up a protostar at the infall rate $\dot{M}(t)$. This mass accumulation continues until it is halted by whatever physical agent terminates protostellar collapse. Each star then contracts on its appropriate pre-main-sequence track to the ZAMS. With both the protostar and pre-main-sequence luminosities provided by theory, we could then trace the development of the full luminosity function. We could also follow the evolution of the individual stellar populations—protostars, pre-main-sequence objects, and main-sequence stars.

In fact, the detailed form of the global birthrate $C(t)$ is *not* known, although observations give us an indication of its overall magnitude in any one example. Let us therefore follow the simple expedient of assuming $C(t)$ to be a temporal constant. Then the actual value of the birthrate is unimportant, and this quantity acts simply as a scale factor multiplying the stellar populations at any time. The temporal behavior of the infall rate $\dot{M}(t)$ is similarly unconstrained. In this case, however, the scaling *is* significant, as it sets the maximum protostar mass at any epoch. Fortunately, equation (2.3) gives us a quantitative prescription for this magnitude.

It would still seem impossible to proceed without further knowledge of infall termination. Suppose, however, that a certain fraction of protostars at any time completes infall and begins pre-main-sequence contraction. This fraction is presumably a function of the stellar mass, so we may denote it as $f(M_*)$. Then $1 - f(M_*)$ must represent the fraction of protostars that continues to gain mass. After adding the mass ΔM_*, the star either stops accreting or not, with a relative probability again determined by $f(M_*)$.

Specification of the function $f(M_*)$ would thus allow us to calculate the evolution of the stellar populations and the bolometric luminosity function. In particular, we would know the final distribution by mass of all stars as they arrive on the ZAMS. But numerous observations of visible open clusters and OB associations show that this distribution is well matched by the field-star initial mass function (e.g. Brown et al. 1994). Thus, we may turn the reasoning around. After *assuming* that the final population is distributed in mass according to the empirical mass function, we may *deduce* the compatible $f(M_*)$. We then read off the prior history of our generic cluster.

Figures 8 and 9 display the evolution of the bolometric luminosity function, as obtained numerically (Fletcher & Stahler 1994b). Each panel includes both the time and the fraction of the cluster's total luminosity contributed by protostars, pre-main-sequence objects, and main-sequence stars. Here, $C(t)$ was taken to be a constant from $t = 0$ to 10^7 yr, after which it was set to zero. The earliest time marks the onset of gravitational collapse in dense cores within the larger cloud. The second, cutoff time signifies the end of the star formation process, an event generally attributed to gas dispersal by massive stars. Finally, the infall rate $\dot{M}(t)$ was set to $1 \times 10^{-5}\,M_\odot\,\mathrm{yr}^{-1}$.

Early on, the luminosity function has two distinct peaks. The one at the lower value of L_* represents pre-main-sequence stars, while the other is the contribution from accreting protostars. With the passage of time, this latter portion fades in significance. The vast majority of the cluster then consists of pre-main-sequence stars. There is also a steadily rising fraction of main-sequence objects. Since more massive stars contract faster, the ZAMS is initially filled in at high L_*. Thus the main-sequence contribution to the aggregate luminosity rises quickly, starting about $t = 1 \times 10^6$ yr. Notice that the protostar contribution to the total is also high at this time. Although the relative number of protostars is very small, their radiative output is dominated by O and B stars that are still accreting. Since, as we noted, massive stars may not undergo steady accretion, this contribution is the most uncertain.

By $t = 10^7$ yr, the luminosity function has started to take on a smooth, featureless appearance. The limiting form is simply obtained by convolving ZAMS luminosities with the field-star initial mass function. This is just the "original luminosity function,"

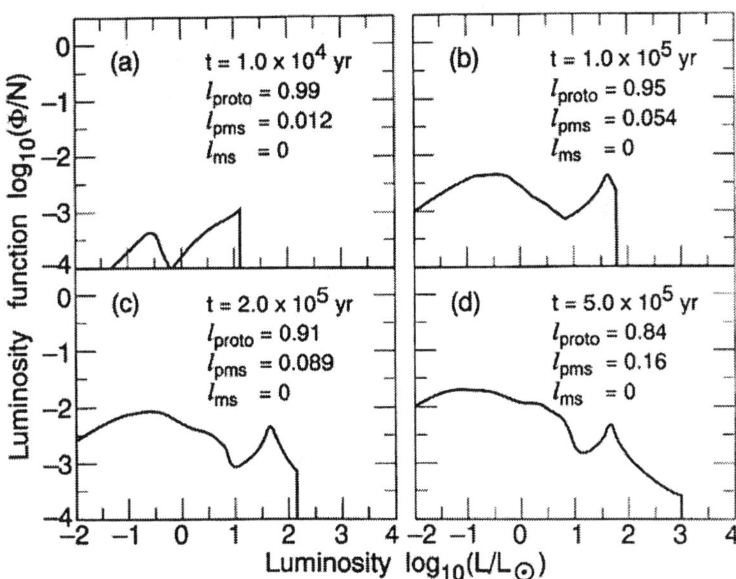

FIGURE 8. Early evolution of the luminosity function within an idealized stellar cluster (Fletcher & Stahler 1994b). Each panel includes, in addition to the time, the fraction of the total cluster luminosity in protostars, pre-main-sequence, and main-sequence stars.

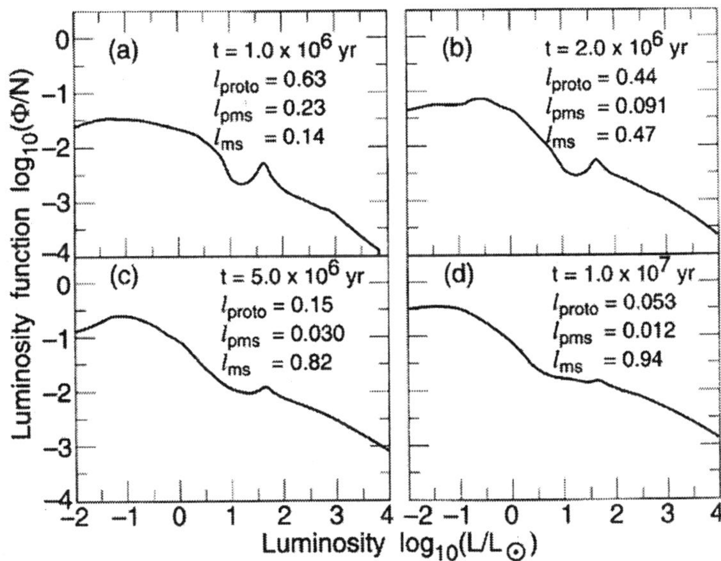

FIGURE 9. Later evolution of the cluster luminosity function (Fletcher & Stahler 1994b). The various symbols are identical to Figure 8.

first constructed by Salpeter (1955).† The evolutionary sequence we have delineated thus represents the prehistory of a stellar group that will ultimately arrive on the main sequence with the canonical mass distribution.

† Salpeter's reasoning was the reverse of ours. That is, he first derived the luminosity function, and then used the ZAMS luminosities to infer the distribution by mass.

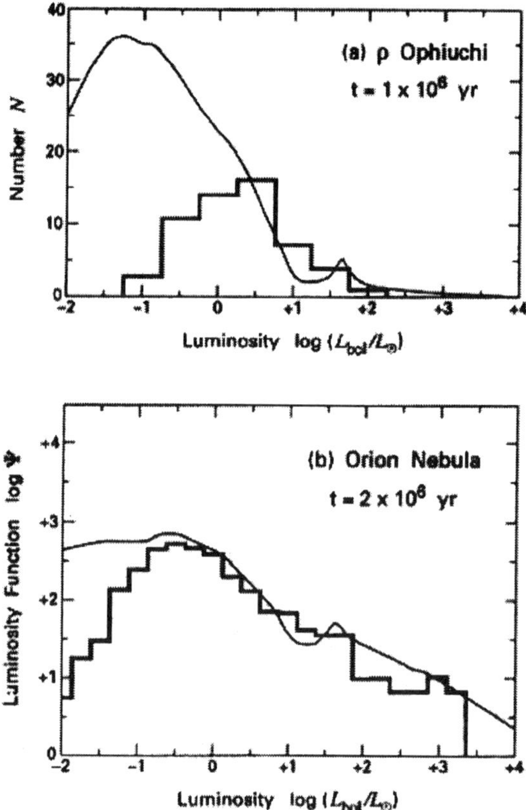

FIGURE 10. Bolometric luminosity functions for (a) the central cluster in ρ Ophiuchi (Wilking et al. 1989), and (b) the Orion Nebula Cluster (Hillenbrand 1997). The smooth curves in both panels are the theoretically predicted luminosity functions for the indicated times (Fletcher & Stahler 1994b).

Comparison of this theoretical prediction with data from very young clusters is in its infancy, owing to the lack of long-wavelength data for the most embedded members. Nevertheless, observers have managed to combine satellite and groundbased fluxes for a limited number of stars to obtain the full L_*. Figure 10 (upper panel) shows one such case, the L1688 cluster in the ρ Ophiuchi cloud complex. The histogram includes data for about 50 cluster members, out of the several hundred known to be present (Wilking et al. 1987; Greene & Young 1992). The solid curve represents the best-fit theoretical prediction, corresponding to a cluster age of 2×10^6 yr. Almost all the "theoretical" stars still lacking measured bolometric luminosities are pre-main-sequence objects.

More convincing agreement is obtained for the Orion Nebula Cluster centered on the Trapezium (Figure 10; lower panel). Here, the best-fit age is 2×10^6 yr, the total number of stars observed is about 10^3, and the empirical luminosity function is consequently much more complete (Hillenbrand 1997). The protostellar bump predicted by theory is still not evident in either ρ Ophiuchi or this example. If Class 0 sources indeed represent protostars, then their inclusion will help in this regard.

As we noted before, the large number of bolometric luminosities in Orion were *not* obtained from direct integration, but by extrapolation from I-band data (Hillenbrand 1997). This procedure was feasible for about half the stars in the cluster, since much of

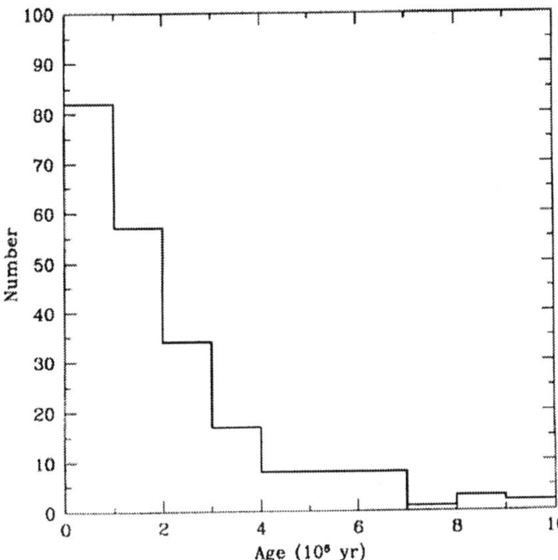

FIGURE 11. Distribution of pre-main-sequence ages in the Orion Nebula Cluster (Palla & Stahler 1998). Ages were obtained using the data of Hillenbrand (1997).

the gas has been cleared away by the few massive members near the center. Narrowband optical spectra are also available for the larger population, from which one may derive effective temperatures. Thus, one may place the entire group in the conventional HR diagram (recall Figure 5), and read off ages and masses from evolutionary tracks.

The histogram in Figure 11 shows the resulting distribution of track ages (Palla & Stahler 1998). This figure effectively depicts the star formation history of the region. It is apparent that, while a few members formed as long ago as 10^7 yr, vigorous formation did not begin until much more recently, and has peaked within the last 2 to 3×10^6 yr. The age obtained from the luminosity function is thus a mean value, heavily weighted by this interval.

Since much of the gas in the central region has already dispersed, the current star formation rate must be quite low. We may plausibly attribute this abrupt falloff to the recent birth of O and B stars, and the rapid spread of their HII regions. In the near future, it should prove most instructive to derive analogous histories for nearby T associations. These will doubtless show a lower overall rate, owing to the reduced population. Nevertheless, there are already indications of a similar acceleration from the past (Palla & Galli 1997). If this is confirmed, we will need to invoke a different kind of truncation mechanism to avoid the overproduction of T Tauri stars compared with their actual numbers.

5. Unanswered questions

Throughout our discussion, the initiation of the pre-main-sequence phase has been a frequently recurring topic. Specifically, we still need to understand what physical event marks the termination of dynamical, protostellar collapse. The derivation of the birthline sidestepped this question by simply assuming that the critical event is suitably brief. We saw, however, that the relevant interval may *not* be small for many of the stars still surrounded by residual gas and dust. More fundamentally, the issue of infall termination is central in our quest to understand the empirical distribution of stellar masses.

One possibility we have mentioned is that stars themselves are the responsible agents, through the momentum of their outflowing winds. This idea is motivated by the nearly universal association of optical jets and molecular outflows with embedded stellar objects (Bachiller 1996). Now outflows are thought to be cloud gas entrained by the high-velocity wind, where the latter is channeled into a narrow jet (e.g. Stahler 1994). On the other hand, we have no clear example of a cloud being decimated by its internal jet. Such well observed objects as the Bok globule B335 show both high column densities in molecular gas and vigorous, well-collimated flows (Hirano et al. 1988). The available evidence is that winds and dense cores coexist for some period. The proposition that the former eventually destroys the latter in all cases remains to be demonstrated.

An alternative viewpoint is that stars accrete all the nearby material available to them. Now the typical masses of cloud cores, as obtained from molecular line observations, are larger by nearly an order of magnitude than the average mass of a newborn star (Myers & Benson 1983). Assuming that the densest, inner portion of the cloud first collapses, the implication is that the outermost envelope is somehow prevented from joining onto the central star.

The most likely means of preventing, or at least forestalling, collapse is through the embedded magnetic field. Measurements of Zeeman splitting have established a well-ordered field of typical magnitude $10\,\mu\text{G}$ in the lower-density gas surrounding cores (Heiles et al. 1993). Unfortunately, this observational technique fails within the core itself, but we may assume that the field penetrates this inner region. The narrow linewidths of NH_3 and other tracers show that the core gas is relatively quiescent. Thus, the dominant expansive force aiding thermal pressure is tension from a smooth field, rather than the pressure associated with MHD waves. By the same token, the broad lines observed outside the cores indicate a strong wave influence throughout the larger cloud volume.

Although bending of field lines can effectively resist gravity, it cannot do so indefinitely. Neutral matter slowly drifts across the field through the process of ambipolar diffusion (Mestel & Spitzer 1956). Thus we may picture, at least crudely, the collapsing inner core tugging on the ambient field until tension halts rapid infall onto the central star. This event marks the end of the protostar phase. Further mass addition and depletion of the core would occur much more slowly, over the diffusive time scale (Safier et al. 1997). The stars receiving this matter would be young, pre-main-sequence objects with strong emission at long wavelengths, perhaps the observed Class I sources.

Other vexing problems remain in the study of pre-main-sequence evolution. Chief among them is the issue of angular momentum. We remarked before that disks represent cloud material of excessive angular momentum that misses the star during collapse. It would seem to follow that the stars themselves must rotate close to breakup. In fact, the observed rotation speeds of T Tauri stars are very slow, typically under 10 percent of the breakup value (Hartmann et al. 1984). Furthermore, there is no clear indication of spinup due to contraction along the evolutionary tracks, except very close to the ZAMS (Stauffer et al. 1985).

It has long been accepted that the torquing action of magnetic winds provides an effective brake on stellar rotation (Schatzmann 1962). Given a field strength and geometry, theory can predict with confidence the rate of angular momentum loss from the stellar surface (e.g. Weber & Davis 1967). Less well understood is the interior response of the star to this braking action. Nor can theory yet specify what field is generated by a given stellar model. Until these gaps are closed, we will be unable to provide a convincing history of pre-main-sequence rotation.

These problems are fundamental, involving both dynamo theory and the stability of rotating fluids. According to current thinking, regions that are stable against convection may still undergo a kind of anisotropic mixing that serves to transport angular momentum outward (Spiegel & Zahn 1992). This same process would also affect the star's internal chemical composition. Thus, the angular momentum problem relates to yet another outstanding issue, the depletion of lithium and other light elements (Martín 1997). In the future, the pre-main-sequence theory outlined here will continue to provide the basic framework for these and other investigations.

REFERENCES

ADAMS, F. C., LADA, C. J., & SHU, F. H. 1987 *ApJ* **312**, 788.

BACHILLER, R. 1996 *ARAA* **34**, 111.

BERILLI, F., CORCIULO, G., INGROSSO, G., LORENZETTI, D., NISINI, B., & STRAFELLA, F. 1992 *ApJ* **398**, 254.

BLITZ, L. 1993. In *Protostars and Planets III* (eds. E. H. Levy and J. I. Lunine. p. 125. U. of Arizona).

BONNELL, I., MARTEL, H., BASTIEN, P., ARCORAGE, J.-P., & BENZ, W. 1991 *ApJ* **377**, 553.

BONNELL, I. A., BATE M. R., & ZINNECKER, H. 1998 *MNRAS* **298**, 93.

BRANDNER, W. & ZINNECKER, H. 1997 *AA* **321**, 220.

BROWN, A. G. A., DE GEUS, E. J., & DE ZEEUW, P. T. 1994 *AA* **289**, 101.

CASEY, B. W., MATHIEU, R. D., VAZ, L. P. R., ANDERSEN, J., & SUNTZEFF, N. B. 1998 *AJ*, **115**, 1617.

CATALA, C. 1989. In *Low-Mass Star Formation and Pre-Main-Sequence Objects* (ed. B. Reipurth). p. 471. ESO.

COX, J. P. & GIULI, R. T. 1968. *Principles of Stellar Evolution and Nucelosynthesis*. Gordon and Breach.

DI FRANCESCO, J., EVANS, N. J., HARVEY, P. M., MUNDY, L. G., & BUTNER, H. M. 1994 *ApJ* **432**, 710.

DUQUENNOY, A. & MAYOR, M. 1991 *AA* **248**, 485.

EZER, D. & CAMERON, A. G. W. 1967 *Can. J. Phys.* **45**, 3429.

FISCHER, D. & MARCY, G. W. 1992 *ApJ* **396**, 178.

FLETCHER, A. B. & STAHLER, S. W. 1994a *ApJ* **435**, 313.

FLETCHER, A. B. & STAHLER, S. W. 1994b *ApJ* **435**, 329.

GHEZ, A. M., NEUGEBAUER, G., & MATTHEWS, K. 1993 *AJ* **106**, 2005.

GREENE, T. P. & YOUNG, E. T. 1992 *ApJ* **395**, 516.

HARTMANN, L., HEWETT, R., STAHLER, S., & MATHIEU, R. 1986 *ApJ*, **309**, 275.

HAYASHI, C. 1961 *PASJ* **13**, 450.

HAYASHI, C. 1966 *ARAA* **4**, 171.

HEILES, C., GOODMAN, A. A., MCKEE, C. F., & ZWEIBEL, E. G. 1993. In *Protostars and Planets III* (eds. E. H. Levy and J. I. Lunine). p. 279. U. of Arizona.

HENYEY, L. G., LELEVIER, R., & LEVEE, R. D. 1955 *PASP* **67**, 154.

HERBIG, G. H. 1960 *ApJS* **4**, 337.

HILLENBRAND, L. A. 1997 *AJ* **113**, 1733.

HILLENBRAND, L. A., STROM, S. E., VRBA, F. J., & KEENE, J. 1992 *ApJ*, **397**, 613.

HIRANO, N., KAMEYA, O., NAKAYAMA, M., & TAKAKUBO, K. 1988 *ApJ* **327**, 69.

IBEN, I. 1965 *ApJ* **141**, 993.

IBEN, I. & TALBOT, R. J. 1966 *ApJ* **144**, 968.

JOY, A. H. 1945 *ApJ* **102**, 68.

KENYON, S. J. & HARTMANN, L. W. 1995 *ApJS* **101**, 117.

KENYON, S. J., YI, I., & HARTMANN, L. 1996 *ApJ* **462**, 439.

LADA, C. J. 1987. In *Star Forming Regions* (eds. M. Peimbert and J. Jugaku). p. 1. Reidel.

LARSON, R. B. 1969 *MNRAS* **145**, 271.

LARSON, R. B. 1972 *MNRAS* **157**, 121.

LARSON, R. B. 1990. In *Physical Processes in Fragmentation and Star Formation* (eds. R. Capuzzo-Dolecetta et al.). p. 389. Kluwer.

MARTÍN, E. L. 1997 *Mem. Soc. Astron. It.* **68**, 905.

MESTEL, L. & SPITZER, L. 1956 *MNRAS* **116**, 505.

MYERS, P. C. & BENSON, P. J. 1983 *ApJ* **266**, 309.

NAKANO, T. 1989 *ApJ* **345**, 464.

O'DELL, C. R., WEN, Z., & HU, X. 1993 *ApJ* **410**, 696.

PALLA, F. & GALLI, D. 1997 *ApJ* **476**, L35.

PALLA, F. & STAHLER, S. W. 1991 *ApJ* **375**, 288.

PALLA, F. & STAHLER, S. W. 1993 *ApJ* **418**, 414.

PALLA, F. & STAHLER, S. W. 1998 *ApJ*, submitted.

SAFIER, P. N., MCKEE, C. F., & STAHLER, S. W. 1997 *ApJ* **485**, 660.

SALPETER, E. E. 1955 *ApJ* **121**, 161.

SCHATZMANN, E. L. 1962 *Ann. d'Astrophys.* **25**, 1.

SHU, F. H. 1977 *ApJ* **214**, 488.

SPIEGEL, E. A. & ZAHN, J.-P. 1992 *AA* **265**, 106.

STAHLER, S. W. 1983, *ApJ* **274**, 822.

STAHLER, S. W. 1988 *ApJ* **332**, 804.

STAHLER, S. W. 1988 *PASP* **100**, 1474.

STAHLER, S. W. 1989 *ApJ* **347**, 950.

STAHLER, S. W. 1994. In *Astrophysical Jets* (ed. D. Burgarella, M. Livio, & C. P. O'Dea). p. 183. Cambridge U. Press.

STAHLER, S. W. & WALTER, F. M. 1993. In *Protostars and Planets III* (eds. E. H. Levy and J. I. Lunine). p. 405. U. of Arizona.

STAHLER, S. W., PALLA, F., & HO, P. T. P. 1998. In *Protostars and Planets IV* (eds. V. Mannings, A. P. Boss, and S. S. Russell). U. of Arizona, in press.

STAHLER, S. W., SHU, F. H., & TAAM, R. E. 1980 *ApJ* **241**, 637.

STAUFFER, J. R., HARTMANN, L. W., SODERBLOM, D. R., & BURNHAM, N. 1984 *ApJ*, **280**, 202.

THÉ, P. S., PÉREZ, M. R., & VAN DEN HEUVEL, E. P. J. (eds.) 1994 *The Nature and Evolutionary Status of Herbig Ae/Be Stars*. ASP.

WALTER, F. M. 1986 *ApJ* **306**, 573.

WEBER, E. J. & DAVIS, L. 1967 *ApJ*, **148**, 217.

WELTY, A. D. 1995 *AJ* **110**, 776.

WILKIN, F. P. & STAHLER, S. W. 1998 *ApJ* **502**, 661.

WILKING, B. A., LADA, C. J., & YOUNG, E. T. 1989 *ApJ*, **340**, 823.

WOLFIRE, M. G. & CASSINELLI, J. P. 1987 *ApJ* **319**, 850.

ZINNECKER, H., MCCAUGHREAN, M. J., & WILKING, B. A. 1993. In *Protostars and Planets III* (eds. E. H. Levy and J. I. Lunine). p. 429. U. of Arizona.

Low-mass young stellar objects

By J O A N   N A J I T A[1]

[1]Space Telescope Science Institute and National Optical Astronomy Observatories

I briefly review the properties of young low-mass stars, and describe how these properties fit into our current understanding of how such stars acquire their initial masses and angular momenta, the starting point for subsequent stellar evolution.

1. Introduction

In the context of this meeting, the goal of star formation is, in some sense, to explain the origin of initial stellar masses and angular momenta, i.e. the initial conditions for subsequent stellar evolution. In this brief review, I hope to give a flavor of our progress to date in the case of stars such as the Sun. I will begin with a summary of the observed properties of young, low-mass stars, focusing on recent developments, and then describe how these properties fit into our current understanding of how such stars acquire their initial masses and angular momenta. Finally, I will highlight some of the unsolved problems associated with basic aspects of this picture and with other related issues in star formation. Due to the limited time available, I will focus primarily on the observational aspects of the subject, and even so will only be able to touch on a few of the highlights in this rich and rapidly developing field. Interested readers in search of additional information and opinions on both observational and theoretical aspects of the subject may wish to consult the comprehensive reviews presented in *Protostars and Planets IV* (Mannings, Boss, & Russell 1999). The review by Bodenheimer (1995) on the more specific topic of the angular momentum evolution of young stars is also a valuable reference.

2. Observed properties of young stars

2.1. *Infall*

As is well-known, molecular clouds are the reservoirs of mass and angular momentum from which stars form. But while it has been appreciated for some time that star formation likely involves a phase of dynamical collapse (rather than quasi-static contraction, for example), the actual detection of infalling motion associated with the star formation process has only been made convincingly in the last few years. Since the mid-1970s, the many molecular line studies of molecular cores, the centrally concentrated regions within molecular clouds from which stars form, have typically revealed evidence for gas engaged in *out*flowing rather than infalling motion. The now apparent ubiquity of stellar outflows has remained one of the primary obstacles to the detection of infall due to their often significant impact on the dynamics of the gas surrounding the star.

As a result, the successful detection of infalling motion has hinged on the choice of appropriate targets and suitable line diagnostics. The very young, so-called "Class 0" sources, have turned out to be particularly fruitful sources for study. This is in part because the outflows in these sources are less well-developed than in older sources and are often directed in the plane of the sky, thereby minimizing their contribution to line profiles.

The signature of dynamical infall seen in these sources is a double-peaked line profile with the blue peak stronger than the red (e.g. Zhou et al. 1993; see also Fig. 1). This

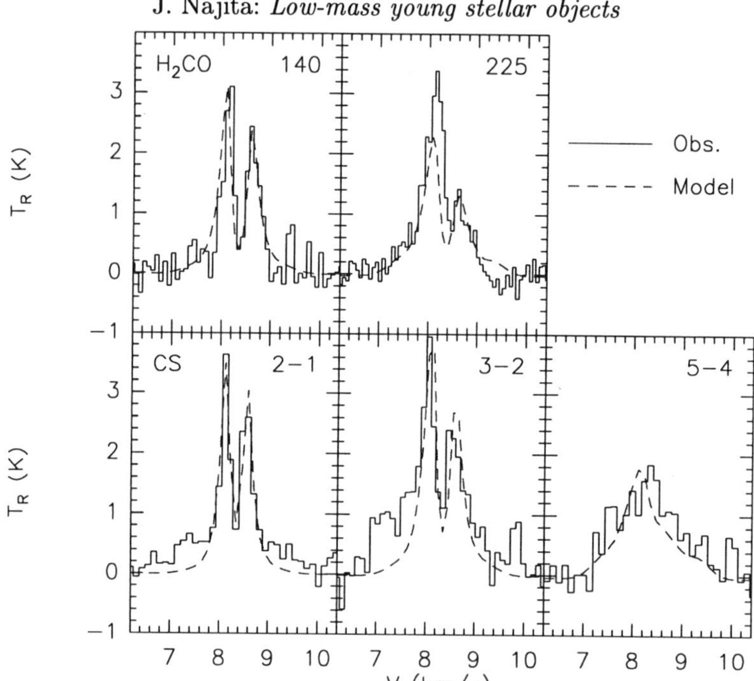

FIGURE 1. Rotational transitions of H_2CO (140 and 225 GHz) and CS ($J = 2 - 1$, $3 - 2$, and $5 - 4$) observed toward the very young Class 0 source B335 (*histogram*) and the best-fit model spectra (*dashed line*) under the assumption of dynamical "inside-out" collapse (Choi et al. 1995). The close agreement between the data and the theoretical model provides strong evidence for a phase of dynamical collapse in the process of star formation.

characteristic profile arises from the centrally peaked density and temperature distributions of the infalling gas, which produce an outwardly decreasing excitation gradient. In the simple but illustrative case of spherical infall, the central absorption is produced by low velocity, low excitation foreground gas in the core which is seen in absorption against the emission from more centrally located, higher excitation gas. Because in general low excitation, red-shifted gas is seen in absorption against higher excitation, red-shifted gas further along the line of sight, while the situation is reversed for the blue-shifted gas (higher excitation gas lies in front of lower excitation gas), the red side of the profile is preferentially weakened relative to the blue side (see, e.g. Fig. 8 in Zhou et al. 1993). A key factor in the detection of this dynamical signature is the choice of appropriate line diagnostics. These must be sufficiently excited in regions where the excitation gradient is large, and have optical depths sufficient to manifest the effect.

As described by Zhou et al. (1993) in their multi-line transitional study of the Class 0 source B335, the theoretical solution of spherically symmetric dynamical collapse (Shu 1977), when coupled with a radiative transfer analysis, provides an excellent fit to the suite of observed CS and H_2CO line profiles (see also Choi et al. 1995; Fig. 1). Because physical situations other than infall can also produce a double-peaked profile (e.g. broad emission plus an unrelated foreground absorber; alternatively, outflow), multi-transitional studies are important in confirming infall and in ruling out other possibilities. For example, in regions undergoing infall, higher excitation lines are predicted to be broader than lower excitation lines, an effect which would not be observed in the absence of systematic motions. Additionally, stronger blueward line asymmetries are predicted for more optically thick lines in the case of infall, an effect which is observed.

Since the infall velocity gradient is measurable, albeit only on small size scales ($\sim 20''$ for the nearest sources), spatial maps of line emission from molecular envelopes have been obtained in order to confirm that the gas is undergoing dynamical infall. Such studies find larger linewidths and line asymmetries closer to the forming star (e.g. Myers et al. 1995), and can even be used to diagnose the presence of rotational motion in the infalling gas (Zhou et al. 1996). Line profiles observed in this latter case are in good agreement with theoretical models of dynamical collapse that include the effects of rotation (Terebey et al. 1984).

From the modeling of multiple line profiles, it is possible then to extract physical and dynamical information about the infalling gas, such as the mass accretion rate, the mass of the star + disk, and the time period over which collapse has been occuring (e.g. Zhou et al. 1993; Choi et al. 1995; Zhou et al. 1996). In a recent development, infalling motions have been detected in a molecular cloud core that is not known to have a star embedded within it (Tafalla et al. 1998). This development opens up the possibility of studying the very early stages of the star formation process using the above techniques. The detail with which the infall process can now be studied, and the close agreement between observations and theoretical predictions, together appear to provide the long-sought observational evidence that stars build up their masses and angular momenta through molecular cloud collapse and a dynamical accretion process.

2.2. *Outflow*

As described above, outflows are a ubiquitous and unexpected signature of star formation. At the very early stages of stellar evolution, when the star is still embedded in the molecular cloud from which it formed, the outflow process manifests itself in several ways. On the largest size scales (parsecs), are molecular outflows which are typically detected in molecular emission lines at millimeter wavelengths (e.g. Lada 1985; Bachiller 1996). The outflows commonly display a bipolar morphology with physically separated lobes of cold red- and blue-shifted gas (e.g. L1551—Snell & Schloerb 1985). The low velocities (~ 10 km s^{-1}) and large masses ($\gtrsim 1$ M$_\odot$) of the outflows indicate that they are swept-up molecular cloud material. This interpretation is sometimes supported by molecular line observations which show that the outflowing gas is distributed in a shell-like structure that surrounds a lower-density cavity (e.g. L1551—Moriarty-Schieven et al. 1987). Since the large momenta of the outflows suggest that they may reverse the motion of the infalling gas and thereby limit the mass of the star, the origin of outflows is of considerable interest.

Support for the idea that molecular outflows are swept-up by an energetic wind originating close to the star comes from the morphology of reflection nebulae within ~ 1000 AU of the star which also imply a cleared polar region (e.g. L1551—Davis et al. 1995). Often collinear with the lobes of molecular outflows are highly collimated emission line jets (e.g. L1551—Davis et al. 1995) which are detected at optical and now IR wavelengths (see e.g. Edwards et al. 1993; Stanke et al. 1998). The jets have terminal velocities of hundreds of km s^{-1}, as indicated by both radial velocities and the proper motions of jet features. Recent wide-field imaging of star forming regions has revealed that jets can extend over very large size scales, comparable to the extent of molecular outflows (e.g. Reipurth et al. 1997). On much smaller size scales, Hubble Space Telescope observations of the emission line jets associated with the nearby sources HH30 and DG Tau (Burrows et al. 1996; Kepner et al. 1993) demonstrate that the flows are collimated on solar system size scales, within 40 AU of the star.

At the smallest size scales, as probed by high resolution spectroscopy, are dense, high-velocity (100s km s^{-1}) winds which produce the strong blueshifted absorption features in the profiles of lines such as Hα and Na I D (e.g. L1551—Mundt et al. 1985; see also Mundt 1984). These wind signatures are observed not only among the young, embedded population, but also among more evolved, optically revealed young stars (§2.4). The densities and temperatures required to excite the hydrogen and Na I D lines strongly favor formation within $\sim 10R_*$ of the star. When combined with the large velocities of the broad blueshifted absorption features, the implication is that winds are accelerated to high speed within a few stellar radii of the stellar surface. For the most active of the optically revealed young stars, winds are inferred to be cool (~ 6000 K) and energetic ($\sim 10^{-7}M_\odot$ yr^{-1}) (e.g. Giovanardi et al. 1991; Najita et al. 1996a). The momenta contained in the wind appears sufficient to drive the associated molecular outflows. For example, the dynamical lifetime of molecular outflows (ratio of linear extent to outflow velocity) is typically 10^5 yr. A wind with the above properties blowing for this length of time would provide a momentum input of a few M$_\odot$ km s^{-1} which is comparable to the momentum contained in the associated molecular outflows (cf. Levreault 1988). We will return in §3 to the origin of winds and jets and their role in the solution of the angular momentum problem for star formation.

2.3. *Circumstellar disks*

Due to the finite angular momentum of molecular cloud cores (Goodman et al. 1993), cloud collapse is expected to result in the formation of a circumstellar disk with a characteristic size comparable to that of the solar system. In addition to their role as the reservoir of mass and angular momentum from which stars accrete a substantial fraction of their mass, the likelihood that disks are also the environments in which planets form has led to long-standing interest in their existence and physical properties.

One of the earliest studied and most commonly observed signatures of disks is the infrared excesses that disks contribute to the spectral energy distribution of the young star system due to the processes of active accretion (e.g. Lynden-Bell & Pringle 1974) and passive reprocessing of stellar light (e.g. Adams et al. 1987). Simple modeling of disks as a radial sequence of blackbodies weighted by emitting area provides a good fit to spectral energy distributions and implies disk temperatures of 1000s K at a few R_* and temperatures of a few 10s K at 100 AU. If infrared excesses are taken as an indicator of the presence of a disk, disk dispersal times in nearby star forming regions are found to be a few Myr, although with a large dispersion among the sources in a given region (e.g. Kenyon & Hartmann 1995; Hillenbrand & Meyer 1999; Meyer et al. 1999; Fig. 2).

While the agreement between spectral energy distributions and the predictions of simple physical models strongly suggests the existence of disks, we need to turn to techniques with greater spatial or spectral resolution in order to confirm the flattened, centrifugally supported nature of circumstellar disks. The excellent spatial resolution of HST has been used to provide perhaps the best evidence for a flattened, disk-like morphology. WFPC2 observations of systems such as HH 30 (Burrows et al. 1996) strongly suggest an edge-on disk system where the outline of a flared disk is revealed by scattered stellar light. The silhouette disks in the Orion nebula (McCaughrean & O'Dell 1996) also provide dramatic evidence for cool, dusty, disk-like structures 50–1000 AU in size that are seen in absorption against the bright emission-line background of the nebula.

An alternative approach is the use of millimeter and submillimeter interferometry to study the structure and kinematics of disks on solar system size scales (~ 50 AU). The sub-arcsecond angular resolution now achieved by interferometers (e.g. the CSO-JCMT Interferometer, the BIMA Array, and the VLA at 7 mm) is well suited to the angular scales of solar system sized disks at the distances of nearby star forming regions (~ 150 pc distant). Multiwavelength continuum measurements are used to study the dust

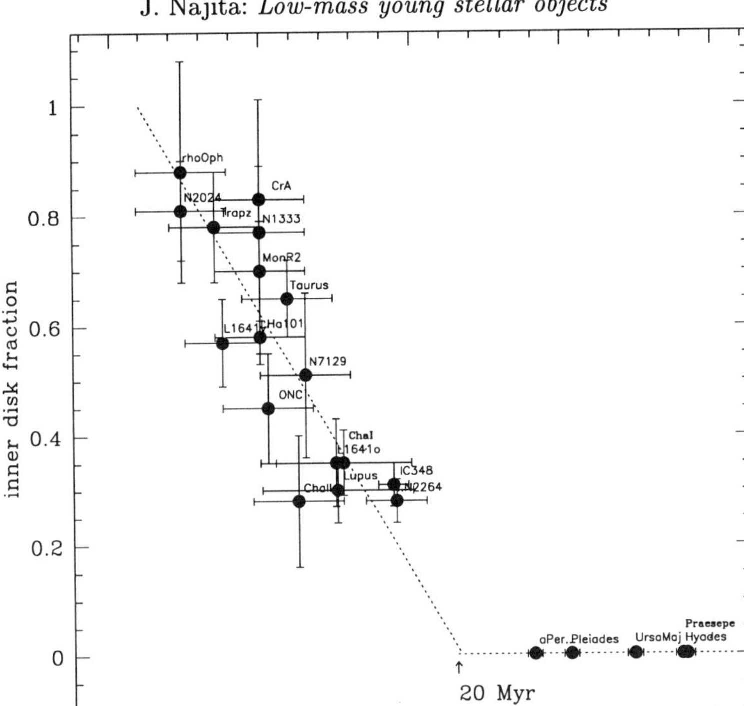

FIGURE 2. A preliminary result from Hillenbrand & Meyer (1999) which shows the fraction of stars that have disks, as diagnosed by near-infrared excesses, in clusters of different ages. The results appear to indicate that disks disperse on a timescale of < 10 Myr.

emissivity and column density distribution of the disk as a function of radius, from which disk masses can be estimated. For example, using the 40 milliarcsecond resolution of the VLA at 7 mm, Wilner et al. (1999) showed that the dust surface density distribution of the disk associated with HL Tau has a shallow slope (flatter than $\propto r^{-1}$) over the range $\sim$ 5–100 AU (Fig. 3). The inferred mass of the disk is $\sim$ 0.6 $M_\odot$ (e.g. Mundy et al. 1996).

Disk kinematics at solar system size scales can be probed by interferometric studies of molecular line emission at millimeter wavelengths. Such studies typically detect molecular line emission on size scales ($\sim$ 500 AU) larger than that of the continuum emission. The kinematics of the emitting gas sometimes show relatively clear evidence for rotation (e.g. Koerner & Sargent 1995), but the signature of radial inflow (e.g. Hayashi et al. 1993) and/or outflow can often dominate the kinematics making the detection of rotational motion difficult.

At much smaller size scales (r < 5 AU), disks are currently studied using infrared spectroscopy, where high spectral resolution is used in lieu of high spatial resolution to probe disk properties. Due to the high densities and warm temperatures (100–5000 K) of disks within 5 AU, molecules are abundant in the gas phase, and their infrared vibrational/rotational transitions sufficiently excited that molecular spectroscopy in the near- and mid-IR is a powerful probe of disk structure and dynamics. For example, R = 20,000 spectroscopy of 2.3μm CO overtone emission from young stars reveals a spectral shape that strongly suggests a disk origin (Fig. 4; Carr et al. 1993; Najita et al. 1996b; see also Chandler et al. 1993). These observations currently provide some of the best evidence for the existence of centrifugally supported structures around young stars.

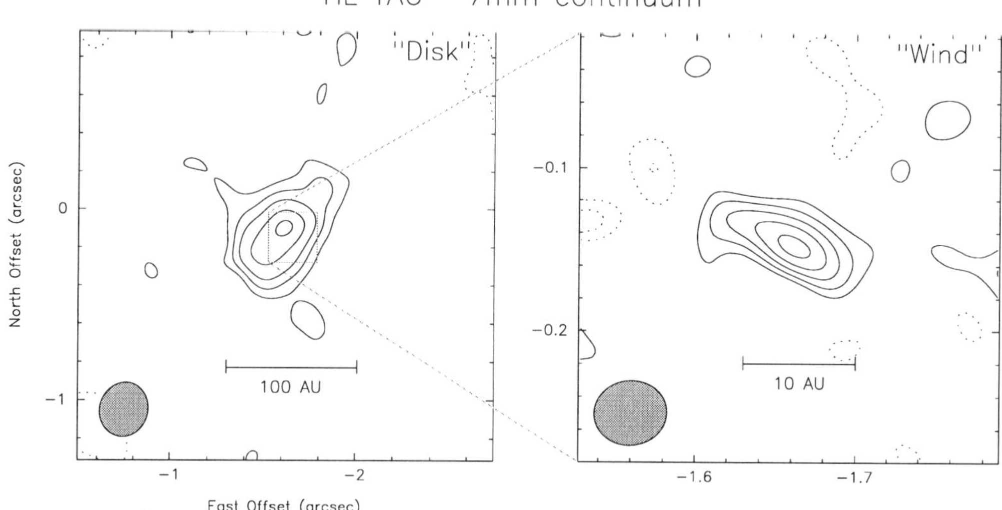

FIGURE 3. Images of the T Tauri star HL Tau obtained with the VLA at 7 mm with 40 milliarc-second resolution (Wilner et al. 1999). The right panel shows the reconstructed image obtained from data taken at baselines longer than $1000k\lambda$ which reveals jet-like emission from an ionized wind. Given the angular scale of the image, this result suggests that jets are collimated very close to the star. The left panel shows a lower resolution map obtained after subtracting the wind component, which reveals dust emission from a disk. The absence of a steep increase in surface brightness toward the star argues for a surface density distribution that is flatter than $\sim r^{-1}$.

2.4. *Disk accretion rates and stellar inflows*

Disk accretion rates can be measured from both the infrared and the optical/UV excesses that the accretion process contributes to the composite spectral energy distribution of the young stellar system. As described in the previous section, the infrared excesses arise in part from the dissipation of gravitational potential energy as disk matter spirals in toward the inner disk boundary (Lynden-Bell & Pringle 1974). While infrared excesses have frequently been used to infer the mass accretion rates of embedded sources, its utility as a measure of the disk accretion rate is compromised by the fact that other physical processes, such as the reprocessing of stellar light by the disk (e.g. D'Alessio et al. 1998) and other circumstellar matter (e.g. Natta 1995), also produce infrared excesses which must be accounted for in order to extract an accurate accretion rate.

In comparison, the UV excesses arise from the dissipation of additional gravitational potential energy as disk matter makes its way from the inner disk boundary onto the star. Since the origin of the emission is more uniquely associated with the accretion process, UV excesses can be regarded as a more robust measure of disk accretion rates (e.g. Bertout et al. 1988; Calvet et al. 1999). Because UV excesses are difficult to measure except in the case of optically visible accreting stars, i.e. the population known as the classical T Tauri stars, this section and the next focus more specifically on the properties of these stars. The classical T Tauri stars are in a later phase of evolution than the embedded sources among which molecular outflows are stellar jets are more common (see §2.2), and are characterized by lower levels of mass accretion and mass loss.

Until the early 1990s, it was believed that circumstellar disks extended essentially up to stellar surfaces and that UV excesses arose from the dissipation of rotational energy as disk matter spun down in a boundary layer to join the slowly rotating star (i.e. a star rotating much below breakup). The hot boundary layer was then believed to

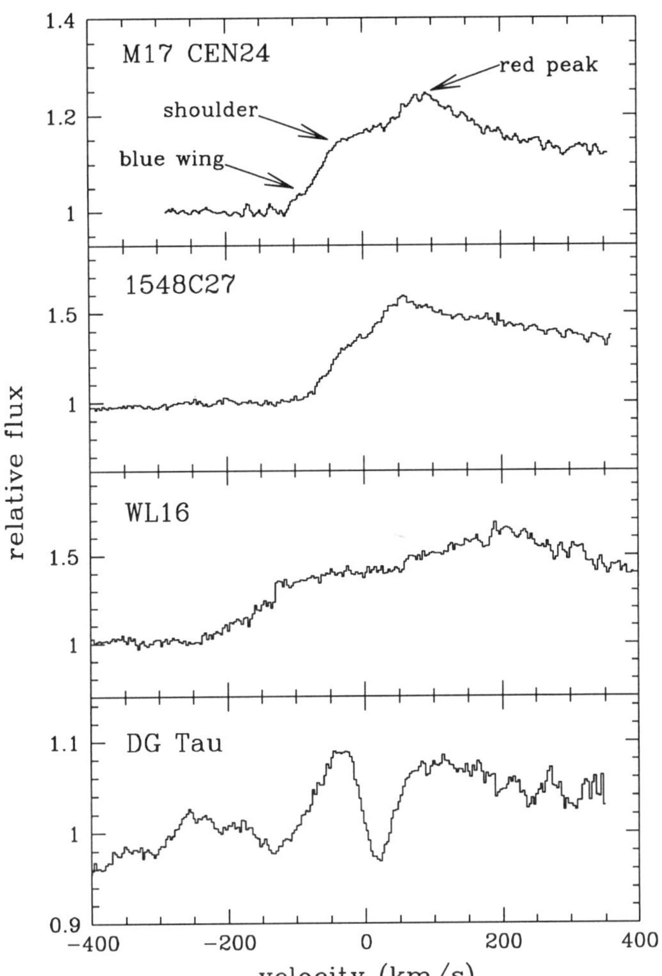

FIGURE 4. The spectroscopic signature of bandhead emission from a rotating disk (a blue wing, shoulder, and red peak) is commonly observed in young stars over a range of masses, from the T Tauri star DG Tau ($\sim 1\ M_\odot$), WL 16 ($\sim 2\ M_\odot$), and 1548C27 ($\sim 4\ M_\odot$), to a high mass star ($\sim 10\ M_\odot$) in the M17 cluster (Carr et al. 1993; Najita et al. 1996b; Carr et al. 1999; Najita et al. 1999). In the DG Tau spectrum, the signature is less clear due to strong stellar CO bandhead absorption.

produce the Balmer continuum emission and much of the rich emission line spectrum that is characteristic of T Tauri stars. However, the origin of these spectral features has recently been reinterpreted within a physical picture that offers an explanation for both the slow rotation rates of T Tauri stars and the origin of their winds.

A variety of observations now support the view that accreting disk matter joins the star not through a boundary layer, but through accretion along stellar magnetic field lines. Evidence for stellar inflows within a few stellar radii of the star come from the inverse P Cygni profiles of the upper Balmer lines of some T Tauri stars (Walker 1972) which indicate infall velocities of several hundred $km\,s^{-1}$. Further evidence comes from the almost ubiquitous presence of blueward asymmetries in the upper Balmer and Brackett line profiles (Edwards et al. 1994; Najita et al. 1996a) which are interpreted as arising in gas infalling onto the star, a situation which results in preferential absorption on the red side of the line profile (cf. §2.1). The interpretation that these line profile properties are signatures of magnetic accretion is supported by radiative transfer models of idealized magnetic accretion flows which reproduce many of the observed line profile features and trends (Calvet & Hartmann 1992; Muzerolle et al. 1998a,b).In this revised picture, the post-shock gas at the base of the magnetic accretion flow is identified as the source of the optical/UV excess formerly attributed to emission from a boundary layer (Königl 1991, see also Bertout et al. 1988).

Several other observations support the idea of a close dynamical relationship between stellar inflows and winds. Evidence for stellar inflows in the upper Balmer lines is often observed simultaneously with the signature of mass loss (i.e. P Cygni profiles) at Hα (cf. §2.2; Edwards et al. 1994). Since excitation considerations require that the hydrogen lines are formed close to the star, these observations provide evidence for the simultaneous presence of winds and stellar inflows within a few stellar radii of young stars.Additional evidence comes from synoptic observations of hydrogen line profiles in T Tauri stars. In the T Tauri star SU Aur, the depths of both a blueshifted absorption component (seen in Hα and Hβ) and a redshifted absorption component (seen in Hβ) are observed to vary at the 3-day rotation period of the star but are 180° out of phase (Johns & Basri 1995). These observations can be explained as the result of the interaction between the disk and a dipole stellar magnetic field that is tilted with respect to the stellar rotation axis (cf. Fig. 15 in Johns & Basri 1995). Because stellar fieldlines bend inward toward the star, favoring inflow, in one region of the magnetosphere, and bend outward, favoring outflow, 180° away in azimuth, the rotation of the magnetosphere at the rotation period of the star can account for the observed period and phase difference.

Yet further evidence, of a more statistical nature, comes from the correlation between wind mass loss rates and stellar and disk accretion rates (Edwards et al. 1993; Hartigan et al. 1995). Although the errors associated with the measurements are large, current estimated values for these mass loss and mass accretion rates appear comparable within an order of magnitude. Gullbring et al. (1998) have reexamined stellar accretion rates implied by the optical/UV spectrum of T Tauri stars in the context of the magnetic accretion scenario, and find a broad distribution of accretion rates with a median value of $\sim 10^{-8}$ M$_\odot$ yr^{-1}. Estimates of wind mass loss rates based on the profiles and fluxes of traditional diagnostics such as the hydrogen and Na I D lines (e.g. Calvet 1997; see also Giovanardi et al. 1991; Najita et al. 1996a) and optical forbidden line emission (e.g. Hartigan et al. 1995) also result in a broad distribution averaging $\sim 10^{-9}$ M$_\odot$ yr^{-1} (e.g. Calvet 1997), although there is considerably larger uncertainty associated with these values compared to the estimates for stellar accretion rates. A hypothesis for the simultaneous presence of mass accretion and loss, and their comparable mass flux rates, is presented in §3.

2.5. *Stellar rotation*

The rotation rates of young stars are measured both from the rotational broadening of stellar photospheric absorption lines and from the observation of periodic photometric variability. The latter is attributed to rotational modulation of the stellar flux by the presence of hot or cool spots on the stellar surface. While cool spots are believed to be the equivalent of large sunspots, the hot spots are now interpreted as arising from the non-axisymmetric components of stellar inflows (Bertout et al. 1988; Bertout et al. 1996; see §2.4). A multitude of rotational broadening and photometric variability studies have shown that the rotation rates of classical T Tauri stars are low, typically much less than the breakup rotation rate, $v \sin i \sim 15$ km s^{-1} or $\Omega_* \sim 0.1$ $\Omega_{\mathrm{breakup}}$ (e.g. Vogel & Kuhi 1981; Hartmann & Stauffer 1989; Bouvier et al. 1993). This result might be considered surprising since these stars are actively accreting, and the accretion of high angular momentum material from a disk might be expected to spin the star to breakup speed.

An important clue to the reason for the slow rotation rates of classical T Tauri stars came from the discovery that these stars, which possess circumstellar disks, rotate more slowly than stars of similar spectral type and age (the weak T Tauri stars) that show no evidence for circumstellar disks (Edwards et al. 1993; Bouvier et al. 1993; Choi & Herbst 1996). In these studies, classical T Tauri stars were found to have rotational periods in a fairly narrow range about ~ 8 days, while weak T Tauri stars were found to have a mean period 2–4 times smaller. These results were interpreted as evidence that in the T Tauri phase, stellar angular momenta are regulated by the presence of a disk, presumably through a kind of magnetic coupling between the stellar field and the inner circumstellar disk (Königl 1991; Shu et al. 1994; Cameron & Campbell 1996; Armitage & Clarke 1996).

While the different theories describing this process disagree in their treatment of the radial extent of the coupling region and the nature of the angular momentum redistribution that allows continued slow stellar rotation, all of the theories share the idea that through disk coupling the star continues to rotate slowly even as it accretes some high angular momentum material from the disk and contracts toward the main sequence. In the context of this picture, the timescale for disk dispersal, and the consequent cessation of disk coupling, is an important factor in setting initial stellar angular momenta. Some degree of variation in initial stellar angular momenta would then appear to be expected, even at ages of a few Myr, given the known dispersion in disk dispersal times, as based, for example, on the intermixed distributions of classical and weak T Tauri stars in the HR diagram for the Taurus star forming region (Kenyon & Hartmann 1995).

These ideas have interesting implications when placed in the context of studies of stellar angular momentum evolution over Gyr timescales. For example, in Figure 5 from Bouvier et al. (1997), the rotation rates of classical and weak T Tauri stars in Taurus are compared with those of post-T Tauri stars in Taurus (10–40 Myr old), and stars in the αPer (50 Myr), Pleiades (100 Myr), and Hyades clusters. If interpreted as a pure evolutionary trend, these results might be taken to indicate that on average stellar rotation increases with age due to stellar contraction until ~ 100 Myr, then decreases with age due to the increasing importance of angular momentum loss in a magnetic solar-type wind (e.g. Bouvier, Forestini, & Allain 1997). The large breadth of the rotational distribution that is seen at all ages can then be interpreted largely as the result of a dispersion in disk dispersal times (e.g. Bouvier et al. 1997), i.e. a spread in initial stellar angular momenta rather than dispersion introduced through subsequent stellar evolution. (See Keppens, MacGregor, & Charbonneau [1995] for a somewhat different view.)

One of the potential difficulties with this picture is that in order to explain the population of slow rotators at ages of ~ 50 Myr, disks must persist much beyond 10 Myr, and stars must couple effectively to them, in a fair fraction of systems. This seems rather unlikely given that disk dispersal times for T Tauri stars appear to be much shorter (§2.3). Other questions for the future include issues such as whether the initial conditions imposed by molecular clouds (e.g. cloud angular momenta, infall accretion rate) and the effects of early dynamical evolution (e.g. development of close companions) are sufficiently homogenized during the epoch of disk coupling that observations of different clusters and associations can be thought to represent a purely time-dependent sequence.

As a final caveat, it is interesting to note that the study of rotational periods in other star forming regions (e.g. NGC 2264) may end up altering somewhat our picture of rotation rates at early times. Using synoptic observations that are sensitive to a broad

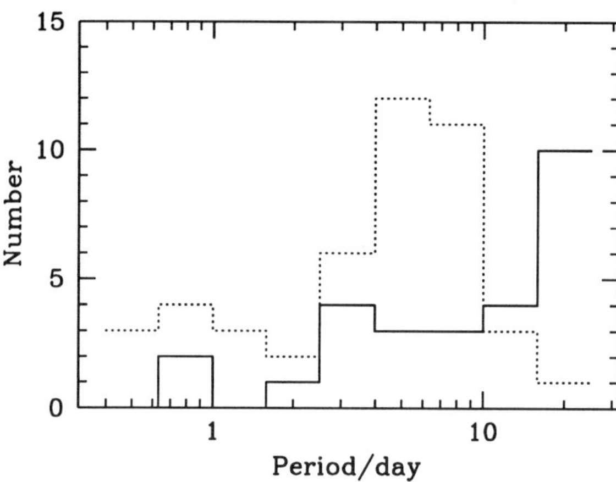

FIGURE 5. Distribution of rotation periods for stars in the young cluster NGC 2264 (Makidon et al. 1999). Although stars with disks (*solid line*) generally rotate slower than stars without disks (*dotted line*), the distribution of both populations is quite broad.

range of rotation periods, Makidon et al. (1999) find a broad period distribution with a significant population of both slow ($P_{\rm rot} \sim 30$d) and fast ($P_{\rm rot} < 1$d) rotators at ages of ~ 3 Myr. While they confirm the trend that stars with disks rotate slower on average than stars without disks, the period distribution for *both* populations is quite broad (Fig. 5). Whether and how such observations fit into our understanding of the rotational evolution of stars is an interesting issue for the future.

2.6. *Stellar magnetic fields*

As indicated in the previous sections, the processes of stellar accretion and angular momentum loss appear to be closely tied to the strength and geometry of stellar magnetic fields at early evolutionary times. Magnetic activity among the T Tauri stars has long been suspected based on a variety of observations including strong X-ray and non-thermal radio emission, and photometric evidence for large starspots. One of the most exciting recent developments in this field is the detection of hard X-ray emission from the much younger protostellar population (Koyama et al. 1996). In this case, hard X-ray emission was used to penetrate the large column densities toward protostars ($A_v \sim 100$) in order to probe the region close to the stellar surface. This result indicates that stellar magnetic activity begins very early in the evolution of low mass stars.

Exciting progress has also been made in the direct measurement of stellar field strengths and geometries. Measurements of stellar magnetic field strengths in T Tauri stars are complicated by the rotational broadening of the photospheric lines (§2.5) which typically exceeds Zeeman broadening at optical wavelengths. For this reason, studies in the optical have ignored line profiles and instead used the enhancement in equivalent width that results from Zeeman broadening as a more indirect measure of stellar field strength. To date, field strengths $\sim 1 - 2.5$ kG have been measured for a few T Tauri stars using this approach (e.g. Basri et al. 1992). The more direct detection of kilogauss strength fields through the study of Zeeman broadened line profiles can be made by going to longer wavelengths where Zeeman broadening ($\propto \lambda^2$) dominates over rotational broadening ($\propto \lambda$). Using this approach, a clear signature of Zeeman broadening has been detected in the Zeeman-sensitive Ti I $\lambda 2.2\mu$m in the spectrum of the classical T Tauri star BP Tau (Johns-Krull, Valenti, & Koresko 1999; Fig. 6). The average field strength implied by the line profile is 3.3 kG which is in excellent agreement with stellar field values required by

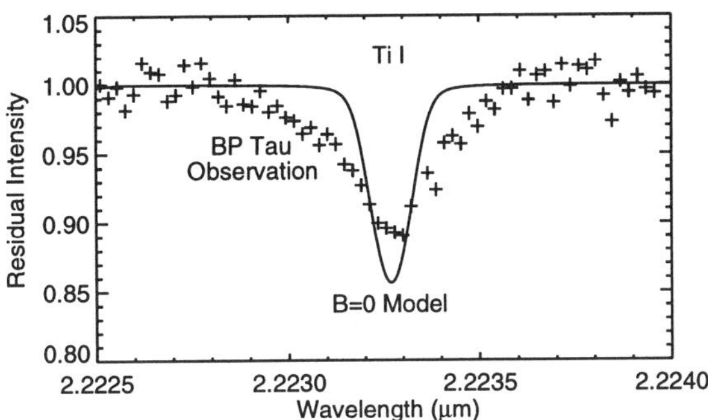

FIGURE 6. Detection of Zeeman broadening in the Ti I $\lambda2.2\mu m$ line in the spectrum of the classical T Tauri star BP Tau (Johns-Krull, Valenti, & Koresko 1999). The Ti I line (*plus signs*) is found to be significantly broader than a Zeeman-insensitive photospheric line (*solid line*). Modeling of the Ti I line indicates a distribution of field strengths on the stellar surface (1–10 kG) with a mean field of ~ 3.3 kG.

theories of magnetic accretion (e.g. Königl 1991; Shu et al. 1994; Cameron & Campbell 1996; see §2.4).

Several other observations support the existence of an organized component to these strong fields. For example, our ability to measure photometric variability due to the rotational modulation of hot or cool spots (§2.5) indicates the existence and long-term stability of an organized magnetic field component. An exciting recent development in the study of the magnetic field geometry of young stars comes from spectropolarimetric studies of certain T Tauri star emission lines. In the classical T Tauri star BP Tau, the He I $\lambda5876$ emission line is strongly polarized and the wavelength shift between the left- and right-circularly polarized components of the line profile implies a field strength ~ 2.4 kG (Johns-Krull, Valenti, Hatzes, & Kanaan 1999; Fig. 7). Since the He I line is believed to originate at the base of the magnetic accretion flow onto the star, these observations provide clear evidence for a strong, organized field that is associated with accretion onto the stellar surface. Such strong, globally organized fields are necessary if the star is to couple magnetically to a disk several stellar radii away and participate in the processes of stellar accretion and angular momentum regulation.

3. Origin of initial stellar masses and angular momenta

Remarkably, a large fraction of the observations described in the previous sections support and/or find a common explanation within our current understanding of how low-mass stars acquire their initial stellar masses and angular momenta. A possible synopsis of the story might be as follows. Mass and angular momentum are accreted from molecular clouds in a phase of dynamical infall. (Following an extended observational search, compelling evidence for this phase of evolution has been obtained; §2.1). The finite angular momentum of infalling cloud material leads to the formation of a circumstellar disk. (Following another observational search, compelling morphological and kinematic evidence for centrifugally supported structures has been obtained; §2.3). As a result of disk angular momentum transport, disk matter spirals in toward the star. The star accretes mass and angular momentum from the disk, producing a rapidly-rotating star. Rapid rotation coupled with the fully convective stellar interior leads to the production of a strong stellar magnetic field. (We now have good evidence for strong, organized stellar fields and the early onset of magnetic activity; §2.6.)

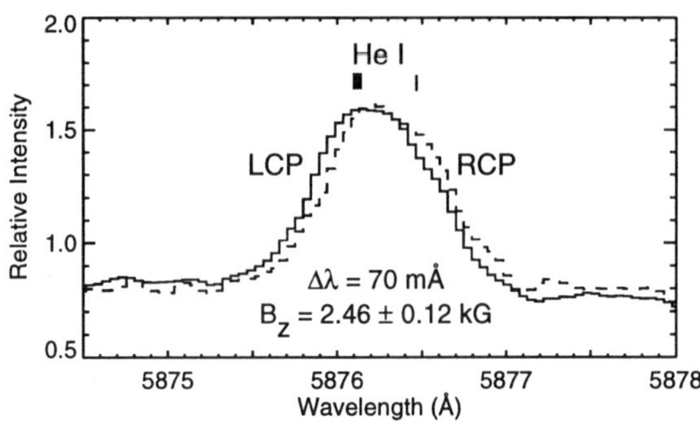

FIGURE 7. Detection of circular polarization in the He I λ5876 emission line in the spectrum of the classical T Tauri star BP Tau (Johns-Krull, Valenti, Hatzes, & Kanaan 1999). The wavelength shift between the right- and left-circularly polarized components indicates a field strength of 2.46 kG. Since the He I line is believed to form in the base of the magnetic accretion flow, these observations argue that organized fields are associated with accretion onto the stellar surface.

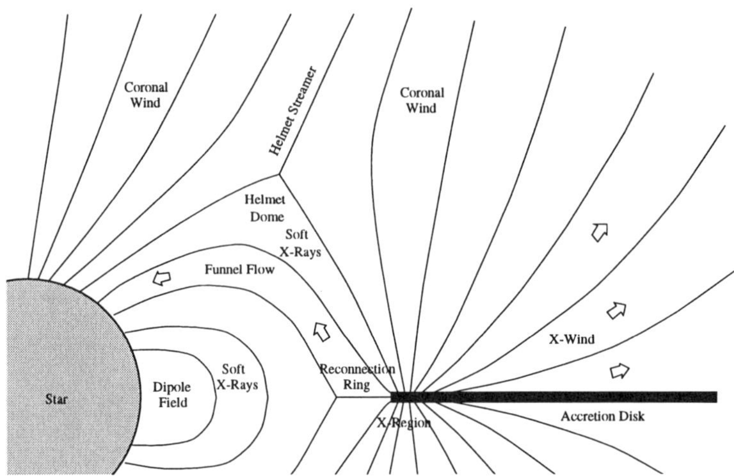

FIGURE 8. Schematic drawing of the X-wind model (Shu et al. 1994, 1999). In the interaction between a stellar dipole magnetic field and a surrounding accretion disk, the disk is truncated and the magnetic configuration of the star (*thin lines*) is altered, allowing mass inflow on to the stellar surface in magnetic accretion flows, and energetic mass loss in a magnetocentrifugal wind.

In the picture of Shu et al. (1994; see also Shu et al. 1999; Königl 1991; Arons 1986; Camenzind 1990; Shu et al. 1988) that describes the interaction of the stellar magnetic field with the disk, the stellar field truncates and couples to the disk at a distance, typically a few stellar radii, that depends on disk accretion rates and stellar parameters including the magnetic field strength (Fig. 8). As a result of the magnetic disk coupling, the stellar rotation rate is slowed and regulated so that it equals the rotation rate at the disk truncation radius (cf. Shu et al. 1997). For typical T Tauri disk accretion rates, stellar parameters, and the mean (~ 8 day) period of classical T'Tauri stars, stellar field strengths of a few kilogauss are required, in good agreement with measured values.

In the Shu et al. picture, disk accretion pinches the stellar magnetic field at the truncation radius, producing a magnetic geometry that favors dynamically the generation of magnetic accretion flows and winds from the truncation region. Thus, disk matter reaching the inner disk radius can be channeled into either an accretion flow onto the star or into an outflowing wind, providing an explanation for the observed simultaneous presence of stellar accretion flows and winds within a few stellar radii of the star (§2.4). The back torque of the wind and magnetic accretion flows on the magnetic field footpoints in the disk truncation region act to maintain the pinched magnetic field in the truncation region.

The magnetocentrifugal action resulting from the combination of strong stellar fields and the rapid rotation of the disk (or star) are capable of driving cool, massive winds with terminal velocities and mass loss rates that are in good agreement with the properties of winds from T Tauri stars (and embedded protostars), as described by Najita & Shu (1994). The magnetocentrifugally-driven winds play an important role in the solution of the angular momentum problem for star formation. They remove significant angular momentum from the young star system, which allows the star to continue to rotate slowly despite ongoing accretion. For T Tauri stars, the observed kilogauss-strength fields are capable of removing sufficient angular momentum that the wind mass loss rate is a comparable fraction of the disk and stellar accretion rates ($\dot{M}_W/\dot{M}_D = 0.1$–$0.3$; $\dot{M}_* = \dot{M}_D - \dot{M}_W$) in reasonable agreement with observationally derived ratios (§2.2).

At distances of tens of parsecs, the wind collimates magnetically into a jet embedded in a wider angle wind (Shu et al. 1995), in good agreement with HST observations of jet collimation distances. The wind and jet together sweep up the surrounding molecular cloud material, creating a molecular outflow (§2.2). At the same time, the outflowing material acts to halt further infall, thereby limiting the mass of the star. At some later time, the disk disperses. The star and disk are consequently decoupled, and the star is free to spin up as it contracts. This picture, as it currently stands, provides a viable framework for understanding the origin and interrelation of many of the observed phenomena associated with young stars and their role in the origin of the initial masses and angular momenta of low-mass stars.

4. Unsolved problems

Despite the success of the above picture, there are numerous issues that remain to be addressed, including several basic aspects of this picture. For example, it is as yet unclear theoretically whether the magnetic interaction between the star and disk is likely to be time steady or not. Various recent simulations (Hayashi et al. 1996; Goodson et al. 1997; Miller & Stone 1997) suggest that the interaction is much less quiescent than in the Shu et al. (1994) picture. Observational evidence for time dependent behavior may be uncovered in synoptic observations of T Tauri line profiles which are notoriously variable.

Another important issue is the origin of angular momentum transport in disks. Gravitational instabilities are efficient when the circumstellar disk mass is a fair fraction of the stellar mass, a situation which may be relevant at early times. In the T Tauri phase, where disk masses appear to be too low to support significant gravitational instabilities, magnetic instabilities such as the Balbus-Hawley instability (Hawley et al. 1995) appear to be the most promising.

Our understanding of the stellar rotational properties of young stars is still in its early stages, despite several conclusions having already been drawn from the existing data. From an observational perspective, observations of large numbers of stars will be needed to study rotation as a function of mass and age, and to average over the differences that may be inherent in trying to compare stars with different formation histories (e.g. molecular cloud initial conditions, etc.). Various groups are currently engaged in expanding the original work on young stars in the Taurus and Orion regions to a larger number of stars and clusters. These results are eagerly awaited.

As discussed above, the longevity of inner accretion disks may be an important issue for the understanding of the origin of slow rotators and the spread in stellar angular momenta that exists in 10–100 Myr old clusters. The factors that control disk dispersal are presently unclear, although some of the relevant processes have been identified. For example, since disk structure is known to be affected by the presence of stellar companions (e.g. Mathieu 1994), the longevity of inner disks may similarly be affected by the presence of close stellar or planetary companions.

Another generic issue is the extent to which this picture, developed to explain the formation of isolated, low-mass stars, applies to stars in clusters and/or to stars of higher masses. It may be necessary to consider other physical processes (e.g. fragmentation, mergers of stars and/or clouds) in order to understand the origin of stellar masses and angular momenta in dense clusters. Further work is needed to understand these processes.

In terms of the applicability of this picture to high mass stars, a topic that will be covered more ably by the next speaker, I would like to offer some additional observational evidence that similar physical processes may be involved. High resolution spectroscopy of the CO overtone emission from a high mass star in the young cluster M17 has revealed the spectral shape of emission from a rotating disk, providing kinematic evidence for disks around young high mass stars (Najita et al. 1999; Fig. 3). This indicates that stars may evolve from low to high mass by accretion through disks (e.g., Palla & Stahler 1993). These observations also support the view that radiation pressure is not a barrier to the formation of high mass stars and that consequently the high end of the stellar initial mass function should be independent of metallicity, as is observed (Massey et al. 1995).

I am very grateful to Neal Evans, Lynne Hillenbrand, Michael Meyer, David Wilner, Steve Strom, Russ Makidon, Jeff Valenti, Chris Johns-Krull, and Frank Shu for providing illustrations for this review.

REFERENCES

ADAMS, F. C., LADA, C. J., & SHU, F. H. 1987 *ApJ* **312**, 788.

ARMITAGE, P. J. & CLARKE, C. J. 1996 *MNRAS* **280**, 458.

ARONS, J. 1986. In *Plasma penetration into magnetospheres*, (eds. N. Kylafis, J. Papamastorakis, & J. Ventura), p. 115. Crete University Press.

BACHILLER, R. 1996 *ARAA* **34**, 111.

BASRI, G., MARCH, G. W., & VALENTI, J. A. 1992 *ApJ* **390**, 622.

BERTOUT, C., BASRI, G., & BOUVIER, J. 1988 *ApJ* **330**, 350.

BERTOUT, C., HARDER, S., MALBET, F., & MENNESSIER, C. 1996 *AJ* **112**, 2159.

BODENHEIMER, P. 1995 *ARAA* **33**, 199.

BOUVIER, J., CABRIT, S., FERNANDEZ, M. & MARTIN, E. L. & MATTHEWS, J. M. 1993 *AA* **272**, 176.

BOUVIER, J., ET AL. 1997 *AA* **318**, 495.

BOUVIER, J., FORESTINI, M., & ALLAIN, S. 1997 *AA* **326**, 1023.

BURROWS, C. J., ET AL. 1996 *ApJ* **473**, 437.

CALVET, N. & HARTMANN, L. 1992 *ApJ* **386**, 239.

CALVET, N. 1997 In *Herbig-Haro Flows and the Birth of Low Mass Stars* (eds. B. Reipurth & C. Bertout). IAU Symp. No. 182. p. 417. Reidel.

CALVET, N., HARTMANN, L., & STROM, S. 1999. In *Protostars & Planets IV* (eds. V. Mannings & S. Russell), in press.

CAMENZIND, M. 1990. In *Reviews in modern astronomy 3* (ed. G. Klare). p. 259. Springer.

CAMERON, A. C. & CAMPBELL, C. G. 1996 *AA*, **274**, 309.

CARR, J. S., TOKUNAGA, A. T., NAJITA, J., SHU, F. H., GLASSGOLD, A. E. 1993 *ApJ* **411**, L37.

CARR, J. S. ET AL. 1999, in preparation.

CHANDLER, C. J., CARLSTROM, J. E., SCOVILLE, N. Z., DENT, W. R. F., & GEBALLE, T. R. 1993 —it ApJ **412**, L71.

CHOI, M., EVANS, N. J. II, GERGERSEN, E. M., & WANG, Y. 1995 *ApJ* **448**, 742.

CHOI, P. I. & HERBST. W. 1996 *AJ* **111**, 283.

D'ALESSIO, P., CANTÓ, J., CALVET, N., & LIZANO, S. 1998 *ApJ* **500**, 411.

DAVIS, C. J., MUNDT, R., EISLÖFFEL, J., & RAY, T. P. 1995 *AJ* **110**, 766.

EDWARDS, S., RAY, T., & MUNDT, R. 1993 In *Protostars & Planets III* (eds. E. H. Levy and J. I. Lunine). p.567. Univ. of Arizona Press.

EDWARDS, S., HARTIGAN, P., GHANDOUR, L. & ANDRULIS, C. 1994 *AJ* **104**, 105.

GIOVANARDI, C., GENNARI, S., NATTA, A., & STANGA, R. 1991 *ApJ* **367**, 173.

GOODMAN, A. A., BENSON, P. J., FULLER, G. A., & MYERS, P. C. 1993 *ApJ* **406**, 528.

GOODSON, A. P., WINGLEE, R. M., & BOEHM, K.-H. 1997 *ApJ* **489**, 199.

GULLBRING, E., HARTMANN, L., BRICEÑO, C., & CALVET, N. 1998 *ApJ* **492**, 323.

HARTIGAN, P., EDWARDS, S., & GHANDOUR, L. 1995 *ApJ* **452**, 736.

HARTMANN, L., & STAUFFER, J. R. 1989 *AJ* **97**, 873.

HAWLEY, J. F., GAMMIE, C. F., & BALBUS, S. A. 1995 *ApJ* **440**, 742.

HAYASHI, M., OHASHI, N., & MIYAMA, S. M. 1993 *ApJ* **418**, L71.

HILLENBRAND, L. & MEYER, M. R. 1999, in preparation.

JOHNS, C. & BASRI, G. 1995 *ApJ* **449**, 341.

JOHNS-KRULL, C. M., VALENTI, J. A., & KORESKO, C. 1999 *ApJ*, in press.

JOHNS-KRULL, C. M., VALENTI, J. A., HATZES, A. P., & KANAAN A. 1999 *ApJ* **510**, L41.

KENYON, S. J. & HARTMANN, L. 1995 *ApJS* **101**, 117.

KEPNER, J., HARTIGAN, P., YANG, C., & STROM, S. 1993 *ApJ* **415**, L119.

KEPPENS, R., MACGREGOR, K. B., & CHARBONNEAU, P. 1995 *AA* **294**, 469.

KOERNER, D. W. & SARGENT, A. I. 1995 *AJ* **109**, 2138.

KÖNIGL, A. 1991 *ApJ* **370**, L39.

KOYAMA, K., HAMAGUCHI, K., UENO, S., KOBAYASHI, N., & FEIGELSON, E. 1996 *PASJ* **48**, L87.

LADA, C. J. 1985 *ARAA* **23**, 267.

LEVREAULT, R. M. 1988 *ApJ* **330**, 897.

LYNDEN-BELL, D. & PRINGLE, J. E. 1974 *MNRAS* **168**, 603.

MAKIDON, R. ET AL. 1999, in preparation.

MASSEY, P., JOHNSON, K. E., & DEGIOIA-EASTWOOD, K. 1995 *ApJ* **454**, 151.

MANNINGS, V., BOSS, A. P. & RUSSELL, S. 1999 eds., *Protostars & Planets IV*, in press.

MATHIEU, R. D. 1994 *ARAA* **32**, 465.

McCAUGHREAN, M. J. & O'DELL, C. R. 1996 *AJ* **111**, 1977.

MEYER, M. R., BECKWITH, S. V. W., STAUFFER, J. R., SCHULTZ, B. 1999, in preparation.

MILLER, K. A. & STONE, J. M. 1997 *ApJ* **489**, 890.

MORIARTY-SCHIEVEN, G. H., SNELL, R. L., STROM, S. E., SCHLOERB, F. P., STROM, K. M., & GRASDALEN G. L. 1987 *ApJ* **319**, 742.

MUNDT, R. 1984 *ApJ* **280**, 749.

MUNDT, R., STOCKE, J., STROM, S. E., STROM, K. E., & ANDERSON, E. R. 1985 *ApJ* **297**, L41.

MUNDY, L. G. ET AL. 1996 *ApJ* **464**, L169.

MUZEROLLE, J., HARTMANN, L. & CALVET, N. 1998b *AJ* **116**, 455.

MUZEROLLE, J., CALVET, N. & HARTMANN, L. 1998a *ApJ* **492**, 743.

MYERS, P. C., BACHILLER, R., CASELLI, P., FULLER, G. A., MARDONES, D., TAFALLA, M., & WILNER, D. J. 1995 Graviational infall in the dense cores L1527 and L483 *ApJ* **449**, L65.

NAJITA, J. & SHU, F. H. 1994 *ApJ* **429**, 808.

NAJITA, J., CARR, J. S., TOKUNAGA, A. T. 1996a *ApJ* **456**, 292.

NAJITA, J., CARR, J. S., GLASSGOLD, A. E., SHU, F. H., & TOKUNAGA, A. T. 1996b *ApJ* **462**, 919.

NAJITA, J., GREENE, T. P., & LADA, C. J. 1999, in preparation.

NATTA, A. 1995 *RevMexAA* **1**, 209.

PALLA, F. & STAHLER, S. 1993 *ApJ* **418**, 414.

REIPURTH, B., BALLY, J., & DEVINE, D. 1997 *AJ* **114**, 2708.

SHU, F. H. 1977 *ApJ* **214**, 488.

SHU, F. H., LIZANO, S., RUDEN, S., & NAJITA, J. 1988 *ApJ* **328**, L19.

SHU, F. H., NAJITA, J., OSTRIKER, E., WILKIN, F., RUDEN, S., & LIZANO, S. 1994 *ApJ* **429**, 781.

SHU, F. H., NAJITA, J., OSTRIKER, E. C., & SHANG, H. 1995 *ApJ* **455**, L155.

SHU, F. H., SHANG, H., GLASSGOLD, A. E., & LEE, T. 1997 *Science* **277**, 1475.

SHU, F. H., NAJITA, J., SHANG, H., & LI, Z.-Y. 1999 In *Protostars & Planets IV* (eds. V. Mannings, A. P. Boss & S. Russell), in press.

SNELL, R. L. & SCHLOERB, F. P. 1985 *ApJ* **295**, 490.

STANKE, T., McCAUGHREAN, M. J., & ZINNECKER, H. 1998 *AA* **332**, 307.

TAFALLA, M., MARDONES, D., MYERS, P. C., CASELLI, P., BACHILLER, R., & BENSON, P. J. 1998 *ApJ* **504**, 900.

TEREBEY, S., SHU, F. H., & CASSEN, P. 1984 *ApJ* **286**, 529.

VOGEL, S. N. & KUHI, L. V. 1981 *ApJ* **245**, 960.

WALKER, M. 1972 *ApJ* **175**, 89.

WILNER, D. J., HO, P. T. P., & RODRIGUEZ, L. F. 1999 *ApJ/*, in preparation.

ZHOU, S., EVANS, N. J. II, KÖMPE, C., WALMSLEY, C. M. 1993 *ApJ/* **404**, 232.

ZHOU, S., EVANS, N. J. II, & WANG, Y. 1996 *ApJ* **466**, 296.

Massive star formation: The role of bipolar outflows

By ED CHURCHWELL

Washburn Observatory, University of Wisconsin

1. Introduction

One of the most durable problems in all of astrophysics is how stars are formed. Since the time of LaPlace, astronomers have postulated numerous scenarios, but to the present, this problem has resisted solution. This is especially true for massive stars. Massive star formation has received much less observational and theoretical attention than low mass star formation and as a consequence, much less is known about the physics of their formation. Indeed it could be argued that we know too little to ask the right questions. Some, however, are obvious. For example, the mechanism(s) of cluster formation is still a mystery. In particular, what determines the initial mass function (IMF)? What are the roles of fragmentation, differential galactic rotation, and supersonic turbulence in regulating the rate, and possibly the IMF, of star formation in molecular clouds? Are bipolar outflows the mechanism for shedding angular momentum from accretion disks? How does matter actually get from an accretion disk onto a protostar? Are massive stars formed entirely by accretion processes or a combination of accretion and stellar mergers? When are stellar winds established, what is their relationship to bipolar outflows, and what role do they play in the formation process, if any?

A few things are known. For example, massive stars typically, perhaps always, form in conjunction with an associated cluster of lower mass stars. They form in massive molecular cloud cores which are generally heavily obscured at optical and UV wavelengths. Consequently, they generally can only be observed at infrared and radio wavelengths. The velocity, density, and temperature structure of the circumstellar matter around a forming massive star is expected to change by several orders of magnitude over distances of ≤ 1000 AU. Although little is known about the detailed distribution and kinematics of gas and dust at distances within several hundred AU of forming massive stars nor the physical process by which they form, there are some critical size scales that we can approximately identify from present observations and theoretical considerations. Stellar winds are expected to accompany massive star formation with wind velocities up to a few thousand km s^{-1}, the effects of which may be detectable out to 1000 AU from the protostar in a hot wind-shocked cavity which supports observed shell-like and cometary untracompact (UC) HII regions. Dense and massive molecular outflows extending out to ~ 1 pc are presumably driven by a central massive protostar. Although the mechanism for initiating and driving these outflows is still unknown, ample observational evidence confirms their existence. UC HII regions surround young massive stars out to radii of about 10^{17} cm typically. See Churchwell (1990, 1991, 1995, 1997a) for reviews of the properties of UC HII regions. During formation, it is expected that a massive accretion disk forms which feeds the central protostar from in-falling ambient cloud core material; the properties of these disks, if they exist, may have radii of a hundred AU or more and masses containing a significant fraction of the central protostar mass. The protostar itself may have a diameter of 10^{12} cm or more depending on the stage of evolution.

UC HII regions are sign posts of massive star formation. The size, morphology, and internal velocity distribution of an UC HII region reflects the interaction of a newly formed massive star with its environment. In some models the lifetime of an UC HII region is also integrally related to the nature of the ambient cloud core in which the nebula is embedded. These nebular properties are not independent; they depend in an integral way on properties of the central protostar, the ambient medium, and the motion of the star relative to the ambient medium. Understanding circumstellar nebular properties involve many important problems of gas dynamics as well as issues associated with the star formation process.

In this review, I will focus on molecular outflows driven by massive protostars and their implications for massive star formation. The evidence for outflows associated with massive star formation is discussed in §2. Properties of outflows in massive star formation regions are given in §3, and issues raised by massive outflows are discussed in §4. In §5 possible mechanisms to explain the large outflow masses are addressed, and in §6 a brief summary is given of the role of molecular outflows in massive star formation.

2. The frequency of outflows from massive star formation regions

By 1995 several massive star formation regions had been observed to have molecular outflows. Among these were the Orion IRC2 outflow (Kwan and Scoville 1976; Zuckerman, Kuiper, Rodriguez Kuiper 1976; Downes et al. 1981; Erickson et al. 1982; Wright et al. 1983; Allen and Burton 1993, and others); G5.89 (Harvey and Forveille 1988; Cesaroni et al. 1991; Choi, Evans, Jaffe 1993); DR21 (Garden et al. 1991a,b; Garden and Carlstrom 1992); NGC 6334I (Bachiller and Cernicharo 1990); G34.26 and G10.62 (Cesaroni et al. 1991); W75N (Hunter et al. 1994); W3IRS 5, GL490, GL2591, S140, and CephA (Choi, Evans, Jaffe 1993); and several low-mass and high-mass outflows are reported by Cabrit and Bertout (1992) and Lada (1985).

In an effort to better determine the frequency of molecular outflows in massive star formation regions, Shepherd and Churchwell (1995; hereafter SC95) observed and analyzed CO(1-0) line profiles with high signal-to-noise toward 94 UC HII regions. The NRAO 12m telescope was used (HPBW∼ 60″). They found that fully 90% of this sample had line wing emission in excess of a gaussian profile fit to the core of the line. These results indicate gas motions in excess of thermal plus turbulent motions, but they do not necessarily imply bipolar outflows as rotation, or expansion, or contraction, or different velocity components along the line of sight could also produce excess line wings. To determine if the excess line wings are due to outflows Shepherd and Churchwell (1996; hereafter SC96) mapped a selection of ten of the sources observed to have excess line wings by SC95 using the NRAO 12m telescope. They found five to have well defined outflows, two had multiple velocity components along the line of sight and were not outflows, and three had either S/N ratios that were too low to confidently map or weak CO emission was in the reference position such that the images were unusable. They conclude that at least 50%, and possibly as high as 80%, of the selected sample had bipolar outflows.

3. Molecular bipolar outflow properties

In Table 1, I have tabulated the sources toward which massive outflows have been detected. These are listed in order of increasing RA. Along with the source names and their positions, I also give the distances from the Sun D, the HII luminosities L_{HII} as determined from radio free-free flux densities or as given by the reference in Table 2, and

SOURCE	RA(1950)	DEC(1950)	D Kpc	L_{HII} $L_\odot$	L_{IR} $L_\odot$	Refs	Notes
W3 (IRS5)	02 21 53.1	61 52 22	2.3		2.00E+05	1	1
GL490	03 23 39.2	58 36 36	0.9		1.7–2.4E3	1,2	
Orion IRC2	05 32 46	−05 24 00	0.5	>1.0E4	2.10E+05	2,16	5
G173.58+2.45	05 36 06.2	35 39 06	1.8	3.00E+03	1.80E+03	3,5,14	
G192.16–3.82	05 55 20.0	16 31 44	2.0	3.00E+03	3.26E+03	3,5,14	
Mon R2 (IRS1)	06 05 20.0	−06 22 39	1.0	4.00E+04	5.00E+04	4,5,6	
G240.31+0.07	07 42 45.1	−24 00 24	6.4	1.26E+04	5.01E+04	7	
NGC 6334I	17 17 32.3	−35 44 04	1.7	3.00E+04	8.00E+04	8	
G5.89–0.39	17 57 26.8	−24 03 55	3.0	1.20E+05	3.00E+05	4,5	
G9.62+0.19F	18 03 16.09	−20 31 59.7	5.7			13	2
G10.62–0.38	18 07 30.7	−19 56 29	6.5	2.10E+05	1.24E+06	4,5	
G25.65+1.05	18 31 40.1	06 02 06	2.8	8.90E+03	1.76E+04	9,5	
G34.26+0.15	18 50 46.1	01 11 12	3.7	1.40E+05	5.93E+05	4,5	
G35.2N (IRS1)	18 59 14.1	01 09 03	2.5	9.10E+04	1.23E+05	4,5,14	
G45.07+0.13	19 11 00.4	10 45 43	8.3	1.40E+05	1.42E+06	4,5	
G45.12+0.13	19 11 06.2	10 48 25	8.3	5.80E+05	1.66E+06	4,5	
IRAS20126+4104	20 12 41.0	41 04 20.5	1.7	5.01E+03	6.31E+04	7,15	
G75.78+0.34NE	20 19 52.7	37 27 22	5.5	4.80E+04	?	3,5	
GL2591	20 27 35.8	40 10 14	2.0	2.00E+04	9.00E+04	1,5	
W75NBb	20 36 50.0	42 26 57	2.0	5.50E+03	1.40E+05	10,5	
DR21	20 37 14.4	42 08 55	3.0	4.00E+05	∼6E+5	11,5	
G98.04+1.45	21 41 21.3	54 42 32	7.4	2.30E+05	1.39E+05	3,5,14	3
S140 (IRS1)	22 17 41.2	63 03 45	0.9		0.5–9.0E4	1,2	1
CEPH A	22 54 19.0	61 45 47	0.7	3.98E+03	1.58E+04	7	
NGC 7538 (IRS9)	23 11 52.8	61 10 59	2.7		.00E+04	12	1
G111.25–0.77	23 13 57.9	59 39 00	3.5	2.00E+04	9.80E+03	3,5,14	4

References:
1. Choi et al. (1993)
2. Cabrit & Bertout (1992)
3. Shepherd & Churchwell (1996)
4. Wood & Churchwell (1989)
5. Panagia (1973)
6. Thronson et al. (1980)
7. MacLeod et al. (1998)
8. Bachiller & Cernicharo (1990)
9. Shepherd & Churchwell (1995)
10. Hunter et al. (1994)
11. Garden & Carlstrom (1992)
12. Mitchell & Nasegawa (1991)
13. Hofner et al. (1995)
14. Shepherd (private communication)
15. Cesaroni et al. (1997)
16. Menten & Reid (1995)

Notes:
1. The IR source does not coincide with a detected HII region.
2. No cm emission is detected, mm emission probably due to warm dust. FIR emission is confused.
3. Two UC HII regions are present, neither of which is likely powering the outflow.
4. There may be multiple outflows toward this UC HII region.
5. The outflow source is not the radio source I, as its luminosity is much too small.

TABLE 1. HII regions where massive outflows have been observed

the total luminosity of the HII region and associated cluster L_{IR} as determined from the integrated infrared flux. The distances are those given by the references listed in Table 2 and the outflow properties were derived on the basis of this distance.

In table 2, I have tabulated the properties of bipolar outflows thought to be driven by massive protostars. These are from single dish observations which do not suffer from "missing flux" problems and therefore provide more reliable global properties of the outflows such as mass, extent, and energetics but little information on detailed morphologies. I have reported values as given by the authors without insisting that the data be analyzed in a uniform and systematic way. As a consequence, there are disagreements among different authors on some flow properties perhaps at the factor of two level. Several sources, such as Orion IRC2, have been observed by many observers with numerous techniques.

In these cases, I have generally tabulated the most complete set of published values or in a few cases the latest values when the techniques or sensitivity were better. Several sources are included for which no outflow properties are given, because observations show that they are clearly bipolar outflows (e.g. G10.62 and G34.26) but no outflow properties were derived in the original publication where they were reported. The headings in Table 2 are, from left to right: the HII region or IR source toward which the outflow is observed, the projected separation of the outflow lobes R_f, the dynamical time scale of the outflow ($\tau_d = R_f/2v_f$), the total mass in the outflow M_f in solar masses, the rate of mass outflow dM_f/dt, the outflow momentum P_f, the momentum supply rate or force of the outflow F_f, the total energy of the outflow E_f, and the mechanical luminosity of the outflow L_f in solar luminosities.

Aside from the usefulness of assembling the list of currently known massive outflows and their properties in one place, the main reason for assembling Table 2 is to provide an overview of massive outflow properties. These outflows are thought to be driven by massive protostars as implied by Fig. 2 and the fact that massive outflows only occur in massive star formation regions as demonstrated by direct near infrared imaging of massive star formation regions (McCaughrean 1993; McCaughrean and Stauffer 1994; Aspin and Walther 1990; and others); by optical imaging of evolved HII regions which contain star clusters; and, indirectly by Kurtz, Churchwell, and Wood (1994) who showed that UC HII regions have 100 μm-to-2 cm flux density ratios well in excess of those produced by single stars of the same luminosity. In all cases where IR imaging has been carried out, a dense cluster of stars or protostars has been found to be associated with the UC HII region toward which outflows have been observed. Single dish spatial resolutions are not good enough to identify the source of the outflow with a particular star or protostar. Identification of the driving source is exacerbated by the fact that massive star formation occurs in distant (typically ≥ 1 kpc), dense stellar clusters. Only radio interferometry in combination with high spatial resolution NIR imaging can isolate the sources of bipolar molecular outflows. In a few cases it appears that an outflow may be driven by the star responsible for ionizing an UC HII region, but frequently outflows are clearly offset from the nearby UC HII region.

Massive stars are formed deeply embedded in molecular clouds where they are generally observable only at radio and infrared wavelengths. Thus they have not benefitted from observations at optical and UV wavelengths. A consequence of large distances and crowded star fields is that the ionizing star of only one UC HII region (the relatively lightly extincted nebula G29.96) has been identified (Watson et al. 1998) to the present time and no star has been unambiguously identified as the driving source for a massive outflow.

Interferometric studies of G45.12, G75.78, and NGC 2024 by Hunter, Phillips, and Menten (1997); Shepherd, Churchwell, and Willner (1997); and Chernin (1996) have shown that these regions harbor two or more outflows when observed with high spatial resolution. We also know that multiple outflows are present in DR21, even from single dish observations. The mass of each of the outflows (with the exception of NGC 2024) is still quite large (in the 10s to 100s of solar masses). The interferometric images have also shown that massive outflows do not appear to be well collimated. The Orion IRC2 outflow, traced by high resolution shocked H_2 emission (Allen and Burton 1993), is shown in Figure 1. This outflow is certainly not well collimated nor is it a smooth gaseous flow. It does not come to a point at its origin but rather appears to originate from a large surface area, most of which is well away from the central position of the outflow. Allen and Burton (1993) referred to it as an explosive event. The "fingers" of this outflow which reach out into the ambient interstellar medium define a jagged interface not a smooth

Source	D kpc	R_f pc	τ_d 10^4 yr	M_f $M_\odot$	dM_f/dt $M_\odot$ yr^{-1}	P_f $M_\odot$ km s^{-1}	F_f $M_\odot$ km s^{-1} yr^{-1}	E_f 10^{45} ergs	L_f $L_\odot$	Refs.	notes
W3 (IRS5)	2.3		0.3	$\sim$8	4.8×10^{-5}	110		16		1	1
GL490	0.9	1.2	1.8	15.4	8.6×10^{-4}	493	7.6×10^{-3}	158	19	2,1	2
Orion IRC2	0.5	0.1	0.075	8.2	1.1×10^{-2}	533	0.2	346	1000	2	3
G173.58+2.45	1.8	0.6	5.3	32	6.0×10^{-4}	170	3.1×10^{-3}	9.4	1.4	3	4
G192.16−3.82	2.0	2	27	58	2.2×10^{-4}	210	7.0×10^{-4}	9.4	0.2	3	4
Mon R2 (IRS1)	0.95	$\geq$3.2	15	202	1.3×10^{-3}	$\sim$2000	1.6×10^{-2}	290	16	4,5,6,7	
G240.31+0.07	6.4	0.76	2.3	8.3	3.6×10^{-4}	93	4.0×10^{-3}	11	3.9	8	7
NGC6334F	1.74	0.36	0.23	>2.3	1.0×10^{-3}	92	4.0×10^{-2}	43	170	9	
G5.89−0.39	3.0	0.1	0.3	77	2.6×10^{-2}	990	0.33	500	1300	10,11,1	5,2
G9.62−0.19F	5.7		0.8	$\sim$50					20		
G10.62-0.38	6.5									11	6
G25.65+1.05	2.8	1.1	1.8	7.6	4.2×10^{-4}	136	7.6×10^{-3}	27	12.4	8	7
G34.26+0.15	3.7									11	6
G35.2N (IRS1)	2.0	1.4	4.6	73	1.6×10^{-3}	1100	2.7×10^{-2}	164	51	2	
G45.07+0.13	8.3	0.3	3.2	50	1.4×10^{-3}	631	1.3×10^{-2}	36	$\sim$11	12	
G45.12+0.13	8.3	1.8	20	$\sim$4800	$\geq1.3\times10^{-3}$	$\sim$31000	$\sim1.3\times10^{-2}$	5600	10–100	12	8
IRAS20126+4104	1.7	0.3	0.58	>90–310	1.6×10^{-2}	540		80	13		
G75.78+0.34NE	5.5	0.8	3	72	2.4×10^{-3}	970	3.2×10^{-3}	130	36	3	
GL2591	2		0.2	33	6.8×10^{-5}	269		23		1	1
W75N	2.0	0.7	$\geq$4.4	48	1.2×10^{-3}	$\sim$700	$\leq1.6\times10^{-2}$	110	22	14,15	
DR 21	3.0	$\sim$2.2		>3000				>2000		16,17,18	9
G98.04+1.45	7.4	1.2	9.1	40	4.4×10^{-4}	240	2.7×10^{-3}	15	1.4	3	
S140 (IRS1)	0.9	1.1	11	24.4	9.4×10^{-4}	488	5.3×10^{-3}	98	8.4	2,1	
Ceph A	0.73		0.2	$\sim$8	2.8×10^{-5}	86		10		1	1
NGC7538 (IRS9)	2.7	0.97	$\sim$1.6	51	4.2×10^{-3}	490		100	790	19	
G111.25−0.77	3.5	0.6	3.7	16	4.3×10^{-4}	110	3.1×10^{-3}	8.4	1.8	3	

References:
1. Choi et al. (1993)
2. Cabrit & Bertout (1992)
3. SC96
4. Wolf, Lada, Bally (1990)
5. Loren (1981)
6. Tafalla et al. (1997)
7. Meyers-Rice & Lada (1991)
8. SC95
9. Bachiller & Cernicharo (1990)
10. Acord et al. (1997)
11. Cesaroni et al. (1991)
12. Hunter et al. (1997)
13. Cesaroni et al. (1997)
14. Hunter et al. (1994)
15. Fischer et al. (1985)
16. Garden et al. (1991a)
17. Garden et al. (1991b)
18. Garden & Carlstrom (1992)
19. Mitchell & Hasegawa (1991)
20. Hofner et al. (1995)

Notes:
1. The mass and other reported flow properties assume the mass is swept-up ISM.
2. Ref. 2 only mapped the inner part of the outflows; they find lower values than other observers.
3. Many observers have studied this source; ref. 2 values are reported here.
4. HPBW$\sim$ 60$''$.
5. Values tabulated are from ref. 10, see note 2.
6. Images shown but no outflow properties given.
7. Optical depth unity assumed.
8. Values from ^{12}CO reported rather than the ^{13}CO values.
9. At least 2 outflows and probably more are present.

TABLE 2. Massive bipolar outflows; single dish data

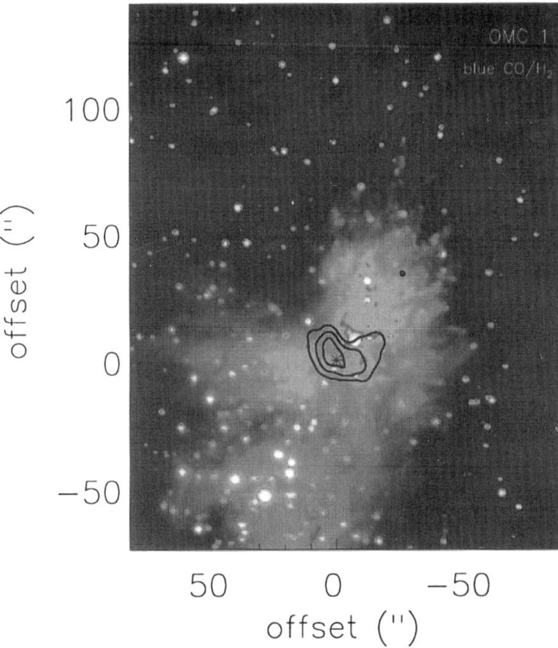

FIGURE 1. The Orion IRC2 outflow as delineated by shocked H_2 (in grey scale) from Allen and Burton (1993) with CO(1-0) emission contours superimposed from Chernin and Wright (1996).

working surface. It has been suggested that the fingers might be jets from individual low-mass protostars at the center of this feature. Although possible, this explanation seems unlikely: one, because it would require many stars to have their jets aligned within about $45°$ of each other; two, NIR photometry does not implicate such stars along the waist of this bowtie feature (although this could be because of large 2 μm optical depths); and three, the outflow looks like a coherent feature, not a montage of many separate features (although this could be a resolution effect).

The outflows found in massive star formation regions have energetics, masses, and outflow rates much larger than those associated with low-mass star formation. This is illustrated quite clearly in Table 3 where the average outflow mass, force, and mechanical luminosity are tabulated from Lada (1985) for star formation regions with bolometric luminosity $L_{bol} > 10^3$ $L_\odot$ versus those with $L_{bol} < 750$ $L_\odot$. If the more recent data given in Table 2 had been used, the mean outflow mass and luminosity from massive star formation regions would be > 130 $M_\odot$ and > 180 $L_\odot$, respectively. The mean properties of low-mass outflows do not seem to have changed significantly excluding the discovery of much larger outflow radii. Thus, massive star formation regions have outflows at least a factor of 10 more massive and more luminous than those from low-mass stars. SC96 have shown that the mass outflow rate $\dot{M}_f$ is a continuously increasing function of stellar bolometric luminosity over the range 1–10^6 $L_\odot$ as shown in Fig. 2. The solid curve is a quadratic least squares fit to the data. Sources G192.16 and G98.04 are believed to be driven by lower luminosity stars than those responsible for ionizing the nearby HII regions and are therefore probably misplotted in Fig. 2. A linear least squares fit to these data imply that

$$\dot{M}_f \approx L_{bol}^{0.7} \propto M_*^{2.1} \ . \tag{3.1}$$

The dependence on stellar mass assumes a mass-luminosity relation $L \propto M^3$ (appropriate for the upper end of the M-L relation). The importance of the tight relation between $\dot{M}_f$

Selected Outflow Parameter	$L_{bol} > 10^3\ L_\odot$ Average	$L_{bol} < 750\ L_\odot$ Average
M_f ($M_\odot$)	44	1.8
F_f ($10^{-3}\ M_\odot$ Km s^{-1} yr^{-1})	8.3	0.8
L_f ($L_\odot$)	8	1.1

TABLE 3. Comparison of mean outflow properties

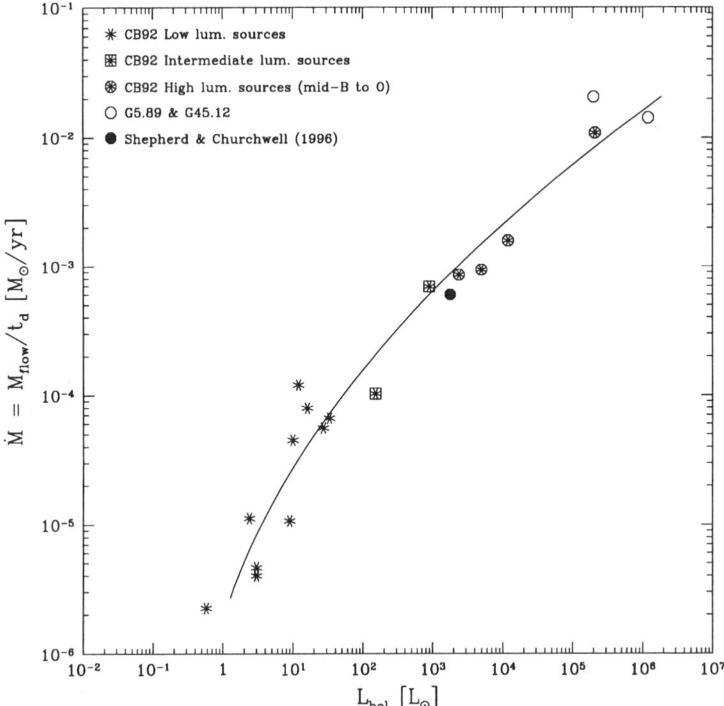

FIGURE 2. The mass outflow rate $\dot{M}_f$ versus the bolometric luminosity L_{bol} inferred from integrated far infrared flux densities. This figure is a compilation of data from several sources as described in SC96. A tight and continuous relationship between $\dot{M}_f$ and L_{bol} exists from 1 to 10^6 solar luminosities.

and L_{bol} is that one can measure $\dot{M}_f$ relatively easily from which it is possible to infer the luminosity of the driving source to relatively high accuracy, even in the absence of radio continuum or IR fluxes. Further, convolution of the $\dot{M}_f$ dependence on L_{bol} shown in Fig. 2 with the stellar luminosity function (Allen 1973) gives the relative contribution of each stellar luminosity (spectral type) to the mass outflow rate. This is plotted in Fig. 3 and indicates that A-F protostars are the major contributors to mass outflow during star formation.

SC96 also plotted the outflow mass versus the dynamical age t_d, which is shown in Fig. 4. Here, the solid lines separate outflows associated with low mass, intermediate mass, and massive stars. One sees from this that within a given luminosity range, the outflow mass increases with age as one would expect and that flows driven by luminous stars are substantially more massive than those driven by lower luminosity stars of the

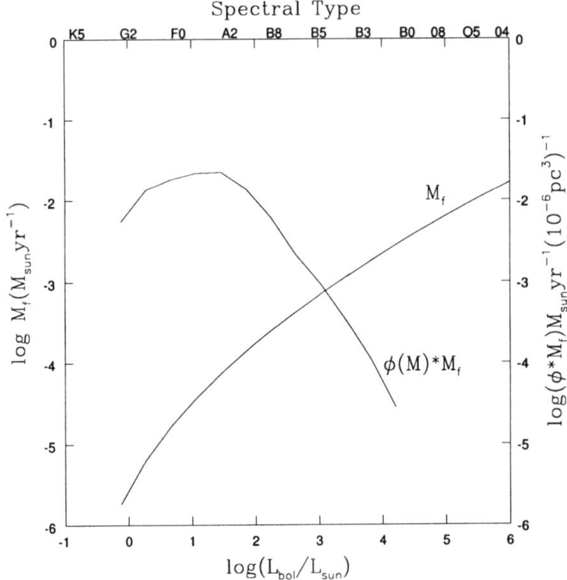

FIGURE 3. The empirical relationship between outflow mass M_f and L_{bpl} (spectral type) is labeled as M_f in this figure. To be completely valid, this relation should be plotted only for outflows with the same age, however, the number of objects for which one could make such a plot would be very small. Also, inspection of Table 2 shows that *most* of the outflows are within about a factor of 10 of the same age. Also in this figure is plotted the product of the stellar luminosity function $\phi(M)$ with the outflow mass $\phi(M)*M_\phi$. To the extent that including sources with a range of ages is valid, this curve implies that stars near F0 are the major contributors to mass outflow during star formation.

same age. For example, the Orion IRC2 outflow has almost 10^3 times as much mass as that of L1448C even though the L1448C outflow is about 10 times older than the Orion IRC2 outflow.

4. Issues raised by massive outflows

The large masses and energetics associated with young molecular outflows from forming massive stars raise some very puzzling questions. Among these are: How can several tens of solar masses of gas be accelerated to supersonic speeds on timescales of $\sim 10^4$ years and remain cold and neutral? What supplies the momentum to the outflows? What is the origin of the mass in the outflows? What is the driving mechanism of the outflows? Why is there no clear evidence of the outflow in the morphologies of UC HII regions? That is, why does one not see obvious disruptions in the shells of cometary or shell-like HII regions where the outflow has broken through the ionized shell? A good example of this is the shell-like UC HII region G5.89 with which a massive and very energetic molecular outflow has been identified. This HII region is shown in Figure 5. Although there is a low intensity extension of the nebula in the SE-NW direction, there is no evidence that the dense, inner, bright shell has been disrupted by a bipolar outflow. One possibility is that the outflow does not originate from the ionizing star of the HII region but from another massive star at an earlier evolutionary stage which has not yet been able to form a detectable HII region because of rapid accretion of matter onto the star. In the following section, I will consider possible origins of the large outflow masses and discuss their implications.

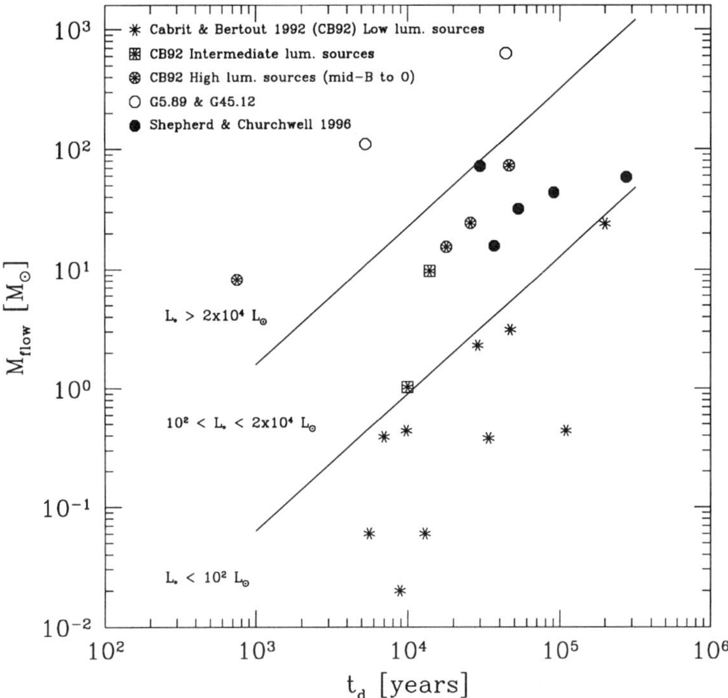

FIGURE 4. A plot of outflow mass versus dynamical age from SC96. The sloped solid lines arbitrarily separate outflows driven by low-mass ($L < 10^2\ L_\odot$), intermediate mass ($10^2\ L_\odot < L < 2 \times 10^4\ L_\odot$), and massive ($L > 2 \times 10^4\ L_\odot$) protostars. This figure shows that the outflows driven by protostars within a given luminosity range have increasing outflow masses as the dynamical age increases and outflows driven by luminous protostars are much more massive that those driven by low luminosity protostars of the same age.

5. Origin of the mass in massive outflows

Churchwell (1997b) noticed that outflow masses are often more massive than the mass of the central protostar believed to be responsible for driving the outflow (inferred from Fig. 2 in conjunction with the mass-luminosity relation). Possible mechanisms to achieve such massive outflows were discussed by Churchwell (1997b) which I will review here with some updating and illustrations. Four possible mechanisms are: accumulated stellar winds; entrained interstellar medium in bipolar jets; swept-up interstellar medium; and, in-falling matter deflected into bipolar outflows.

5.1. *Accumulated stellar winds*

This idea is illustrated in Figure 6 where a high mass-flux, bipolar stellar wind is blown away from the protostar in directions perpendicular to the accretion disk. The accumulated wind material plus whatever ISM has been accelerated by the wind is indicated at the ends of the outflows. The winds from O and B stars are radiation driven (Abbott 1985; Lamers and Groenewegen 1990; Pauldrach et al. 1990; Castor 1993). For stellar winds to supply an outflow of 50 solar masses over a typical period of 10^4 yr, the average mass loss rate would have to be about $5 \times 10^{-3}\ M_\odot\ yr^{-1}$. This is about two orders of magnitude greater than the highest mass loss rate observed toward O-stars and WR stars. A mass loss rate of this magnitude, if it is radiation driven, cannot be greater

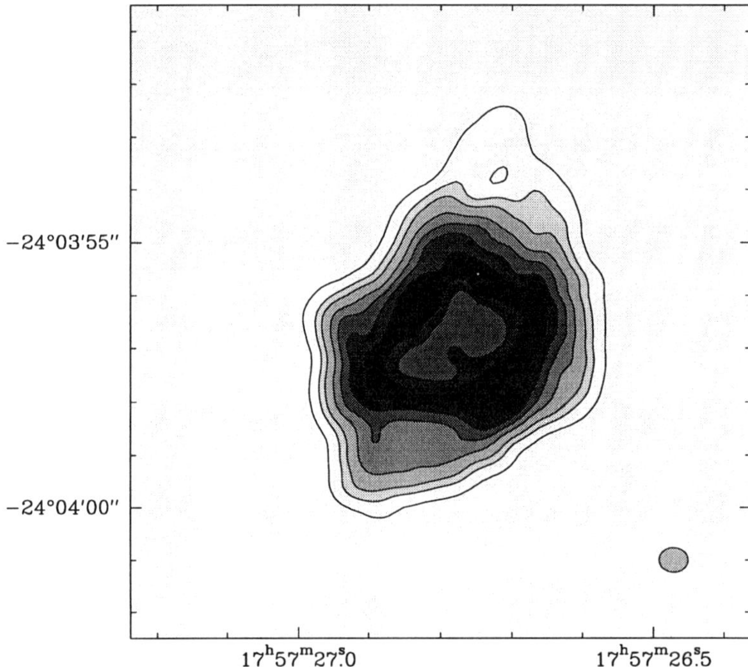

FIGURE 5. A 3.6 cm VLA image of the shell-like UC HII region G5.89–0.39 from Acord, Churchwell, and Wood (1998; ACW98). The synthesized beam (HPBW = $0.57'' \times 0.47''$) is shown in the lower right corner. The coordinates are RA(1950) and Dec(1950) and the contours are similar to those given in Fig. 1 of ACW98.

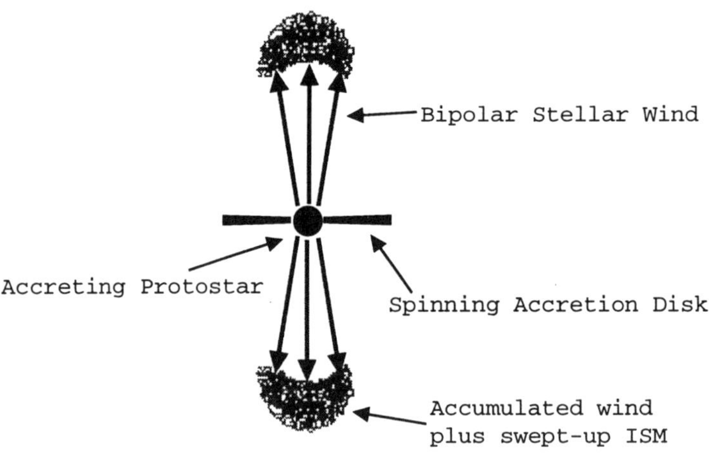

FIGURE 6. A schematic illustrating accumulation of stellar wind material from a polar wind. A segment of the accretion disk is shown edge-on.

than $(L_*/c)/v_\infty$ where L_* is the luminosity of the star, c is the speed of light, and v_∞ is the wind terminal velocity. This requires that the wind terminal velocities must be in the range 0.4 to 4 km s^{-1} for $L_* = 10^{5-6}$ solar luminosities. Such velocities are several times lower than observed molecular outflow velocities.

Mass accumulations could be greater if the dynamical lifetimes (determined from the outflow lobe separation divided by the outflow speed) underestimates the actual ages possibly because of confinement by ambient pressure. Inclination of the outflow axis to the line of sight foreshortens the lobe separation and the flow velocity in the same way so τ_d is independent of inclination effects. The typical pressure applied by massive outflows is 4–5 orders of magnitude greater than that of the ambient ISM. It therefore seems unlikely that the ambient pressure can significantly increase the lifetime beyond that implied by the dynamical age. The effective radiative momentum can be increased, and therefore mass-loss rates increased, if multiple scattering is important. Lucy & Abbott (1993) and Gayley, Owocki, and Cranmer (1995) have shown that multiple scattering can increase the mass-loss rate by as much as a factor of 10 in extreme cases (factors of 3 or less are more typical), but only in a spherically symmetric geometry and only when comoving opacities are large. In a bipolar outflow, multiple scattering is unlikely to be important because of scattering losses perpendicular to the flow axis. Thus, neither increases in τ_d by ambient pressure nor multiple scattering appears to be capable of increasing the accumulated outflow masses by large enough factors to account for the observed values. The fact that $\dot{M}_f \propto L_{\rm bol}^{0.7}$ (Fig. 2) also indicates that a wind cannot drive the outflows since radiatively driven winds have $\dot{M}_w \propto L_{\rm bol}^{1.7}$. The outflow masses are unlikely to be the result of accumulated stellar winds plus swept-up ISM, because they do not have enough radiative momentum.

5.2. *Entrained ISM in stellar bipolar jets*

We consider here a highly collimated, high speed, bipolar jet ejected along the spin axis of a rapidly accreting, massive protostar. It is not known whether the jets originate from the inner disk, the protostar, or both. A mixing layer at the interface of the jet walls with the ambient ISM entrains matter and accelerates the entrained matter in the direction of the jet flow. Also, some ISM will be accelerated ("swept-up") at the leading shock fronts ("working surfaces") of the jets. This scenario is sketched in Figure 7 below.

The issue here is how much ISM can be entrained in a bipolar stellar jet. The entrainment rate per unit area is

$$\dot{M}_e = \varepsilon \rho_0 c_0 \ , \tag{5.2}$$

(Cantó and Raga 1991) where ε is the entrainment efficiency ($\ll 1$), ρ_0 is the ambient ISM density, and c_0 is the speed of sound in the ambient ISM. For T = 20–50 K and $n_{H_2} \approx 10^5$ cm^{-3},

$$M_e = \varepsilon(0.9 - 1.4) \times 10^{-14} \text{ gm cm}^{-2} \text{ s}^{-1} \ .$$

If we take the central flow in G75.78NE as an example (see Shepherd, Churchwell, and Willner 1997), we have a lobe separation of 0.69 pc and an age of 3.7×10^4 yr. To estimate its area, an approximate cone shape is assumed. For a flow mach number in the range 10–20, the jet opening angle is expected to be $\theta = 2/\sigma \approx 0.04 rad = 2.3$ deg from experimental results for a "mixing layer limited" entrainment regime (see fig. 2 in Cantó and Raga 1991). In the above, σ is the "spreading parameter" which is inversely proportional to the opening angle. This gives a total surface area (both lobes) of about 1.5×10^{35} cm^{-2} and an entrained mass

$$M_f = R_e A_{\rm flow} \tau_d \approx \varepsilon(0.8 - 1.2) \text{ M}_\odot \ , \tag{5.3}$$

where $A_{\rm flow}$ is the area of the turbulent mixing layer of the jet and τ_d is the dynamical lifetime of the outflow. In the case of stellar jets with temperature $T_j \approx 10^4$ K and an ambient temperature $T_0 < 100$ K, Cantó and Raga (1991) find that the "mixing layer" limited regime is appropriate and that the entrainment efficiency in this regime is $\varepsilon < c_o/2c_j$ where c_o and c_j are the sound speed in the ambient medium and the jet, respectively. For the above conditions, $\varepsilon < 0.1$ and the total mass of entrained ISM

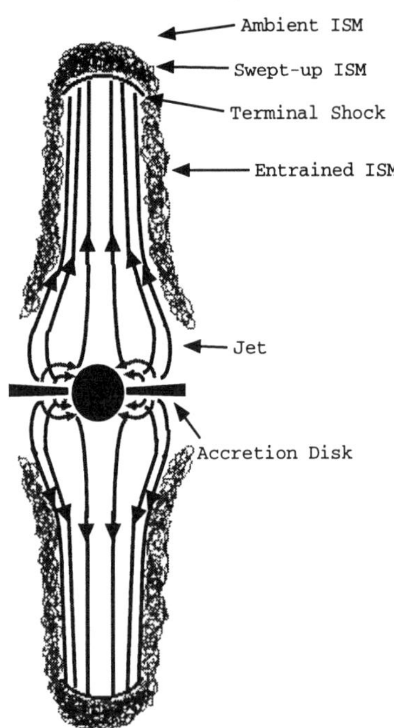

FIGURE 7. A schematic of entrained plus swept-up interstellar matter along the interface of a bipolar jet with the ambient interstellar medium. A segment of the accretion disk is shown edge-on with arrows showing matter going from the disk to the star and some arrows indicating a flow of matter into the jet from both the star and the inner accretion disk. It is not known where the jets actually originate (i.e. disk, star, or both).

would be $M_f < 1$ M$_\odot$. One could increase the entrained mass by assuming that the stellar mass loss occurs not as a narrow jet but as a flow of matter with much larger opening angles. For example, if $\theta = 30$ deg, $M_f \approx \varepsilon(13 - 20)$ M$_\odot \leq 2$ M$_\odot$. Thus, even in the large opening angle case, one can only entrain a few solar masses in outflows; certainly nothing like the several tens of solar masses observed.

Three caveats are in order here. One, the entrained mass is proportional to the third power of the lobe separation. An under-estimate of this parameter would increase the estimated entrained mass. On average this effect would under estimate the lobe separation by a factor of $\sqrt{2}$, which results in an average under estimate of M_f by a factor of $\sim 35\%$, not enough to account for the observed masses. Two, this whole analysis depends on the entrainment efficiency ε being small relative to unity. The conclusion of Cantó and Raga (1991) that ε is small is based on experimental data and theoretical arguments. If, however, the experimental data are not applicable to the conditions in the ISM, then the entrained mass could be substantially greater. Three, there is clearly evidence for entrainment of ISM in low-mass outflows and one will surely find similar evidence in massive outflows when the data become precise enough to measure it. However, the masses involved in low-mass outflows are typically a small fraction of a solar mass. Such masses are easy to account for by entrainment. It is much more problematic to account for 10s–100s of solar masses by entrainment if Cantó and Raga (1991) are correct. However, caution is in order because it is not absolutely clear that the experimental data on which their conclusions rest are applicable to the ISM.

5.3. *Swept-up ISM*

It is also possible to accelerate a large mass of ambient ISM into motion along an outflow axis if: 1) the density of the ISM in the immediate vicinity of a forming star is large; 2) the opening angle of the outflow is large (i.e. has a large working surface); and, 3) the interface of the outflow lobe (working surface) with the ISM is a reasonably continuous surface. Under these conditions, the outflow lobes could act like pistons or snowplows which sweep-up ISM along their leading shock fronts rather than sweeping aside the ISM as narrow jets tend to do. This scenario is sketched in Figure 8 below. ISM densities are large toward massive star formation (Cesaroni et al. 1991; Churchwell, Walmsley, and Wood 1992; Cesaroni, Walmsley, Churchwell 1992; Hofner et al. 1995; and others). Thus the main issues are whether the opening angles are systematically large for massive star formation and if the interfaces of the outflows with the ISM indeed form reasonably impervious surfaces or are leaky pistons. None of the highest resolution data currently available (for G75.78, SCW97; G45.12, HPM97; and G192.16, Shepherd et al. 1998) indicate well collimated outflows. The opening angles of the outflows in all three of these regions seem to be several tens of degrees. Unfortunately, all these sources are ≥ 2 kpc away and the best millimeter interferometer resolution does not match that of the best images toward Orion. Therefore, let us examine the Orion IRC2 shocked H_2 outflow (shown in Fig. 1) as a possible prototype of an outflow from a massive protostar. First, it is clear that this outflow has a wide opening angle. Second, the interface with the ISM is not a smooth surface. In fact, Wiseman and Ho (1996; 1998) have imaged NH_3 (1,1 and 2,2) emission toward the Orion IRC2 outflow and find long filaments (≥ 0.5 pc) of dense ammonia gas elongated along the outflow axis (see Figure 9). These filaments are density enhancements as confirmed by $^{12}C^{18}O$ observations. They find a systematic change in velocity and temperature of the filaments with distance from origin. These results are strongly suggestive that the elongation of the filaments is a result of the outflow stretching the dense molecular core condensations out into long filaments as the flow passes around the dense condensations. It appears that rather than sweeping-up dense ISM components, the flow simply sweeps around them while stretching them out into long filaments aligned with the outflow axis and possibly superimposing a weak velocity gradient on the filaments. Thus, while the data suggest that there is a strong interaction of the outflow with the ISM, it is not a simple "snowplow" type interaction, but rather a very leaky snowplow that is not particularly effective at sweeping up large masses of ISM. Much more work needs to be done before one can make a definitive statement on this issue. From the currently available data one can only get an impression of what might be happening, but there is certainly not enough data to be more quantitative or more certain about how effective wide-angle, massive outflows might be in sweeping-up large amounts of ISM.

5.4. *Accretion driven outflows*

Protostars, or more precisely their accretion disks, must shed angular momentum in order to accrete matter. The discovery of optical jets and radio bipolar molecular outflows associated with YSOs suggests that this is how they shed angular momentum and permit stars to form. The process by which accreting matter is diverted into outflows is not understood, especially for massive stars. Several models have been proposed (Pudritz 1988; Shu 1991, 1995; Shu et al. 1991; Shu et al. 1994; Fiege and Henriksen 1996a,b; Henriksen and Valls-Gabaud 1994; and others). In particular, Fiege and Henriksen (1996a,b) have developed protostellar bipolar outflow models in which in-falling material is diverted into bipolar outflows due to high central pressures produced by radiative and/or dissipative heating. They find that these models are consistent with radio observations of molecular

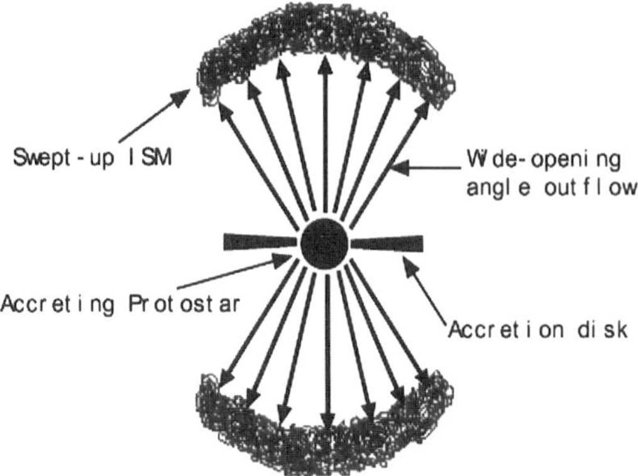

FIGURE 8. A schematic of swept-up interstellar matter due to a wide-angle outflow with a fairly continuous terminal shock surface.

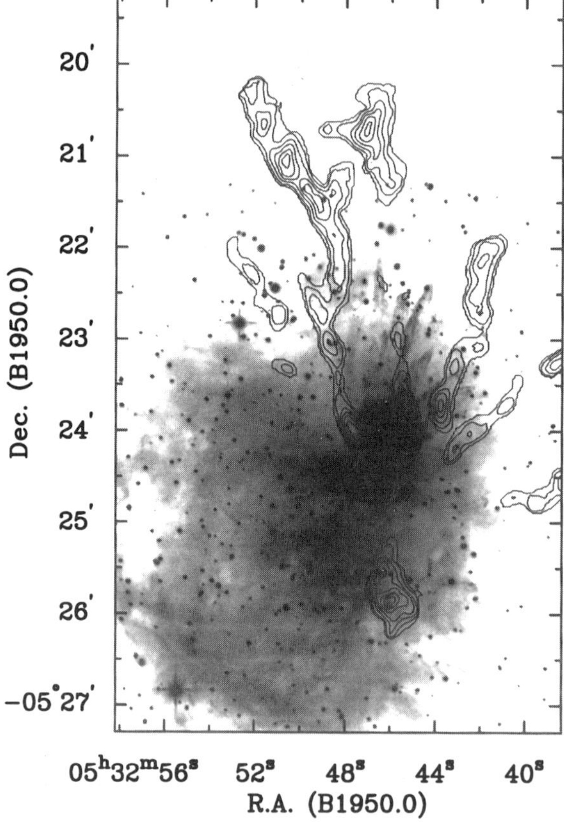

FIGURE 9. An overlay of NH$_3$ streamers (contours) from Wiseman and Ho (1998) onto the Allen and Burton (1993) shocked 2 μm H$_2$ outflow from Orion IRC2. This shows that all the streamers seem to be pointing toward the origin of the outflow and possibly shaped by the outflow flowing past the dense molecular clumps.

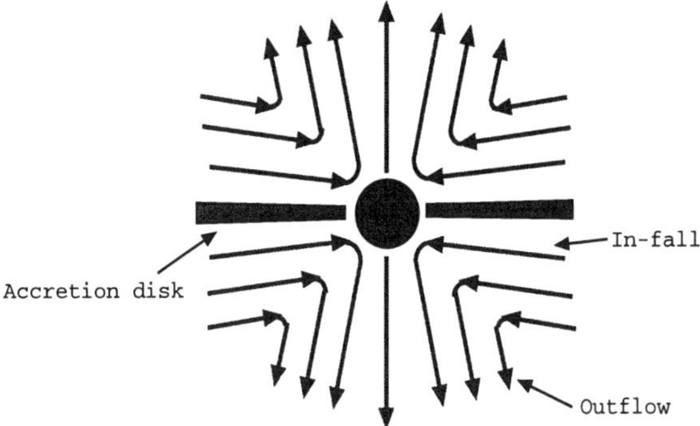

FIGURE 10. A schematic of in-falling matter diverted into a broad bipolar outflow. The physical mechanism for achieving this is not understood, although several models have been proposed (see text).

outflows. Tomisaka (1998) has developed a collapse driven outflow model which achieves a mass outflow rate $\sim 1/3$ the accretion rate for a Class 0 low-mass protostar. Since entrainment and swept-up mass do not appear to be able to account for the very large observed outflow masses, it seems likely that deflection of in-falling matter into bipolar outflows may be an important (perhaps primary) mechanism that feeds outflows of massive YSOs. This is illustrated in Figure 10.

Now, I will examine some of the consequences of this scenario. First, it implies that massive YSOs must go through a very rapid accretion phase with a mass accretion rate

$$\dot{M}_{\mathrm{acc}} > \dot{M}_f = 10^{-2}\ \mathrm{M}_\odot\ \mathrm{yr}^{-1}$$

$\dot{M}_{\mathrm{acc}}$ must be larger than $\dot{M}_f$ since some of the accreting matter ultimately ends up on the central star. $\dot{M}_{\mathrm{acc}}$ also places restrictions on the density structure of the accreting matter. Since the infall cannot exceed the free-fall rate, if gravity is the only inward force, we have:

$$\rho(r) \geq \frac{\dot{M}_{\mathrm{acc}} r^{-3/2}}{4\pi (2GM_r)^{1/2}}\ , \tag{5.4}$$

where r is the distance from the central protostar and M_r is the total mass (protostar plus disk and core material) interior to r. This implies a very rapid increase in density toward accreting protostars, consistent with observations (Hofner et al. 1995; Bronfman, Nyman, and May 1996).

Second, the outflow plus stellar mass gives the *minimum* mass that a molecular cloud core must have to form a star of a given mass. For example, the central outflow in the G75.78NE region is apparently driven by an early B-star (M_* about 10 M$_\odot$) and it has an outflow mass of about 58 M$_\odot$. Thus the fraction of the in-falling matter actually incorporated into the star is only about 15% and about 85% is deflected back outward into the outflow. In this sense, the efficiency of getting matter onto the central star is only about 15% and one concludes that a cloud core with a mass at least 6–7 times that of the final star would be required to form the star. However, we know that massive stars only form in clusters, so the initial cloud mass is likely to be ≥ 100 times that of the most massive star. This may also be part of the reason why massive stars are so rare; typical cloud cores generally do not achieve such large masses.

Third, accretion rates of 10^{-2} to 10^{-3} $M_\odot$ yr^{-1} will result in most of the stellar UV radiation being absorbed close to the central star, thus delaying the formation of a detectable UC HII region until after this phase of star formation. For example, a density greater than 10^8 cm^{-3} of in-falling matter toward an O5 star would result in a Strömgren sphere of radius ≤ 100 AU. This implies that the massive outflows observed to date are likely to be associated with YSOs that have not yet formed a detectable UC HII region. Those appearing to be driven by the central star of a UC HII region, such as G5.89, may either be relics of an earlier rapid accretion phase and are no longer being driven or are misidentified. This has the further implication that detection of massive YSOs in their rapid accretion phase or earlier can only be achieved via their molecular outflows or their thermal dust emission at FIR to mm wavelengths or high excitation molecular line probes, but not from radio free-free emission.

6. Summary

This review has concentrated on molecular outflows associated with massive star formation. The main conclusions are as follows:

1. Molecular outflows are found toward $\geq 50\%$ of massive star formation regions.
2. Outflow properties are:
   Mass outflow rates: 10^{-2} to 10^{-4} solar masses per year ($\dot{M}_f \propto L_{\rm bol}^{0.7}$)
   Outflow masses: 10s to 100s of solar masses
   Outflow ages: typically $\sim 10^4$ yr
   Collimation: poor
   Opening angles: wide (several 10s of degrees)
   Energetics of massive outflows ~ 10 times those of low-mass outflows

3. Massive outflows raise the interesting question of how several 10s of solar masses of material can be accelerated to supersonic speeds (relative to the ambient medium) over time scales $\leq 10^4$ yr and remain neutral (molecular) and cool.

4. Possible origins of the mass in massive outflows associated with massive star formation were considered. It was shown that radiation-driven stellar winds cannot accumulate the momentum nor the mass observed. It does not appear possible to entrain 10s to 100s of solar masses of material into outflows if the Cantó and Raga (1991) analysis is correct. Also, the Orion IRC2 outflow seems to indicate that rather than sweeping-up large masses of ISM, outflows simply sweep by denser condensations of ISM in their paths. These arguments seem to favor the possibility that massive outflows may be accretion driven, but do not by any means constitute proof of such.

5. The issues raised in this review will require much more observational underpinning to achieve further progress in their resolution. In particular details of the outflow morphologies, their velocity structures, and their relationship to accretion disks need to be studied in much more detail.

REFERENCES

ABBOTT, D. C. 1985. In *Radio Stars* (eds. R. M. Hjellming & D. M. Gibson). p. 61. Reidel.

ACORD, J. M., CHURCHWELL, E., & WOOD, D. O. S. 1998 *ApJ* **495**, L107.

ACORD, J. M., WALMSLEY, C. M., & CHURCHWELL, E. 1997 *ApJ* **475**, 693.

ALLEN, C. W. 1973. *Astrophysical Quantities, 3rd ed.* Athlone Press.

ALLEN, D. A. & BURTON, M. G. 1993 *Nature* **363**, 54.

ASPIN, C. & WALTHER, D. M. 1990 *A&A* **235**, 387.

BACHILLER, R. & CERNICHARO, J. 1990 *A&A* **239**, 276.

BRONFMAN, I., NYMAN, L.-Ä., & MAY, J. 1996 *A&AS* **115**, 81.

CABRIT, S. & BERTOUT, C. 1992 *A&A* **261**, 274.

CANTÓ, J. & RAGA, A. C. 1991 *ApJ* **372**, 646.

CASTOR, J. I. 1993. In *Massive Stars: Their Lives in the Interstellar Medium* (eds. J. P. Cassinelli & E. B. Churchwell). ASP. Conf. Proc. 35, p. 297. ASP.

CESARONI, R., FELLI, M., WALMSLEY, C. M., & OLMI, L. 1997 *A&A* **325**, 725.

CESARONI, R., WALMSLEY, C. M., & CHURCHWELL, E. 1992 *A&A* **256**, 618.

CESARONI, R., WALMSLEY, C. M., KÖMPE, C., & CHURCHWELL, E. 1991 *A&A* **252**, 278.

CHERNIN, L. M. 1996 *ApJ* **460**, 711.

CHOI, M., EVANS, N. J. II, & JAFFE, D. T. 1993 *ApJ* **417**, 624.

CHURCHWELL, E. 1990 *A&A Rev.*, **2**, 79.

CHURCHWELL, E. 1991. In *The Physics of Star Formation and Early Stellar Evolution* (eds. C. J. Lada and N. D. Kylafis). p. 221. Kluwer.

CHURCHWELL, E. 1995 *Ast. Sp. Sci.* **224**, 157.

CHURCHWELL, E. 1997a. In *Herbig-Haro Flows and the Birth of Low Mass Stars* (eds. B. Reipurth and C. Bertout). IAU Symp. No. 182, p. 525. Kluwer.

CHURCHWELL, E. 1997b *ApJ* **479**, L59.

CHURCHWELL, E., WALMSLEY, C. M., & WOOD, D. O. S. 1992 *A&A* **253**, 541.

DOWNES, D., GENZEL, R., BECKLIN, E. E., & WYNN-WILLIAMS, C. G. 1981 *ApJ* **244**, 869.

ERICKSON, N. R., GOLDSMITH, P. F., SNELL, R. L., BERSON, R. L., HUGUENIN, G. R., ULICH, B. L., & LADA, C. J. 1982 *ApJ* **261**, L103.

FIEGE, J. D. & HENRIKSEN, R. N. 1996a *MNRAS* **281**, 1038.

FIEGE, J. D. & HENRIKSEN, R. N. 1996b *MNRAS* **281**, 1055.

FISCHER, J., SANDERS, D. B., SIMON, M., SOLOMON, P. M. 1985 *ApJ*, **293**, 508.

GAYLEY, K. G., OWOCKI, S. P., & CRANMER, S. R. 1995 *ApJ* **447**, 296.

GARDEN, R. P., GEBALLE, T. R., GATLEY, I., & NADEAU, D. 1991a *ApJ* **366**, 474.

GARDEN, R. P., HAYASHI, M., GATLEY, I., HASEGAWA, T., & KAIFU, N. 1991b *ApJ* **374**, 540.

GARDEN, R. P. & CARLSTROM, J. E. 1992 *ApJ* **392**, 602.

HARVEY, P. M. & FORVEILLE, T. 1988 *A&A* **197**, L19.

HENRIKSEN, R. N. & VALLS-GABAUD, D. 1994 *MNRAS* **266**, 681.

HOFNER, P., KURTZ, S., CHURCHWELL, E., WALMSLEY, C. M., & CESARONI, R. 1995 *ApJ* **460**, 359.

HUNTER, T. R., PHILLIPS, T. G., & MENTEN, K. M. 1997 *ApJ* **478**, 283 (HPM97).

HUNTER, T. R., TAYLOR, G. B., FELLI, M., & TOFANI, G. 1994 *A&A* **284**, 215.

KURTZ, S., CHURCHWELL, E., & WOOD, D. O. S. 1994 *ApJS* **91**, 659.

KWAN, J. & SCOVILLE, N. Z. 1976 *ApJ* **459**, 638.

LADA, C. J. 1985 *ARA&A* **23**, 267.

LAMERS, H. G. J. L. M. & GROENEWEGEN, M. A. T. 1990. In *Properties of Hot Luminous Stars* (ed. C. D. Garmany). ASP. Conf. Proc. 7, p. 189. ASP.

LOREN, R. B. 1981 *ApJ* **249**, 550.

LUCY, L. B. & ABBOTT, D. C. 1993 *ApJ* **405**, 738.

MACLEOD, G. C., SCALISE, E. JR., SAEDT, S., GALT, J. A., & GAYLARD, M. J. 1998 *AJ*, in press.

MC CAUGHREAN, M. 1993. In *Massive Stars: Their Lives in the Interstellar Medium* (eds. J. P. Cassinelli and E. Churchwell). ASP Conf. Series No. 35, p. 80. ASP.

MCCAUGHREAN, M. & STAUFFER, J. R. 1994 *AJ* **108**, 1382.

MENTEN, K. M. & REID, M. J. 1995 *ApJ* **445**, L157.

MEYERS-RICE, B. & LADA, C. J. 1991 *ApJ* **368**, 445.

MITCHELL, G. F. & HASEGAWA, T. I. 1991 *ApJ* **371**, L33.

PANAGIA, N. 1973 *AJ* **78**, 929.

PAULDRACH, A. W. A., PULS, J., GABLER, R., & GABLER, A. 1990. In *Properties of Hot Luminous Stars* (ed. C. D. Garmany). ASP. Conf. Proc. 7, p. 171. ASP.

PUDRITZ, R. E. 1988. In *Galactic and Extragalactic Star Formation* (eds. R. E. Pudritz and M. Fich). p. 135. Kluwer.

SHEPHERD, D. S. & CHURCHWELL, E. 1995 *ApJ* **457**, 267 (SC95).

SHEPHERD, D. S. & CHURCHWELL, E. 1996 *ApJ* **472**, 225 (SC96).

SHEPHERD, D. S., CHURCHWELL, E. & WILLNER, D. J. 1997 *ApJ* **482**, 355 (SCW97).

SHEPHERD, D. S., WATSON, A. M., SARGENT, A. I., & CHURCHWELL, E. 1998 *ApJ*, in press (SWSC98).

SHU, F. H. 1991. In *The Physics of Star Formation and Early Evolution* (eds. C. J. Lada and N. D. Kylafis). p. 365. Kluwer.

SHU, F. H. 1995 *Rev. Mex. A. A. (Ser. de Conf.)* **1**, 375.

SHU, F. H., NAJITA, J., OSTRICKER, E., WILKIN, F., RUDEN, S., & LIZANO, S. 1994 *ApJ* **429**, 781.

SHU, F. H., RUDEN, S. P., LADA, C. J., & LIZANO, S. 1991 *ApJ* **370**, 31.

TAFALLA, M., BACHILLER, R., WRIGHT, M. C. H., & WELCH, W. J. 1997 *ApJ* **474**, 329.

THRONSON, H. A. JR., GATLEY, P. M., SELLGREN, K., & WERNER, M. W. 1980 *ApJ* **237**, 66.

WATSON, A. M., COIL, A. L., SHEPHERD, D. S., HOFNER, P., & CHURCHWELL, E. 1998 *ApJ*, in press.

WISEMAN, J. J. & HO, P. T. P. 1996 *Nature* **382**, 139.

WISEMAN, J. J. & HO, P. T. P. 1998 *ApJ*, in press.

WOOD, D. O. S. & CHURCHWELL, E. 1989 *ApJS* **69**, 831.

WOLF, G. A., LADA, C. J., & BALLY, J. 1990 *AJ* **100**, 1892.

WRIGHT, M. C. H., PLANBECK, R. L., VOGEL, S. N., HO, P. T. P., & WELCH, W. J. 1983 *ApJ* **267**, L41.

ZUCKERMAN, B., KUIPER, T. B. H., & RODRIGUEZ KUIPER, E. N. 1976 *ApJ* **209**, L137.

Observation and theory of the initial stellar mass function

By BRUCE G. ELMEGREEN

IBM Research Division, T. J. Watson Research Center, P.O. Box 218, Yorktown Hts, NY
10598; bge@watson.ibm.com

Observations of normal galactic star-forming regions suggest there is widespread near-uniformity in the initial stellar mass function (IMF) in spite of diverse physical conditions. Fluctuations may come largely from statistical effects and observational selection. There are also tantalizing, but uncertain reports that the IMF shifts systematically in peculiar regions, giving a low mass bias in quiescent gas, and a high mass bias in active starbursts. Theoretical proposals for the origin of the IMF are reviewed. The theories generally focus on a combination of four physical effects: wind-limited accretion of stellar mass, coalescence of protostellar gas clumps, mass limitations at the thermal Jeans mass, and power-law cloud structure. Hybrid theories combining the best of each may be preferred.

---

1. Observations

1.1. *Hints at a universal IMF in normal star-forming regions*

The initial stellar mass function (IMF) has been observed directly in dozens of clusters in our Galaxy and the Large Magellanic Clouds, and inferred indirectly for our Galaxy and other galaxies from metallicities, supernova rates, ionization levels, luminosities, dynamical masses, colors, and other tracers.

There is some convergence in these observations to an IMF that is approximately a power law above 1 $M_\odot$, having a slope between $x \sim -1$ and ~ -1.6 on a $\log N - \log M$ histogram (see reviews in Massey 1998 and Scalo 1998), with a flattening below this down to ~ 0.1 $M_\odot$ or lower. The power-law part is in rough agreement with the Salpeter (1955) function, which has a slope of $x = -1.35$ on such a plot, and also with the Scalo (1986) determination, which revised the often-used Miller & Scalo (1979) function at intermediate and high mass (the Miller-Scalo function is too steep).

Scalo (1998) recently emphasized the variations in the derived power law slopes from region to region, which amount to about ± 0.5. There are no obvious correlations between these variations and physical properties of the star-forming regions, including the cluster density, galactocentric radius, and metallicity (Scalo 1998). There are also "gaps" in the power law parts of some cluster IMFs, according to Scalo (e.g. NGC 103; Phelps & Janes 1993), and a "ledge" in the Orion IMF by Hillenbrand (1997). These features led Scalo to suggest that physical effects might be involved, making IMF variations larger than Poisson statistics.

An emphasis on the similarities, rather than the differences, in the IMF has been made by Massey and collaborators (see review in Massey 1998). They find an IMF power law slope that is fairly constant, around the Salpeter value, for clusters with a factor of ~ 200 range in densities (Massey & Hunter 1998; Luhman & Rieke 1998) and a factor of ~ 10 range in metallicities (Massey, Johnson & DeGioia-Eastwood 1995a; see also the globular cluster study by Bellazzini et al. 1995). This slope was also found by Kennicutt, Tamblyn & Congdon (1994) in galaxy-wide surveys, and by Bresolin & Kennicutt (1997) for individual HII regions in galaxies, using the equivalent widths of hydrogen emission lines, and it was found by Greggio et al. (1993), Marconi et al. (1995), Holtzman, et al.

(1997), and Grillmair, et al. (1998) from color magnitude diagrams of stars in the LMC and local dwarf galaxies.

Chemical evolution also tends to suggest an invariant IMF, at least at intermediate and high mass, in the stellar halo (Nissen et al. 1994) and disk (Tsujimoto et al. 1997; Wyse 1998), in QSO damped Lyα (Lu et al. 1996) and Lyα forest (Wyse 1998) lines, the intracluster medium (Renzini et al. 1993; Wyse 1997, 1998; however, see Loewenstein & Mushotzky 1996), and elliptical galaxies (see review in Wyse 1998). The observable in this case is the ratio of the iron to the oxygen abundance, which is always about the same. Since this ratio scales with the ratio of intermediate mass to high mass stars, the IMF must be about the same too. A different result was obtained by Heikkilä, Johansson & Olofsson (1998), who conjectured that a low $C^{18}O/C^{17}O$ ratio in the LMC results from a relatively low proportion of high-mass stars, as observed, for example, by Massey et al. (1995b); clearly the chemical and fractionation processes involved with this important isotope ratio have to be investigated further.

Gibson & Matteucci (1997) also obtained a different result from evolution and metallicity models of elliptical galaxies, including infall and wind ejection of gas. They suggested the IMF had a higher proportion of massive stars than the Solar neighborhood, as it appears to in some starburst models (discussed below). Angeletti & Giannone (1997) modeled elliptical galaxies and fit the mass-to-light and [Fe/H] ratios with an IMF that has a mass-dependent high mass slope, different from the Salpeter value.

The IMF in local OB associations might differ from the Salpeter value too. Brown (1998) reviewed observations, including those with distances from the Hipparcos satellite, which suggest a slightly steeper slope, ~ -1.7 to -1.9, for local associations, but the stellar types in these studies were obtained from photometry, and Massey (1998) discussed how such data can artificially steepen the IMF without spectroscopic observations too.

At the low mass end, there is no *systematic* indication yet of a turnover at masses lower than 0.1 M$_\odot$, but there are some regions where such a turnover has been reported, e.g. in nearby field stars (Reid & Gazis 1997a), Orion (Hillenbrand 1997), the Pleiades (Reid 1998), NGC 6231 (Sung, Bessell, & Lee 1998), 30 Dor (Nota 1998), and the globular cluster NGC 6397 (King et al. 1998). There are also regions where the IMF appears to rise below 0.1 M$_\odot$, again for local stars (Méra et al. 1996; Gould et al. 1997). Scalo (1998) suggests that the local field results at low mass contain uncertainties in the star formation rate, pre-main-sequence contribution, and mass-luminosity relation. The older cluster results are uncertain also because of evaporation of low mass members, although Reid (1998) suggests evaporation is not important for the Pleiades. In any case, Scalo (1998) and Reid (1998) both provide evidence that the low mass IMF, below 1 M$_\odot$, varies slightly in shape from region to region, although no sensible correlations with properties of the environment have yet been noticed. Variations in the low mass IMF are also seen for globular clusters, without any obvious dependence on metallicity (Santiago et al. 1996; Piotto et al. 1997; see review in Cool 1998), but there may be some dependence on the cluster evaporation rate and susceptibility to orbit shocking (Cool 1998), since low mass stars leave the cluster most easily. The IMF for the young open cluster mentioned above, NGC 6231 (Sung et al. 1998), is somewhat peculiar, because the turnover mass is around 2.5 M$_\odot$, much higher than in OB associations and other clusters. However, NGC 6231 is a dense cluster ~ 10 My old, and a high turnover mass in the central regions might be expected from rapid thermalization leading to mass segregation (e.g. Scalo 1986).

The low mass IMF is also probed by observations of embedded clusters, where most of the stars are pre-main sequence (Zinnecker et al. 1993; Fletcher & Stahler 1994a,b; Lada & Lada 1995; Lada, Alves & Lada 1996; Megeath 1996; Horner, Lada, & Lada 1997).

Pre-main-sequence (PMS) clusters have not yet evolved kinematically, so differential evaporation and segregation of high and low mass members should not be as much of a problem as it is for older clusters. PMS stars are also much brighter than they will be on the main sequence, so even stars less massive than the hydrogen burning limit can be observed in local clusters (see review in Lada, Lada, & Muench 1998). Lada et al. (1998) combined Herbig's (1998) optical data on the cluster IC 348, which was used to determine the history of star formation in the region, with their own K-band luminosity function to determine the low mass IMF. They found that the IMF in this cluster turns over below 0.25 $M_\odot$.

Recent searches for brown dwarfs (Zuckerman & Becklin 1992; Reid & Gizis 1997b; Reid 1998) suggest these objects are so infrequent that the IMF cannot continue to rise at low mass. The number of possible brown dwarfs is consistent with an extension of the flattening seen between 1 $M_\odot$ and 0.1 $M_\odot$ (Reid 1998).

1.2. *Starburst IMFs and the extreme field*

In spite of these approximate similarities in the IMF from region to region, there have been two clear and consistent deviations that, if real, would have important consequences for star formation. One is the apparent shift toward relatively higher masses in regions of extreme activity, such as starburst galaxies (e.g. Rieke et al. 1980), and the other is the apparent shift toward relatively lower masses in regions of extreme *inactivity*, the so-called "field" population, away from OB associations, as discussed by Massey et al. (1995b). Indeed, there may be a monotonic transition from low-mass weighted IMFs to high-mass weighted IMFs with increasing star formation rate, ISM pressure, radiation field, etc., perhaps because of a change in cloud properties, cloud destruction capabilities of young stars, and other things. The "normal" IMF occupies what seems to be a broad range of star formation rates between these two limits, but this broad range is not so broad in terms of *possible* ISM conditions. It may only be broad in the sense that it is common in main galaxy disks where some type of feedback between star formation and gas stability is maintained for long periods of time.

Reviews of IMFs in starburst galaxies are in Scalo (1990), Zinnecker (1996), and Leitherer (1998). The first observation that suggested a "top-heavy" IMF was that the luminosity of the starburst regions are so high, and the dynamical masses from the rotation curves so low, that there cannot be a full complement of low mass stars, which dominate the mass, to go with the high mass stars, which dominate the light (Rieke et al. 1980, 1993; Kronberg, Biermann, & Schwab 1985; Wright et al. 1988; see reviews in Telesco 1988, Joseph 1991, Rieke 1991, Thuan 1991). Thus the IMF contains a higher proportion of high mass stars than the Solar neighborhood. Doane & Matthews (1993) reached the same conclusion for M82 based on many properties of galactic evolution models, including the remaining gas mass, total mass, star formation rate, and supernova rate. Doyon, Joseph, & Wright (1994) and Smith et al. (1995) found top-heavy IMFs from spectroscopic line ratios for the starburst galaxies NGC 3256 and UGC 8387, respectively. Smith, Herter & Haynes (1998) found the same result from infrared excesses in 20 starburst galaxies.

The distribution of stars was often not resolved in the early starburst studies. Satyapal et al. (1995, 1997) derived a smaller extinction-corrected K band luminosity for the inner $30''$ of the starburst galaxy M82, and suggested the IMF could have the Salpeter slope all the way from 0.1 to 100 $M_\odot$ with only 30% of the dynamical mass in the star formation event. Ho & Filippenko (1996a,b) measured the velocity dispersions in two dense clusters in starburst galaxies, clusters that may evolve into globulars like those in the halo of our Galaxy (Meurer 1995). They found that the young globulars have large masses like old globulars, and that the young clusters would evolve into old clusters with a mass-to-light ratio like the old globulars after a Hubble time. Because of this, they suggested that the IMF might be the same in the young starburst globulars as it is in old globular clusters, and in particular, that it contains low mass stars (e.g. De Marchi & Paresce 1997).

Other considerations have led to revisions of the top-heavy models as well. Devereux (1989) observed 20 nearby starbursts like M82 with infrared photometry, and found no evidence for an unusual IMF in any of these; he pointed out that a revision of extinction corrections for M82 could change the Rieke et al. (1980) result too, as recently confirmed by Satyapal et al. (1997). Schaerer (1996) considered detailed evolutionary models of young stellar populations in starburst galaxies and also found that the IMF was normal. Calzetti (1997) modeled multiwavelength spectroscopy and broad-band infrared photometry of 19 starburst galaxies and found that the Salpeter IMF between 0.1 and 100 $M_\odot$ fits the observations. Stasińska & Leitherer (1996) modeled the emission line spectra of giant HII region and starburst galaxies, having a factor-of-ten range in metallicities, and found a Salpeter IMF up to 100 $M_\odot$; the lower mass limit could not be obtained, however.

The proposed peculiarity of starburst IMFs is therefore not certain, a conclusion reached by Scalo (1990) as well. Maybe only the extreme examples of starbursts require a high proportion of massive stars, while other starbursts, perhaps more "ordinary," do not. The exact form of the peculiarity in a starburst IMF is not known either. It could be "truncated" in the sense that the lower mass limit of the power law, where the power law slope begins to flatten, is higher than usual, perhaps as high as several $M_\odot$ instead of several tenths of a $M_\odot$. Such truncation may be observed already for 30 Dor (Nota 1998). Or, it could have a power-law slope that is shallower than -1.35, or possibly, a higher upper mass limit. These differences are often hard to distinguish, even though all of them affect the mass-to-light ratio. Fortunately, the exact form of the IMF sometimes affects other things as well. For example, Charlot et al. (1993) suggested that a truncated IMF will produce a very red population of red giants, without the corresponding main sequence stars, after the turnoff age reaches the stellar lifetime at the truncation mass. Wang & Silk (1993) suggested that the truncated model gives an oxygen abundance that is too high when the star formation process is over. Presumably these other constraints will help narrow down the possible choices.

The extreme field studied by Massey et al. (1995b) is another example where the Salpeter IMF seems to break down. In regions of the LMC and our Galaxy that are more than 30 parsecs from catalogued OB associations, Massey et al. found O-type stars that probably did not drift there during their short lifetimes. Massey et al. showed with an H-R diagram that there was an approximately steady rate of star formation there for the last 10 My years, and they derived a time-averaged IMF. The result was very steep, with $x \sim 4$ instead of ~ 1.35 for a Salpeter function.

1.3. *Other IMF oddities*

There are several other peculiarities of the IMF that may or may not have a specific physical origin. For example, it has been known for a long time (Larson 1982; Myers & Fuller 1993) that high mass clouds produce high mass stars, and low mass clouds produce low mass stars. Two decades ago, this observation and others were related to the concept of a bimodel IMF (Eggen 1976, 1977; Elmegreen 1978; Güsten & Mezger 1983; Larson 1977, 1986), and to the impression that spiral arms preferentially formed massive stars (Mezger & Smith 1977), presumably because the molecular clouds in spiral arms are more massive, or somehow different from interarm clouds. Today, we know that massive clouds also produce low mass stars in great abundance (Herbig & Terndrup 1986), and

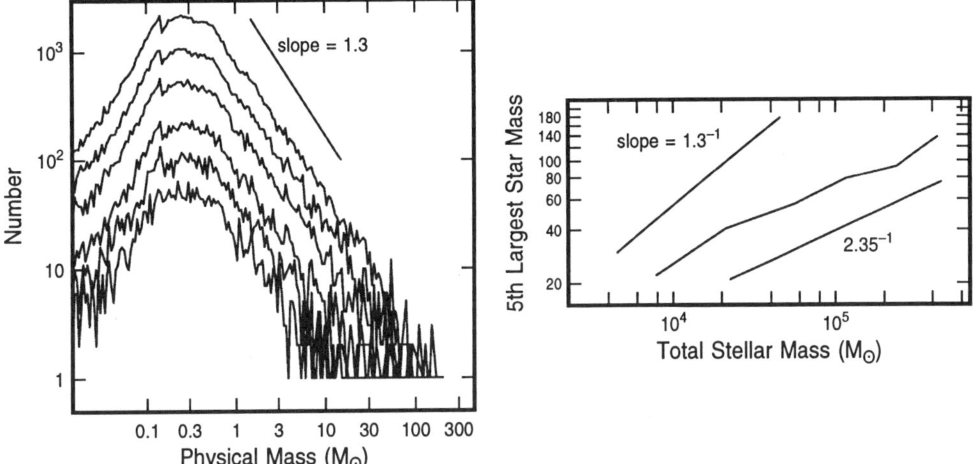

FIGURE 1. Model IMFs with different numbers of stars showing an increase in the maximum stellar mass with increasing number of stars. The figure on the right shows the mass of the fifth largest star versus the total stellar mass, with an expected slope equal to $1/x = 1/1.3$ and a measured slope equal to $\sim 1/2.35$.

there are essentially no local regions that produce high mass stars exclusively. Thus the "observation" that the maximum mass of a star correlates with the mass of the cloud could be only a statistical sampling effect (Larson 1982; Elmegreen 1983; Walter & Boyd 1991; Massey & Hunter 1998): high mass clouds produce more stars of all types, and so they are more likely to sample at the tail of the IMF to produce a few massive stars. In the same sense, high mass clouds should also produce more dwarf stars than low mass clouds.

An IMF model based on random sampling from a hierarchical cloud (Elmegreen 1997a) shows this effect clearly in figure 1. Each curve on the left is from a different cluster, with a number of stars in the clusters equal to 2500, 5000, 10000, 25000, 50000, and 100000. As the number of model stars increases, the mass of the largest star increases too, as shown on the right of the figure. The expected slope of the power law relation between cluster mass, M_{cl}, and maximum star mass, M_{max}, is $1/x$, i.e. $M_{max} \sim M_{cl}^{1/x}$; actually the calculation gives a slightly steeper slope because of storage limitations to the number of hierarchical levels allowed in the simulation (these models use 8 levels with an average of 3 subclumps per clump at each level). This property of increasing maximum star mass with cloud mass should not depend on any specific IMF model or on the star formation process.

To be more precise about the stochastic interpretation of this effect, we can consider an IMF $n(M)dM = n_0 M^{-1-x} dM$, which gives the Salpeter function when $x = 1.35$. The largest star, with mass M_{max}, typically satisfies $\int_{M_{max}}^{\infty} n(M)dM = 1$. This gives $n_0 = x M_{max}^x$. Because the total cluster mass is $M_{tot} = \int_{M_{min}}^{M_{max}} M n(M)dM$, the largest star is related to the total cluster mass as

$$M_{tot} = \frac{x}{x-1} M_{max} \left[\left(\frac{M_{max}}{M_{min}} \right)^{x-1} - 1 \right]. \tag{1.1}$$

For $M_{max} \sim 10 M_{min}$ and $x = 1.35$, this gives $M_{tot} \sim 4.8 M_{max}$, and for $(M_{max}/M_{min})^{0.35} \gg 1$, this gives $M_{tot} \sim 3.9 M_{max}^{1.35}/M_{min}^{0.35}$. Thus, a cluster requires several hundred M_☉ before a small O-type star is likely to appear, and a cluster with 10^3 M_☉ is likely to contain ~ 10 modest O-type stars or a few very massive ones. This is the mass range for the largest open clusters in the Milky Way disk, indicating that when a few O stars appear, a cluster is likely to terminate its formation in an unbound state, presumably as a result of rapid cloud destruction, and make an expanding OB association (Elmegreen 1983).

Larson (1991, 1992) and Khersonsky (1997) interpret this observation in the opposite way, saying that there is a physical effect which tends to produce higher mass stars in clouds of higher mass, and that the IMF partly results from this physical effect. Larson (1992) attributes the physical effect to a larger accretion length in larger clouds for material going into a star. If there is such an effect, and the observation is not entirely statistical as discussed above, then the superposition of IMFs from clouds of different mass would be different from the IMF in any one cloud. That is, clouds of low mass would produce an IMF with a particular slope up to some maximum mass, and clouds of higher mass would produce an IMF, perhaps with the same slope, up to higher masses. The sum of these two IMFs would be an IMF with greater slope than either separate cloud, because the summed IMF contains low mass stars from both clouds, but only high mass stars from the more massive cloud. Larson (1991) proposed his model by saying, in effect, that only one star forms in each cloud fragment at each level in the hierarchy, and that the star mass is always proportional to the square root of the fragment mass. Then the sum over all fragments is the final IMF, and this has a slope of $x = 2$, which is not too bad. Stated this way, there is no problem with the model because each component in the sum is only a single star, not a whole cluster of stars with a separate IMF. However, even in this model, if we add together the stellar populations from many separate clouds with different total masses, then the summed IMF will be steeper than the IMF in each. At this point, it is important to recall that the IMF obtained from large-scale surveys of galaxies and their metallicities is about the same as the IMF observed in individual clusters. Thus the summed IMF from many regions of star formation has to be about the same as the IMF from each separate cluster. This would seem to suggest that the observed increase in maximum star mass with cloud mass is not the result of a physical process that specifically increases the upper stellar mass in larger clouds.

A second peculiarity about star formation is that the high mass stars tend to form after the low mass stars (Herbig 1962a,b; Iben & Talbot 1966). This has been determined most recently for 30 Dor (Massey & Hunter 1998). Again one can think of physical reasons for this, such as a gradual warming of the cloud following the formation of low mass stars, and an increase in the Jeans mass with temperature (Silk 1977; Yoshii & Saio 1985), but statistical effects can explain it too.

The decreasing nature of the IMF at high mass implies that massive stars are likely to form only after a lot of low mass stars have already formed. For a constant star formation rate, the average time between the formation of stars in a logarithmic mass range centered on M is proportional to the rate at which these stars form, which is inversely proportional to the relative number of the stars, or $(Mn[M])^{-1} \propto M^x$. For a constant star formation rate, this is also the average time after star formation begins for the *first* appearance of a star with this mass. Thus the most common, low-mass stars form first, followed by the intermediate and then the high mass stars. In terms of the proportion of all stars formed, and for a constant star formation rate in units of mass per year, the maximum stellar mass increases with time as

$$\int_{M_{min}}^{M_{max}(t)} Mn(M)dM = At \qquad (1.2)$$

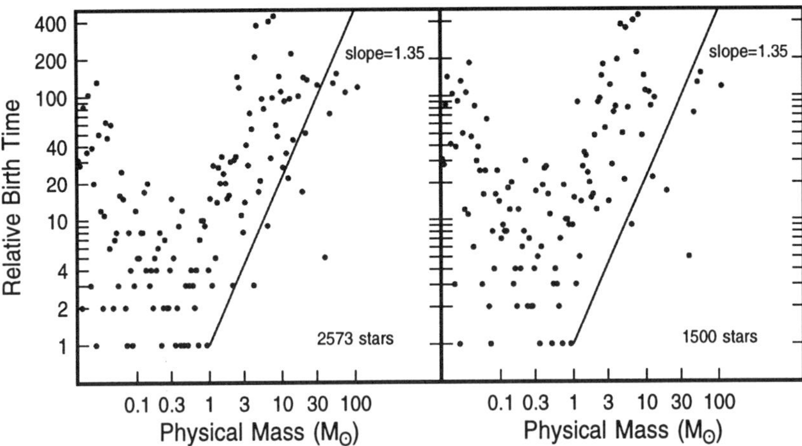

FIGURE 2. The relative birth time for the first appearance of a star with a certain mass is shown as a function of mass for a composite of 11 clusters on the left, containing 2573 stars total, and 6 clusters on the right, containing 1500 stars. The relative birth time is measured in units of the time of first appearance of any star in the model. The solid line has a slope of 1.35.

for star formation rate A. Thus, a star of mass $M(t)$ forms after the proportional time given by

$$\frac{t}{t_{max}} = \frac{\int_{M_{min}}^{M(t)} M n(M) dM}{\int_{M_{min}}^{M_{max}} M n(M) dM}.$$ (1.3)

For the power law portion of the IMF only, with $M_{min} = 1$ M$_\odot$ and $x = 1.35$, this becomes, for masses in M$_\odot$,

$$\frac{t}{t_{max}} = \frac{1 - M(t)^{-0.35}}{1 - M_{max}^{-0.35}}.$$ (1.4)

For example, in a cluster with a maximum stellar mass of 30 $M_\odot$, stars with 10 M$_\odot$ form in the last 20% of the time.

Note that this increase in time with stellar mass is only for the first appearance of a star with that mass. The *average* time of appearance is independent of stellar mass in the stochastic interpretation. This suggests a way to check the stochastic model of birth orders. If the average time of appearance of all stars with mass M is the average age of the cloud, and this is true for any M, and if the first time of appearance of a star with mass M increases with time, as described above, then the birth order is dominated by stochastic effects.

Figure 2 shows the relative birth order of stars in the random sampling IMF model by Elmegreen (1998). Each point marks the earliest time of formation of a model star with that mass, among all of the stars that formed in 11 clusters on the left and 6 clusters on the right. Each cluster contains 200 to 500 stars and is independently generated. The plot shows a distribution of points that traces the inverse of the IMF in log intervals. The fiducial line has a slope of 1.35 for comparison. The increase at low mass occurs because the model uses a Gaussian probability for failure to form a star at low mass, which is an arbitrary assumption at this time (it causes the model IMF to turn over at low mass).

There are no quantitative measures yet of how the earliest appearance of a star with a particular mass increases with time relative to the birth order of all stars, so the predictions of theory cannot yet be checked. In a real star-forming region, the star formation rate is probably not constant in time, so the ordinate in figure 2 should not be time, but star number in its proper birth order; i.e. one should plot, for the first appearance of a star of mass M, its mass versus the number of stars that formed before it.

An interesting implication of the above two discussions is that *massive stars must not be able to severely limit or halt star formation in their primordial clouds*. If they did, then, considering their relatively late appearance, each cloud would produce stars with the same local IMF up to some maximum stellar mass at which point the cloud is finally destroyed. This maximum stellar mass at the time of cloud destruction is presumably larger for more massive clouds, because it is harder to destroy a massive cloud than a low mass cloud. Thus the summed IMF from many regions of star formation would be steeper than the local IMF in each region, for the reasons discussed above, and we could not explain why the integral IMF in a galaxy is the same as the cluster IMF. Perhaps massive stars only shred the low density parts of these clouds, allowing (or perhaps stimulating), most of the cores that have already formed in these regions to continue their evolution toward stars. In view of this, it seems that stars of all mass are probably equally likely to form in any cloud (that contains at least this much mass), regardless of what other stars have formed before them. Moreover, the maximum stellar mass is not physically limited, it is only statistically limited in the sense that extremely massive stars are rare. An exception to this conclusion may arise in the "field" population studied by Massey et al. (1995b), in which the summed IMF of presumed dispersed clusters appears to be steeper than the IMF in each. Maybe massive stars can more easily disrupt their clouds in low pressure "field" environments. On the other hand, maybe the Massey et al. observation can be explained in other ways.

A third peculiarity of the IMF is that massive stars are often closer to a cluster center than low mass stars. This effect has been discussed for a long time (e.g. Sagar et al. 1988), but it was never certain whether the observations reflected a birth position, or just a relaxation effect after the cluster formed. Thermal relaxation makes the high mass stars sink to the center of the cluster, on average, in a thermalization time scale. However, Hillenbrand & Hartmann (1998) and Bonnell & Davies (1998) have recently shown that this peculiar mass distribution, which is observed in the Trapezium cluster, could not result from thermalization because the Orion cluster is too young (see also Jones & Walker 1988). Similar mass segregation has been observed in NGC 2157 (Fischer et al. 1998). Thus it is an effect that should be explained by an IMF theory, and it is likely to result from a physical process during star formation rather than stochastic sampling.

There have been many attempts to explain this relative birth position (see reviews in Larson 1991, and Zinnecker, McCaughrean & Wilking 1993), including excess gas clump collisions in the center of a cloud (Larson 1990; Stahler, Palla, & Ho 1998), excess gas accretion onto protostars in the cloud core (Larson 1978; Larson 1982; Zinnecker 1982; Bonnell et al. 1997), enhanced gas drag for massive protostars (Larson 1990, 1991), late-time formation of high mass stars in a collapsing cloud (Murray & Lin 1996), and hierarchical cloud structure in which the massive trunk of a hierarchical tree for the gas cloud is closer to the center of the tree than the low-mass branches (Elmegreen 1998). A combination of these effects might be involved.

1.4. *Summary*

There are evidently four distinct physical effects that have to be explained by a theory of the IMF:

(1) the power-law slope at intermediate to high stellar mass, with a value close to that found by Salpeter (1955);

(2) the flattening at low mass for clusters and star-forming regions in our Galaxy;

(3) the increase in the transition mass between power-law and flat regions of the IMF in starburst galaxies,

(4) the preferential birth of high mass stars close to cluster centers.

There are apparently three additional observations that can be explained by stochastic sampling effects:

(1) the tendency for the largest mass star in a region to increase with the total number of stars;

(2) the tendency for the largest mass stars to form last, and

(3) the seemingly random variations in the intermediate and high mass IMF slopes from region to region, along with the appearance of gaps, ledges, and other peculiar departures from a power law.

In addition to these observations, there are tentative indications that other IMF features may be present, including:

(1) a turnover at low mass, and

(2) a steepening of the power-law part in field regions.

Future observations should be able to illuminate these uncertain features.

2. What determines a star's mass?

2.1. *A steady stream of theories*

The theory of the IMF depends on both the theory of star formation and the theory of cloud structure. Because of the complexity of these issues, the IMF may not be understood for a long time. Here we review some recent developments. Previous reviews were in Cayrel (1990) and Clarke (1998).

One of the key problems in understanding the origin of the IMF is to determine the processes that limit the pre-stellar gas accretion onto a star and define the star's final mass. There have been several ideas on this.

The final mass could result from the star's own ability to limit the accretion of new gas onto its surface, perhaps because of an intense proto-stellar wind (Larson 1982; Shu, Adams, & Lizano 1987), which is known to exist at this phase of evolution (Lada 1985). This possibility has led to several theories of the IMF. One, proposed by Nakano, Hasegawa, & Norman (1995), considered three mass-limiting agents: protostellar winds, ionization, and depletion of the gas clump in which the star forms. They concluded that under normal conditions, the protostellar wind would limit the accretion. Then they calculated the fraction of the clump that would get into the star, finding that the final star mass scaled with the clump mass to the 7/6 power. To get an IMF slope of $x = 1.7$, which they got from the field-star IMF in Scalo (1986), they required a very steep clump mass spectrum, $n(M_{clump}) \sim M_{clump}^{-3}$ in linear intervals, instead of the usual observation of $M_{clump}^{-\alpha}$ with $\alpha \sim 1.5 - 1.8$ (Blitz 1993; Kramer et al. 1998). They justified this steep clump spectrum by noting that if the linewidths found in a CS survey of Orion clumps (Tatematsu et al. 1993) were assumed to be virial velocities, and the masses calculated accordingly, then the Orion CS clumps would have such a spectrum.

Another IMF model with wind-limited stellar masses was proposed by Adams & Fatuzzo (1996). They conjectured that the stellar luminosity scales with the wind mass loss rate, which, at the time when the accretion stops, is proportional to the direct accretion rate onto the star, independent of what goes onto the disk. This gave them a relation between the stellar properties, i.e. luminosity and mass, and the gas cloud properties, sound speed and angular rotation rate. The angular rotation rate entered because they had to determine the fraction of the accreting gas that goes into the star and not the disk. Finally, they related the stellar mass to the total luminosity of the star, from the sum of the accretion-driven luminosity, which depends on mass, and the luminosity of a main sequence star at intermediate mass, from the standard mass-luminosity relation. The result is a relation between stellar mass and the cloud properties, i.e. sound speed and rotation rate. Taking the mass distribution for clumps with the observed slope α discussed above, and the empirical relationship between clump mass and velocity dispersion from Larson's (1981) laws, they got a distribution function for the clump velocity dispersion, which then led to a distribution function for the final star mass if the angular rate of all the clumps is the same. Adams & Fatuzzo (1996) also discussed more general equations giving the stellar mass from cloud properties, considering random variations in these cloud properties, and derived a log-normal final mass distribution, as in Larson (1973), Elmegreen & Mathieu (1983), Zinnecker (1984), and Elmegreen (1985).

Silk (1995) also got a relation between cloud core mass and turbulent linewidth, different from Adams & Fatuzzo's, considering centrifugally supported cores with luminosities equal to the accretion luminosities. He considered an equality between the turbulent linewidth and the thermal velocity in the core, which entered into the luminosity through the temperature using the usual radiative transfer equation for a star and the Rossland mean opacity. Then, after substituting the empirical correlations between clump rotation rate and size, and between linewidth and size, he got the cloud core mass as a function of the turbulent linewidth. The distribution function for cloud core mass then followed from a distribution function for linewidth, which came from a theory for the time-dependent deceleration of wind-driven bubbles in a cloud. The star mass is then taken to be proportional to the cloud core mass. Silk (1995) made several comments that protostellar outflows limit the mass of a star, but this assumption was not present in any of the theory in his paper, nor in the IMF that resulted, particularly considering that a fixed fraction of the cloud core mass was assumed to go into the star. This was unlike the results of the wind-limited accretion models proposed by Nakano, Hasegawa, & Norman (1995) and Adams & Fatuzzo (1996).

One of the most popular and persistent methods for determining the mass of a star has been with a combination of cloud fragmentation, accretion, and clump collisions. Larson (1978) showed with three-dimensional N-body experiments that gas clouds fragment hierarchically by self-gravity, and the fragments accrete material in competition with each other. The resulting mass distribution was modeled after a fractal, which was a remarkably prescient concept in astronomy considering that Mandelbrot (1977) began to popularize his fractal geometry only a year earlier. Larson showed that the fractal mass function would have a slope of about $x = 1$, in reasonable agreement with observations. Larson also reasoned that the lower mass limit to a star is the thermal Jeans mass in the cloud because smaller condensations are not likely to form in the initial collapse. This experiment was an important departure from standard fragmentation models of the time, since it was previously believed, following Hoyle (1953) and Hunter (1962), that fragmentation decreased the Jeans mass, leading to ever more fragmentation. In Larson's result, the number of clumps formed was about equal to the number of Jeans masses in the initial cloud, with no subfragmentation into smaller pieces. This result was seconded by Tohline (1980), but investigated again by Silk (1982), who concluded that initial cloud fragments, particularly elongated fragments, formed by dynamical collapse could in fact fragment again (see also Bonnell & Bastien 1993, Burkert et al. 1997).

Silk (1977) considered a different picture of cloud fragmentation, using a thermal Jeans mass that increased with time as a result of heating from the stars that already formed. His model of fragmentation was simpler than Larson's, so there was no built-in, power-law mass spectrum from the fragmentation process itself. The power law in Silk's model came from the time increase in the thermal Jeans mass and the identification of this mass with the stellar mass at all times. Yoshii & Saio (1985) followed this model by considering the additional influence of coalescence among the opacity-limited fragments. They showed that the opaque fragments are usually so small that coalescence is unimportant; thus fragmentation and heating alone determine the IMF as a time sequence of increasing thermal Jeans masses, as in the Silk (1977) model. Bastien (1981) obtained a different result by using the Jeans length to determine the fragment collision cross section. The initial Jeans length is much larger than the size of the opaque fragments considered by Yoshii & Saio (1985), so Bastien (1981) found that fragment collisions were important. Subsequent work by Lejeune & Bastien (1986), and Allen & Bastien (1995, 1996) considered time-dependent coalescence, and concluded again that both fragmentation and coalescence were important, with coalescence dominating the formation of massive stars.

Price & Podsiadlowski (1995) used protostar collisions in a different way. They proposed that stars grow by accretion at some more-or-less uniform rate, and this accretion stops when two protostars collide, disrupting the gas reservoirs from each. The final stellar mass was then determined by the product of the accretion rate and the time interval between collisions. The IMF was built up over time as the final stellar mass decreased in the presence of an increasing collision rate that resulted from the continuous formation of more and more protostars.

Another coagulation model was proposed by Murray & Lin (1996). They suggested that fragmentation driven by thermal instabilities leads to clumps that fall in the potential well of a cloud, after which they collide and coalesce with other clumps that have fallen too. A star forms when the clump mass exceeds the thermal Jeans mass, although further coalescence can increase that mass afterwards. The power-law distribution of masses follows from the coagulation process, in a manner similar to that proposed by Silk & Takahashi (1979). There is no assumption about wind-limited accretion here; the star mass is identified with the core mass directly. An interesting aspect of their model is that they considered positionally correlated velocities, as in a turbulent fluid, but they showed this had no important effect on the model IMF.

Numerical SPH simulations of an evolving IMF made of accreting, pseudo-star particle clumps was done by Bonnell et al. (1997). The advantages of numerical solutions like this is that they can treat well the competition for gas among all the nucleated centers. They found that nucleating centers originally close to the center of the whole cloud grow faster and to larger masses because of the larger gas density there.

2.2. *An IMF from random selection of mass in hierarchical clouds*

A different class of theory considers only the statistical aspects of the IMF. The Hoyle (1953) picture of fragmentation led naturally to these ideas, because successive fragmentation produces hierarchical structure and power law mass spectra regardless of many physical details. Larson (1973) considered a modification to these ideas by proposing that only part of each fragment undergoes further fragmentation, and got from this a log-normal mass distribution instead of a power-law. When Miller & Scalo (1979) proposed that the IMF actually had a log-normal form, the statistical implications of this were developed further by Elmegreen & Mathieu (1983), Zinnecker (1984), and Elmegreen (1985), who showed that a variety of processes, acting together, would combine to make a log-normal function, regardless of the distribution functions resulting from each separate process. This point was made again more recently by Adams & Fatuzzo (1996).

The most recent development in this area has been by Elmegreen (1997a, 1998), who considered the random selection of masses from a hierarchically structured cloud. These papers depart from the usual scenarios by asserting that hierarchical cloud structure is independent of star formation—that it is set up long before star formation begins in the diffuse cloud stage, and then continuously reestablished during the molecular cloud stage as a result of supersonic turbulence compression. There is no gravitationally-driven fragmentation at all, and no significant clump coalescence. In fact, the clumps in this model need not even be constant objects, able to move around and coalesce; they can be ever-changing gas compressions and wavepackets in the chaotic turbulent flow. This model is motivated by the pervasive appearance of hierarchical structure and spatially-correlated velocities in interstellar clouds. These features are reminiscent of structures and flows in laboratory turbulence (e.g. see reviews in Sreenivasan 1991, and Falgarone & Phillips 1991).

The basic point of the Elmegreen (1997a, 1998) model is that essentially all local star formation processes that have been considered in detail have a time scale for evolution that varies approximately as the inverse square root of local density, including contraction or collapse from self-gravity, turbulence compression, magnetic diffusion in virialized cores, and coalescence at each level in the hierarchy. This means that in a model like this, stars will appear here and there, randomly, alone or with neighbors, at a rate that depends almost exclusively on the local square root of density. On a large scale, this implies that dense clouds in high pressure regions of galaxies will form all of their stars quickly, while low density clouds in low pressure regions will take much more time. On a small scale, within any one cloud, it means that different parts of the hierarchy form stars at different times. The density always increases at lower levels in a hierarchical structure, and this is where the clumps, contained by other clumps on larger scales, are small and have low mass. Thus the low mass pieces in a cloud are likely to form stars first. As a result, there is slightly less mass for other stars that form later in the same or higher levels. This competition for mass tends to steepen the power in the power-law mass function by several tenths, i.e. from $x = 1$ to $x = 1.35$, as in the Salpeter IMF.

Perhaps a more important difference in this new model is the way in which cloud structure is measured. Typically, the power law index for molecular and diffuse cloud structure is determined from the mass distribution of separate clumps, resolved in large-scale surveys by various telescopes and then separated into discrete objects either by eye or by computer algorithms that do about the same thing as the eye. These mass distributions are always much flatter than the IMF, having slopes α in the range from 1.5 to 1.8, when the IMF has a slope of about $1 + x = 2.35$. To understand the IMF, however, we have to try to view interstellar clouds from the perspective of a forming star. A star does not care about the clumps that our telescopes resolve and our eyes choose to label as discrete, but only about the general distribution and motion of gas in all forms. In pre-star-forming clouds, where this structure is first established, it is largely hierarchical, from scales that are much smaller than mm-wave telescopes can resolve, up to perhaps a galactic scale height.

Hierarchical means that most of the small clouds are contained inside larger clouds, and most small stellar groupings are inside larger stellar groupings (see review in Elmegreen & Efremov 1998). For example, Efremov (1995) and colleagues have estimated that 90% of the OB associations in the Milky Way (Efremov & Sitnik 1988), M31 (Efremov, Ivanov, & Nikolov 1987; Battinelli 1991, 1992; Magnier et al. 1993; Battinelli, Efremov & Magnier 1996), M33 (Ivanov 1987, 1992), and the LMC (Feitzinger & Braunsfurth 1984) are inside larger star complexes. Scalo (1985) has reviewed hierarchical structure for interstellar clouds. It appears as if the hierarchical embedding of interstellar and young stellar structures is nearly all-inclusive.

Emission line surveys leading to "discrete" clump masses do not consider this aspect of cloud structure. If two clumps are close together and part of a larger structure, the algorithms call them two separate objects, and do not tabulate the larger "object" that contains them. For this reason, all emission line or extinction surveys catalogue clouds that are within a factor of ~ 3 to 10 of the angular resolution, regardless of the cloud distance or the wavelength of the observation. This factor of 3–10 for recognized clump size corresponds to a factor of ~ 100 for cloud mass (which scales as the size squared or cubed in CO surveys), so the mass functions look reasonably well sampled, but in fact only a small part of the cloud structure is included. Everything smaller than the telescope resolution is not seen, and everything larger than about 3 to 10 resolution elements is subdivided into its component parts and not called a separate object. This is how every cloud or clump mass function has been evaluated since the beginning of this exercise (i.e. since before Field & Saslaw 1965).

The Elmegreen (1997a, 1998) model takes a different point of view. It considers the structure on all scales and asks for the probability that any particular mass is chosen from anywhere in the whole hierarchy. This is presumably what happens as a result of the combination of physical processes that leads to star formation in a real cloud: because of the self-similar nature of turbulent flows, each level in the hierarchy looks the same as any other level (for masses above the thermal Jeans mass), so each choice of level for a star-formation event would have the same likelihood as any other choice, modulated only by the density-dependent local evolution rate discussed in the previous paragraph. Aside from this variable rate, the instantaneous distribution of masses for all structures, and the instantaneous probability of selecting any particular mass, is proportional to M^{-2} for linear intervals in mass (i.e. $\alpha = 2$ in the notation of the clump spectra given above). Such a distribution for hierarchical structure was also recognized by Larson (1978) and Fleck (1996). The result gives a steeper instantaneous mass function than emission line or extinction surveys, but there is no conservation of mass as in these standard surveys (i.e. the sum of the masses of all possible structures is larger than the mass of the whole cloud, because nearly all of the structures are contained in other structures, and so are multiply counted). This way of viewing cloud structure seems to be closer to what a real star-forming process "sees" before it actually begins the sequence of events that makes a star. The density dependent rate of star formation then steepens the mass function from M^{-2} to $M^{-2.35}$. This result is then identified with the Salpeter IMF.

For the purposes of understanding the IMF, there are probably better ways to measure cloud structures than with clump-finding algorithms. The structure on the edges of clouds has been analyzed in terms of fractals, rather than clumps, for many years (Beech 1987; Bazell & Désert 1988; Scalo 1990; Dickman, Horvath, & Margulis 1990; Falgarone, Phillips, & Walker 1991; Zimmermann & Stutzki 1992, 1993; Henriksen, 1991; Hetem & Lepine 1993; Vogelaar & Wakker 1994; Pfenniger & Combes 1994).

For the structures inside clouds, Stutzki et al. (1998) showed that Fourier transform power spectra of emission line intensity scans across molecular clouds gave power laws, and concluded that the internal structure was scale-invariant, as in a fractal. They reproduced this structure with a random fractal-generation model. Other models that made cloud fractals were in Hetem & Lepine (1993) and Elmegreen (1997b). The concept that the cloud mass distribution function is the result of fractal structure, presumably generated by turbulence, began with papers by Fleck (1996), Elmegreen & Falgarone (1996), and Stutzki et al. (1998). This seems to be a more reasonable explanation than the older collisional-build-up models of clump structure, particularly since off-center supersonic collisions between clumps should not be sticky (Scalo & Pumphrey 1982; Kimura & Tosa 1996; Fujimoto, & Kumai 1997).

The probability of selecting structures for star formation in hierarchical clouds seems to show up more directly in the distribution of masses for open clusters. When a bound cluster forms, a high fraction of the gas mass has to go into stars (see Verschueren 1990 and references therein), so the cluster mass distribution should reflect the mass distribution of cloud structures pretty well. In this case, the result is obviously independent of telescope resolution or cloud-clump recognition bias, because clusters are seen optically, even at great distances. Also, the range of masses for clusters can be reasonably large, exceeding a factor of 100, so a power law can be measured fairly well if it exists. Cluster selection suffers from various other biases, however, such as extinction, age limitations, and a loss of low-mass clusters with increasing distance. Nevertheless, there are some studies that get around these biases, and they confirm the expected result. In two samples of nearby clusters, each with calibrated masses, Battinelli et al. (1994) found power law mass functions with slopes of -2.13 ± 0.15 and -2.04 ± 0.11. These clusters are close enough to the Sun to be relatively free of distance and extinction effects. Also, in the LMC, where ~ 600 clusters are catalogued (Bica et al. 1996), calibrated photometrically for age (Girardi et al. 1995), and all at about the same distance, Elmegreen & Efremov (1997) found $M^{-2}dM$ mass functions for discrete age groups (e.g. 10^8–10^9 yrs; dividing the clusters into age groups is important when there are no direct measures of cluster mass, because the conversion from luminosity to mass depends on age).

The IMF should be steeper than the cluster mass function because of the density-dependent formation rate of individual stars, and the competition for mass in the IMF. These effects are not important for clusters. Open clusters take a high fraction of all the gas mass originally available to them (or else they would not have ended up bound), and the rate at which they do this is not important for their final masses. In the case of stars, new objects forming at one level steal mass away from the objects that form later at a higher level. But clusters do not do this: those which form inside other clusters in the general hierarchy can just merge together to make a single cluster of higher mass, appropriate to the higher level in the hierarchy. And if they do not merge, then they stay with their original masses, which are appropriate for the levels in which they form.

Other IMF models based on fractal or hierarchical cloud structure were developed earlier by Henriksen (1986, 1991) and Larson (1992). Henriksen used the size, L, distribution function of structures in a fractal of dimension D, given as $n(L)d\log L \propto L^{-D}d\log L$ by Mandelbrot (1983), and assumed that the density ρ varied with size too, not as in a fractal (which would be $\rho \propto L^{D-3}$) but as actually observed in self-gravitating clouds, namely $\rho \propto L^{-1}$. This gave a mass function for cloud structure that agreed well with observations if $D \sim 2.7$. Note that if Henriksen used the density dependence for a fractal, in a self-consistent manner with the size distribution, then the resulting mass function would have been independent of the fractal dimension, $M^{-2}dM$, as discussed above for hierarchical clouds in general. Larson (1992) started again with the size distribution for a fractal, and assumed that final stellar mass is directly proportional to the linear size of the cloud structure, suggesting that such structures were filamentary anyway. Then the slope of the IMF in linear intervals became $1 + x = 1 + D \sim 3.3$ for $D = 2.3$, as implied by the fractal dimension of cloud perimeters, namely $D_p \sim 1.3$, added to 1 to account for the higher dimension of the surface. This IMF, with a slope of $1 + x = 3.3$, is significantly steeper than the Salpeter IMF, with a slope of $1 + x = 2.3$, but Larson suggested that maybe stars form in subparts of clouds where D is smaller than 2.3, or perhaps larger structures have larger temperatures, which would break the assumed linear relationship between star mass and scale size.

These models differ from the Elmegreen (1997a, 1998) model, even though both employ "fractal" cloud structure, because the latter uses general cloud shapes, not necessarily filaments or sheets, random sampling from all levels in the hierarchy of structures (giving $M^{-2}dM$ from sampling alone), a density-dependent rate that steepens the IMF by preferred sampling at lower mass (i.e. steepening the result to $M^{-2-0.5+1.5/D} \sim M^{-2.15}$), and a competition for mass, which steepens the IMF further, to $M^{-2.35}dM$, as in the Salpeter IMF. With these different assumptions, the IMF in the Elmegreen model hardly depends on the fractal dimension at all; it enters into the exponent of the IMF power law as approximately $0.5 - 1.5/D$, which is a very small increment for D in the likely range of 2 to 3.

2.3. *Theories for the lower stellar mass limit*

Understanding the lower mass limit to star formation is an important part of any IMF theory. Various assumptions about this have already been mentioned in the review of theories presented above. Three views on this subject seem to pervade most of them. In the wind-limited accretion models, the assumption is made that accretion onto a dense core continues until a protostellar wind develops, which presumably follows the onset of deuterium burning (Shu et al. 1987). Thus low mass objects do not stay that way very long, they just keep accreting until they become active protostars. In most of the older fragmentation-coalescence models, the minimum mass is the mass of a gravitationally bound fragment that is optically thick. In the various scale-free models, based on gravitationally-driven fragmentation or turbulence, the minimum mass is usually the thermal Jeans mass, particularly in the Larson (1992) and Elmegreen (1997a, 1998) models. This assumption is made because, without further fragmentation during the collapse itself, the thermal Jeans mass is the smallest structure that can ever become self-gravitating, even in the presence of turbulence and magnetic fields.

The first and the third of these mass limits equals about the observed limit, ~ 0.1 M$_\odot$. In the first case, it is self-defining: objects accrete until they become star-like, and then they stop accreting, so naturally the lowest possible mass is the mass of a star. This concept has the powerful advantage over the others that the basic star mass, and perhaps even the IMF itself, can be the same everywhere, independent of cloud conditions. In the third case, the minimum theoretical mass just happens to equal the observed minimum in standard cloud conditions. An expression for this mass is the Bonner-Ebert critical mass,

$$M_{BE} = 0.35 \left(\frac{T}{10K}\right)^2 \left(\frac{P}{10^6\,k_B}\right)^{-1/2} \text{ M}_\odot. \tag{2.5}$$

Here we have normalized the result to standard cloud conditions, taking the pressure inside a cloud equal to the typical self-gravitating core pressure, not the boundary pressure (which is sometimes lower by a factor of ~ 30). Our use of this form for the thermal Jeans mass is motivated by the near constancy of T and total P in molecular clouds, according to the scaling relations (Larson 1981). This is much more sensible than writing the thermal Jeans mass in terms of T and density, for example, when the density varies by several orders of magnitude inside a cloud. Our use of the total pressure for P also allows for a decrease of Jeans mass in compressive flows, as in Hunter & Fleck (1982).

Nevertheless, another theory of the IMF in which all of the stellar masses have the local thermal Jeans mass has been advanced by Padoan, Nordlund, & Jones (1997), with variations in mass entirely the result of variations in the density in a turbulent fluid. In this model, the high mass stars form in the low density regions, because the Jeans mass is high there for a given temperature.

The second minimum fragment mass discussed above, from the opacity limit of a self-gravitating condensation, is typically much smaller than the minimum star mass, so coalescence has to bring the star mass up to the observed value in these models.

Larson has looked for signatures of the thermal Jeans mass in several ways, predicting that it should vary with cloud temperature (Larson 1985, 1992). In Larson (1992), he suggested that such variations were observed by Myers (1991) in different molecular clouds. Larson (1995) also suggested that the length scale at a break in the separation distribution for T Tauri stars was the Jeans length (see also Simon 1997; but see Bate, Clarke, & McCaughrean 1998 and Nakajima et al. 1998 for other interpretations.)

Elmegreen (1997a, 1998) considered M_{BE} from a different point of view, stating that most star-forming regions in normal galaxies actually do have nearly constant values of $T^2/P^{1/2}$, with small variations in this quantity possibly leading to the observed variations in cluster IMFs at low mass (see Section 1.1 above). This constancy is a result of the numerator in M_{BE} being approximately proportional to the cooling rate per unit mass of molecular gas (Neufeld, Lepp, & Melnick 1995), and the denominator being approximately proportional to the local surface density of stars and gas in the galaxy, which determine the heating rate per unit mass from cosmic rays and background starlight. As long as cooling roughly balances background heating, this ratio will be about constant. Other, more local variations in T and P tend to cancel out when combined as the quantity $T^2/P^{1/2}$. For example, when P goes up in a triggered region next to a previous generation of star-formation, T generally goes up too because of the enhanced radiation field. Conversely, when P is low, as in low-mass clouds like Taurus, then T is low as well.

Elmegreen (1997a, 1998) also points out that the lower mass limit for stars actually *does* change, perhaps by a factor of 10 in extreme starburst regions (see Sect. 1.1 above) where the average T can be much larger than in local molecular clouds, e.g. 100 K instead of 10 K (cf. Aalto et al. 1995), and the pressure much larger too (e.g. $\times 10^4$). But this variation should be rare in normal galaxies because of feedback processes in the interstellar medium that tend to connect young stellar activity, interstellar pressure, and gas temperature. If we consider T to be a measure of the rate of cooling per unit cloud mass, then T in a star-forming region will be high when there is an unusually large embedded or nearby luminosity of stars per unit gas mass. This occurs when the efficiency for star formation is high in a high mass cluster (the cluster has to be fairly high mass to contain the luminous stars that dominate the total radiation field). A high efficiency in a low mass cluster would not necessarily increase M_{BE} because low mass clusters are not very luminous per unit mass. A low efficiency in a high mass cluster, leading to an OB association, might not change M_{BE} much either, because the radiation luminosity per unit gas mass is not particularly large there either. Thus the formation of globular clusters in regions with only moderately large pressures, not the enormous pressures typical of early galactic halos, might be elevate M_{BE}, but this conclusion is very uncertain.

Figure 3 shows two numerical experiments that made model IMFs from random selections in a hierarchically structured gas cloud, using a rate of selection proportional to the square root of the gas density, as discussed above (Elmegreen 1997a). The models differ only in the value of M_{BE}, which enters into the probability for failure to form a star, taken to be $P_f = e^{-(M/M_{BE})^2}$ in these cases (M_{BE} in the figure is written in terms of the computer-program mass, which is scaled to a physical mass after multiplication by 0.15. The IMFs are identical except for the larger value of the lower mass limit in the case with high M_{BE}.

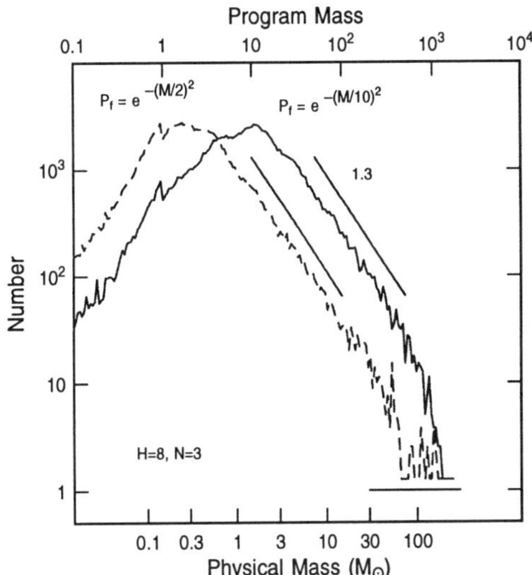

FIGURE 3. Two model IMFs with different lower mass limits, M_{BE}. The entire IMF shifts to the right when M_{BE} increases because the power law portion always has the same slope. The fall-off at high mass is the result of computational limitations. The models have $H = 8$ levels in the hierarchy with $N = 3$ subclumps per clump, on average.

2.4. *A combination of theories?*

The two main contenders for the lower mass limit of star formation, wind-regulation and the thermal Jeans mass, have obvious differences in predicted behaviors of the shape of the IMF at low mass. If this shape varies a lot, with some clusters having clearly higher lower mass limits than other clusters, then wind self-regulation might be ruled out. Moreover, if cluster IMFs in regions with large turnover masses also have larger pre-stellar M_{BE}, then the thermal Jeans mass would seem to be important.

Curiously, both of these models, the wind-regulated IMF, and the M_{BE}–fractal cloud model, have their strong points where the other is weakest. The wind-regulated IMF model seems to have difficulty explaining the nearly universal slope of the IMF without some sensitivity to cloud parameters, yet it seems certain that wind self-regulation must play some role in determining the mass that gets onto a star, at least near the lower mass limit for star formation. Similarly, models that rely on M_{BE} have an uncomfortable susceptibility to changes in cloud properties near the lower mass limit, but get the Salpeter slope above this limit with essentially no free parameters. Perhaps a combination of the two models would be better.

There are several ways this could occur. First, the fractal model supposes that a star's mass can be identified with the mass of some structure in a turbulent cloud, but this is obviously too simple—there is a lot that can happen during star formation that will vary the fraction of the cloud mass that gets into a star. If this fraction is randomly distributed, then the same IMF slope and scatter about that slope result, as shown in connection with figure 5 below, which is discussed in more detail in Elmegreen (1998). If it is not randomly distributed, then the power-law slope could change. But the entire power-law slope is not likely to come from wind-regulation, that would give the slope too great a sensitivity to cloud properties and the randomness of wind-clearing. This means that, perhaps to within a factor of 2 to 5, star mass should really be identified with the mass of the clump in which it forms, i.e. that bigger stars really do form in bigger clumps (Casoli et al. 1986; Myers, Ladd, & Fuller 1991; Myers & Fuller 1993; Pound &

Blitz 1995; Motte, André, & Neri 1998; not to be confused with the statistical sampling statement discussed in section 1.1, that bigger stars form in bigger *whole* clouds). This factor of 2 to 5 may contain all the detailed physics of wind clearing.

Protostellar winds should also play some role in regulating the smallest stars that form. According to the models by Shu and collaborators, as reviewed in Shu et al. (1987), the onset of the wind is triggered by deuterium burning in a young, pre-stellar object. If the object is too small, there will be no wind, and presumably accretion will continue until deuterium burning begins.

This concept combines well with the arguments based on a thermal Jean mass. For example, if M_{BE} is much smaller than the deuterium burning limit, then a large number of small brown dwarfs or Jupiters should form, and these could either leave the cloud in that state, forever drifting as brown dwarfs, or they could accrete more mass over time and end up finally able to start a wind. In the latter case, the effective lower mass limit to the IMF would be the deuterium burning mass, not M_{BE}, and the larger masses would follow the fractal mass distribution as before, giving the Salpeter function. On the other hand, if there are a lot of free brown dwarfs, which does not appear to be the case in the Solar vicinity (cf. Sect. 1.1), and if there are also a lot of regions where M_{BE} is low, which does not appear to be the case locally either, then a model with accretion up to a wind-limited mass would not seem to work.

If M_{BE} is much larger than the deuterium burning limit, then large, self-gravitating clumps will form in the cloud, but the rapid onset of a wind in these clouds might limit the gas mass that actually gets into the star to only a small fraction of what is available. Then all of the excess mass between the deuterium limit and M_{BE} would get dispersed back into the cloud for recycling into other stars, and the lower mass limit would be the deuterium limit. In this second case, the mass fraction that goes into each star will be small, but if it is about the same fraction for all clumps, then the higher mass stars will still map out the fractal properties of the cloud and produce the Salpeter IMF. Again the details of how the IMF gets established in a cloud depend strongly on unknown processes during the final collapse and dispersal phase of star formation, but the basic shape of the IMF could be relatively independent of these details.

2.5. *Reflections on the various theories of the IMF*

There have been some fundamental changes lately in how we view molecular clouds and star formation, and these changes inevitably affect the various models for the IMF.

For example, we are beginning to think that star formation is relatively rapid in cloud cores, occurring in only one or two crossing times regardless of scale (Elmegreen & Efremov 1996; Efremov & Elmegreen 1998). This means that we cannot generally wait for magnetic diffusion to occur as a cloud core slowly accretes across field lines. Diffusion typically takes about 10 free fall times for cosmic ray ionization (Shu et al. 1987), longer for ionization by starlight (Myers & Khersonsky 1995), and longer still if the gas is clumpy (Elmegreen & Combes 1992). Indeed, Nakano (1998) suggested that all star-forming cloud cores are magnetically *super*critical; i.e. they would collapse dynamically across the field lines if turbulent motions were not present.

There are two important implications of this change in thinking, if correct. First, there would no longer be a strong motivation for IMF theories that consider bimodel star formation in the sense that low mass stars come from subcritical cores, and high mass stars come from magnetically supercritical cores (Lizano & Shu 1988). The IMF is so remarkably uniform anyway, it does not seem possible to have widely different modes of star formation at high and low mass. Any change from low mass dominance to high mass dominance in the IMF could more easily be accommodated by an upward shift in the thermal Jeans mass (cf. Section 2.3).

Secondly, this time scale for star formation may be too short to allow multiple interactions between protostellar clumps. This would then rule out a broad class of coalescence and slow accretion models. Such coalescence is unlikely anyway for clumps that move supersonically relative to each other; they will fragment or disperse upon collision, rather than stick together. Clump collisions may trigger star formation when they occur (Bhattal et al. 1998), but multiple collisions are probably not the source of the cloud or star mass distributions.

Another observation that suggests the same thing is the extremely high stellar densities in young globular clusters, which form today in starburst regions. As suggested in Section 1.1, these globulars look like they will evolve into objects similar to our Milky Way's halo globulars, and the old Milky Way globulars have an apparently normal IMF. If coalescence is important in local "normal" embedded clusters, it would seem to be devastating in globular clusters, where the stellar density can be 10^4 times higher. Conversely, if these globulars have normal IMFs, then protostellar coalescence must be unimportant nearly everywhere. A similar constraint would come from the IMF in small regions, where only a few stars form. There, protostar interactions are not likely to be important either.

A second major shift in our view of molecular clouds is that most astronomers now think that pre-stellar structure comes from turbulence, not gravitationally-driven fragmentation. This is because the same character of structure is observed in both self-gravitating and non-self-gravitating clouds, and both types of clouds have correlated motions reminiscent of laboratory turbulence. A related point is that the cloud cores in which clusters form are not obviously collapsing as a whole: there are no inverse P-Cygni profiles indicative of collapse motions for whole cores, as there are in some individual protostars. The stars are apparently forming inside more-or-less stable cloud cores. Once again, the general collapse models seem untenable.

The presence of supersonic turbulence also seems to rule out models involving thermal instabilities. Turbulent motions in cluster-forming cores are generally much larger than thermal, and so the dominant forces that structure the gas are the turbulent forces, not the thermal. This means that turbulence causes the observed structures in pre-stellar clouds, not thermal instabilities. Thermal processes are probably important on the smallest scales, but these are at and below M_{BE}.

Turbulence in pre-stellar clouds also appears to generate structure that is much lower in mass than even the smallest stars, perhaps as low as 10^{-4} $M_\odot$ (Heithausen et al. 1998; Kramer et al. 1998). This means that star formation occurs in the middle range of all cloud structures, not at the bottom end. It also means that, while turbulence may generate a cascade of structures down to smaller and smaller scales, the end result of this cascade is not the formation of a star. Turbulence makes cloud structures independent of star formation. Self-gravity, sonic wave generation, magnetic reconnection, and many other aspects of interstellar cloud dynamics are likely to be important too by the time star formation begins.

2.6. *Implications of a nearly uniform IMF*

The remarkable near-uniformity of the IMF in regions with a wide range of stellar densities, cluster masses, metallicities, ages, and galactic types strongly suggests there is a *unifying process in star formation* that makes the relative probability of forming various masses somewhat independent of physical parameters. For example, the IMF seems independent of how star formation begins, i.e. whether the region is triggered by some ex-

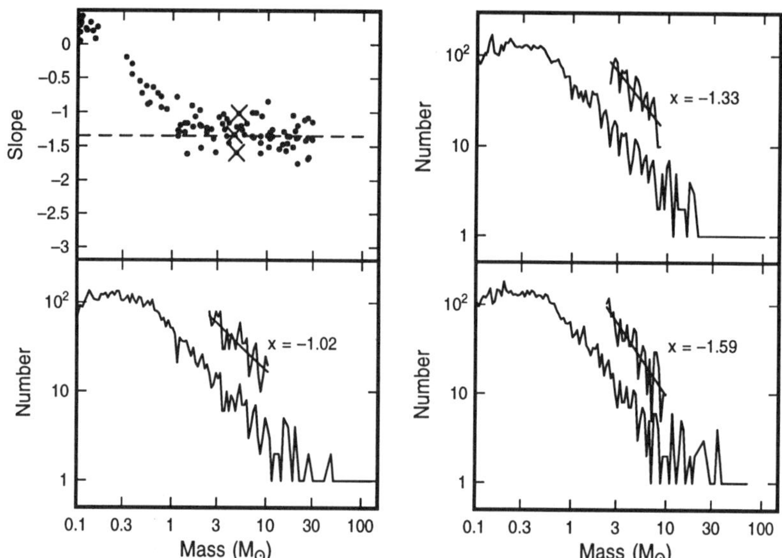

FIGURE 4. (top left) IMF slopes in 100 models, plotted as a function of the average logarithm of the mass. Each IMF slope is fit using 200 stars, with the largest star in the fit taken to be smaller by 20 filled mass bins than the largest star produced in the model. The Salpeter value of -1.35 is shown by a dashed line. The three values indicated by crosses have their complete IMFs shown in the other panels. The fitted portions of these IMFs are indicated by the offsets.

ternal pressure or forms quiescently (e.g. see Parker et al. 1992 and the contrary opinion by Oey & Massey 1995), independent of the general cloud shape (shell, layer, filament, etc.), magnetic field strength, ionization level, presence of other stars, and so on. This is remarkable because these physical variables do control the *rate* of star formation in some regions, and they certainly control *where and when* star formation begins. They just do not appear to influence the IMF.

The near-uniformity of the *slope* of the power law also implies several other things. (1) There is a *universal aspect of cloud structure and evolution that produces the power-law portion of the IMF independently of the temperature, pressure, and metallicity*. This means that the power-law probably does not depend on the protostellar collapse process, e.g. on the fragmentation or collapse of isothermal clumps, the accretion of gas in thermal equilibrium, or the thermal Jeans mass. For all of these processes, thermal temperature and magnetic diffusion are probably important at some stage (see reviews in Nakano 1984; Mouschovias 1991; Shu, Adams, & Lizano 1987). (2) The IMF power-law is also likely to be a *highly reduced average over many physical processes that either all give about the same result, or are combined so finely in every region that the same proportion of each process is always present*.

The variations in slope of the IMF from region to region (Scalo 1998) seem at first to suggest something different, that the IMF is not in fact uniform but depends sensitively on physical conditions. But are these variations statistically significant? Rarely do studies of cluster IMFs contain more than several hundred stars.

Figure 4 shows sample IMFs from the model of random selection in a hierarchical cloud (Elmegreen 1998). The panel in the upper left shows IMF slopes for 100 different models, each with a different number of stars and therefore different upper mass limit. In all cases, the range of stars chosen for the power-law slope determination is such that there are 200 stars in the fit, starting from the histogram bin that is twenty filled bins (a

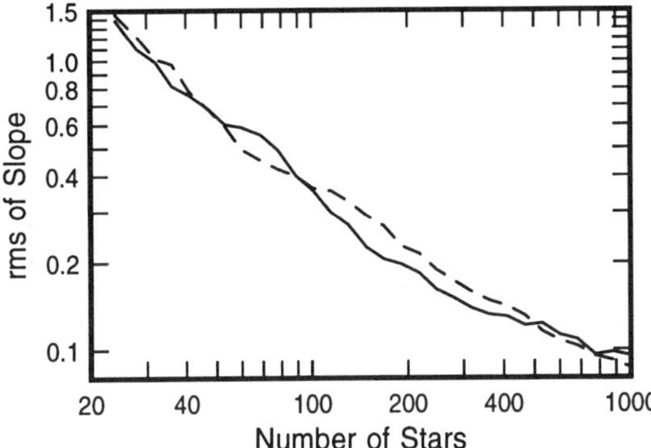

FIGURE 5. The rms deviation in the fitted slopes of the model IMFs for the 1–10 $M_\odot$ range, plotted versus the number of stars that are included in the fit. The solid line is for the case shown in Fig. 4, and the dashed line is for another 100 independent IMFs that have an additional randomness in the ratio of the star mass to the clump mass.

factor of 3 in mass) away from the most massive star, and going to lower mass bins. This starting point for the fit ensures that there are enough stars to get an accurate slope, and it coincides with an astronomer's decision to avoid fitting the IMF to the scarce few highest-mass stars in the cluster. We pick a different number of total stars for each IMF, and use only the highest mass parts of each model, because this is what an astronomer would do as well. Generally an observer catalogues only the brightest stars in a cluster and misses the low mass members. If a cluster has an IMF made only from low mass stars, then that cluster is generally so small that there are no high mass stars at all.

Evidently the IMF slope varies a lot around the Salpeter value, shown in the figure by the dashed line at $x = 1.35$. This average slope is nearly independent of physical cloud properties in this model (see the discussion in section 2.1). The other three panels in figure 4 show particular IMFs and the mass ranges and fits used for the "X" marks in the top left panel.

Figure 5 shows the rms deviations in the IMF slopes around the mean values for power-law fits that include different numbers of stars in the same 100 IMFs (from Elmegreen 1998). Each plotted value, connected by a smooth line, represents the rms deviation for the 100 models, calculated in the mass range from 1 to 10 $M_\odot$ (the characteristic physical mass in the model comes from M_{BE}, which is taken to be 0.35 $M_\odot$ from equation 2.5). The rms deviation around the Salpeter slope decreases systematically as more and more stars are included in the fit. The deviation can be as high as ±0.5 when only 80 stars are used, decreasing to ~ 0.1 when 1000 stars are used. This implies that an IMF calculated with, say, 150 stars, and having a slope of -1.15 instead of the Salpeter slope of -1.35, is in fact statistically consistent with this Salpeter slope in a "universal" IMF. The fluctuations found by Scalo (1998) in his review correspond to ±0.5 around the value $x \sim 1.35$. These fluctuations could be statistical, considering the number of stars included in typical IMF surveys and the likely presence of measurement and star selection errors in the real data.

The dashed line in figure 5 is for another 100 models, independent of the first, and evaluated with a large ($\times4$) and random fluctuation in the ratio of the star mass to the clump mass. Such variations do not affect the average IMF slope or the statistical fluctuations around it.

3. Summary

The IMF has been observed directly and indirectly for many years, with many different results, but there seems to be some convergence now in the basic form of the IMF, and also some tantalizing indications that this form changes a little when the physical conditions for star formation change a lot. The main observational results were summarized in section 1.4.

The theory of the IMF seems to be all over the map, even in recent years. This variation reflects our ignorance in the processes that determine a star's mass and in the processes that cause the complex spatial structures seen in interstellar clouds. The common aspects in many of these models, including clump and protostellar interactions, wind erosion of protostellar gas concentrations, minimum masses from the lack of self-gravity, and hierarchical cloud structure, are all likely to play some role in generating the IMF. The worry is that the IMF is such a highly reduced average over many physical processes that each process does not a have clear signature in the final result.

In view of the observations, we should perhaps consider "acceptable" those models that do not have much sensitivity to either the local aspects of cloud structure, or the large scale aspects of time and place in the Universe. This drives the recent appeal towards models based on common turbulence in one way or another, considering that turbulent structures are likely to be robust. Unfortunately, we do not understand compressible MHD turbulence much either.

There are several key observations that would help. One is the determination of star formation time scales, always in units of the cloud or clump crossing time. There are three important timescales: the rise time of the star formation rate in a cloud, the duration of star formation, and the decay time scale. If the rise time is less than a crossing time, and the duration only one or two crossing times, then IMF models based on multiple clump interactions would seem to be ruled out.

There is also a need to know the mass dependence of the mass fraction of clump gas that goes into a star. This might be determined from the ratio of luminosity to gas mass for Type 0 sources, plotted as a function of gas mass for the dense cores where the stars are actually forming.

Theory should tell us what gas structures and evolutionary timescales are expected from self-gravitating MHD turbulence in the supersonic-subAlfvénic regime of molecular clouds. If turbulence only makes moving waves and wavepackets, and star formation actually occurs in such regions, then star formation has to be relatively quick in terms of the local crossing time. It may be that in the absence of strong self-gravity, turbulent clumps are transient and amorphous, but when self-gravity becomes important, the clumps get some integrity and persist for a relatively long time.

Thanks to Gerald Gilmore for pre-publication copies of reviews from the IMF conference that he organized with I. Parry, and S. Ryan.

REFERENCES

AALTO, A., BOOTH, R. S., BLACK, J. H., & JOHANSSON, L. E. B. 1995 *A&A* **300**, 369.

ADAMS, F. C. & FATUZZO, M. 1996 *ApJ* **464**, 256.

ALLEN, E. J. & BASTIEN, P. 1995 *ApJ* **452**, 652.

ALLEN, E. J. & BASTIEN, P. 1996 *ApJ* **467**, 265.

ANGELETTI, L. & GIANNONE, P. 1997 *A&A* **321**, 343.

BASTIEN, P. 1981 *A&A* **93**, 160.

BATE, M. R., CLARKE, C. J., & McCAUGHREAN, M. J. 1998 *MNRAS* **297**, 1163.

BATTINELLI, P. 1991 *A&A* **244**, 69.

BATTINELLI, P. 1992 *A&A* **258**, 269.

BATTINELLI, P., BRANDIMARTI A.. & CAPUZZO-DOLCETTA R. 1994 *A&AS* **104**, 379.

BATTINELLI, P., EFREMOV, YU. N., & MAGNIER, E. A. 1996 *A&A* **314**, 51.

BAZELL, D. & DÉSERT, F. X. 1988 *ApJ* **333**, 353.

BEECH, M. 1987 *Astrophys. Sp. Sci.* **133**, 193.

BELLAZZINI, M., PASQUALI, A., FEDERICI, L., FERRARO, F. R., & FUSI PECCI, F. 1995 *ApJ* **439**, 687.

BHATTAL, A. S., FRANCIS, N., WATKINS, S. J., & WHITWORTH, A. P. 1998 *MNRAS* **297**, 435.

BICA, E., CLARIA, J. J., DOTTORI, H., SANTOS, J. F. C., JR., & PIATTI, A. E. 1996 *ApJS* **102**, 57.

BLITZ, L. 1993 In *Protostars and Planets III* (ed. E. H. Levy and J. I. Lunine). p. 125. Tucson: University of Arizona Press.

BONNELL, I. A. & BASTIEN, P. 1993 *ApJ* **406**, 614.

BONNELL, I. A., BATE, M. R., CLARKE, C. J., & PRINGLE, J. E. 1997 *MNRAS* **285**, 201.

BONNELL, I. A. & DAVIES, M. B. 1998 *MNRAS* **295**, 691.

BRESOLIN, F. & KENNICUTT, R. C., JR. 1997 *AJ* **113**, 975.

BROWN, A. G. A. 1998 In *The Stellar Initial Mass Function* (eds. G. Gilmore, I. Parry & S. Ryan). p. 45, ASP.

BURKERT, A., BATE, M. R., & BODENHEIMER, P. 1997 *MNRAS* **289**, 497.

CALZETTI, D. 1997 *AJ* **113**, 162.

CASOLI, F., DUPRAZ, C., GERIN, M., COMBES, F., & BOULANGER, F. 1986 *A&A* **169**, 281.

CAYREL, R. 1990. In *Physical Processes in Fragmentation and Star Formation* (eds. R. Capuzzo-Dolcetta, C. Chiosi & A. Di Fazio). p. 343. Kluwer.

CHARLOT, S., FERRARI, F., MATTHEWS, G. J., & SILK, J. 1993 *ApJ* **419**, L57.

CLARKE, C. 1998. In *The Stellar Initial Mass Function* (eds. G. Gilmore, I. Parry, & S. Ryan). p. 189. ASP.

COOL, A. M. 1998. In *The Stellar Initial Mass Function* (eds. G. Gilmore, I. Parry, & S. Ryan). p. 139. ASP.

DE MARCHI, G. & PARESCE, F. 1997 *ApJ* **476**, L19.

DEVEREUX, N. A. 1989 *ApJ* **346**, 126.

DICKMAN, R. L., HORVATH, M. A. & MARGULIS, M. 1990 *ApJ* **365**, 586.

DOANE, J. S. & MATTHEWS, W. G. 1993 *ApJ* **419**, 573.

DOYON, R., JOSEPH, R. D., & WRIGHT, G. S. 1994 *ApJ* **421**, 101.

EFREMOV, YU. N. 1995 *AJ*, **110**, 2757.

EFREMOV, YU. N., IVANOV, G. R., & NIKOLOV, N. S. 1987 *Astrophys. Sp. Sci.*, **135**, 119.

EFREMOV, YU. N. & SITNIK, T. G. 1988 *Sov. Astr. Lett.*, **14**, 347.

EFREMOV, YU. N. & ELMEGREEN, B. G. 1998 *MNRAS*, in press.

EGGEN, O. J. 1976 *Quart. J. Roy. Astron. Soc.*, **17**.

EGGEN, O. J. 1977 *PASP* **89**, 187.

ELMEGREEN, B. G. 1978 *The Moon and the Planets* **19**, 159

ELMEGREEN, B. G. 1983 *MNRAS* **203**, 1011.

ELMEGREEN, B. G. 1985. In *The Birth and Infancy of Stars*, (eds. R. Lucas, A. Omont, & R. Stora). p. 257. New Holland.

ELMEGREEN, B. G. 1997a *ApJ* **486**, 944.

ELMEGREEN, B. G. 1997b *ApJ* **477**, 196.

ELMEGREEN, B. G. 1998 *ApJ*, submitted

ELMEGREEN, B. G. & MATHIEU, R. 1983 *ApJ* **202**, 305.

ELMEGREEN, B. G. & COMBES, F. 1992 *A&A* **259**, 232.

ELMEGREEN, B. G. & FALGARONE, E. 1996 *ApJ* **471**, 816.

ELMEGREEN, B. G. & EFREMOV, YU. N. 1996 *ApJ* **466**, 802.

ELMEGREEN, B. G. & EFREMOV, YU. N. 1997 *ApJ* **480**, 235.

ELMEGREEN, B. G. & EFREMOV, Y. N. 1998 In *The Orion Complex Revisited*, (eds. M. J. McCaughrean & A. Burkert), in press. ASP Conference Series.

FALGARONE, E. & PHILLIPS, T. G. 1991. In *Fragmentation of Molecular Clouds and Star Formation* (eds. E. Falgarone, F. Boulanger, & G. Duvert). p. 119. Kluwer.

FALGARONE, E., PHILLIPS, T., & WALKER, C. K. 1991 *ApJ* **378**, 186.

FEITZINGER, J. V., & BRAUNSFURTH, E. 1984 *A&A* **139**, 104.

FIELD, G. B. & SASLAW, W. C. 1965 *ApJ* **142**, 568.

FISCHER, P., PRYOR, C., MURRAY, S., MATEO, M., & RICHTLER, T. 1998 *AJ* **115**, 592.

FLECK, R. C., JR. 1996 *ApJ* **458**, 739.

FLETCHER, A. B. & STAHLER, S. W. 1994a *ApJ* **435**, 313.

FLETCHER, A. B., & STAHLER, S. W. 1994b *ApJ* **435**, 329.

FUJIMOTO, M., & KUMAI, Y. 1997 *AJ* **113**, 249.

GIRARDI, L., CHIOSI, C., BERTELLI, G., & BRESSAN, A. 1995 *A&A* **298**, 87.

GIBSON, B. K. & MATTEUCCI, F. 1997 *MNRAS* **291**, L8.

GREGGIO, L., MARCONI, G., TOSI, M., & FOCARDI, P. 1993 *AJ* **105**, 894.

GRILLMAIR, C. J., ET AL. 1998 *AJ* **115**, 144.

GOULD, A., BAHCALL, J. N., & FLYNN, C. 1997 *ApJ* **482**, 913.

GÜSTEN, R. & MEZGER, P. G. 1983 *Vistas Astr.* **26**, 159.

HEIKKILÄ, A., JOHANSSON, L. E. B., & OLOFSSON, H. 1998 *A&A* **332**, 493.

HEITHAUSEN, A., BENSCH, F., STUTZKI, J., FALGARONE, E., & PANIS, J. 1998 *A&A* **331**, 65.

HENRIKSEN, R. N. 1986 *ApJ* **310**, 189.

HENRIKSEN, R. N. 1991 *ApJ* **377**, 500.

HERBIG, G. H. 1962a *Adv.A&A* **1**, 47.

HERBIG, G. H. 1962b *ApJ* **135**, 736.

HERBIG, G. H. 1998 *ApJ* **497**, 736.

HERBIG, G. H. & TERNDRUP, D. M. 1986 *ApJ* **307**, 609.

HENRIKSEN, R. N. 1991 *ApJ* **377**, 500.

HETEM, A., JR., & LEPINE, J. R. D. 1993 *A&A* **270**, 451.

HILLENBRAND, L. A. 1997 *AJ* **113**, 1733.

HILLENBRAND, L. A. & HARTMANN, L. 1998 *ApJ* **492**, 540.

HO, L. C. & FILIPPENKO, A. V. 1996a *ApJ* **466**, L83.

HO, L. C. & FILIPPENKO, A. V. 1996b *ApJ* **472**, 600.

HOYLE, F. 1953 *ApJ* **118**, 513.

HOLTZMAN, J. A., ET AL. 1997 *AJ* **113**, 656.

HORNER, D. J., LADA, E. A., & LADA, C. J. 1997 *AJ* **113**, 1788.

HUNTER, C. 1962 *ApJ* **136**, 594.

HUNTER, J. H., JR., & FLECK, R. C., JR. 1982 *ApJ* **256**, 505.

IBEN, I., JR., & TALBOT, R. J. 1966 *ApJ* **144**, 968.

IVANOV, G. R. 1987 *Astrophys. Sp. Sci.* **136**, 133.

IVANOV, G. R. 1992 *MNRAS* **257**, 119.

JONES, B. F. & WALKER, M. F. 1988 *AJ* **95**, 1755.

JOSEPH, R. 1991. In *Massive Stars in Starbursts* (eds. C. Leitherer, N. R. Walborn, T. M. Heckman, & C. A. Norman). p. 259. Cambridge University Press.

KENNICUTT, R. C., JR., TAMBLYN, P., & CONGDON, C. W. 1994 *ApJ* **435**, 22.

KHERSONSKY, V. K. 1997 *ApJ* **475**, 594.

KIMURA, T. & TOSA, M. 1996 *A&A* **308**, 979

KING, I. R., ANDERSON, J., COOL, A. M., & PIOTTO, G. 1998 *ApJ* **492**, L37.

KRAMER, C., STUTZKI, J., RÖHRIG, R., & CORNELIUSSEN, U. 1998 *A&A* **329**, 249.

KRONBERG, P. P., BIERMANN, P., & SCHWAB, F. R. 1985 *ApJ* **291**, 693.

LADA, C. J. 1985 *ARAA* **23**, 267.

LADA, E. A. & LADA, C. J. 1995 *AJ* **109**, 1682.

LADA, E. A., ALVES, J., & LADA, C. J. 1996 *AJ* **111**, 1964.

LADA, E. A., LADA, C. J., & MUENCH, A. 1998. In *The Stellar Initial Mass Function* (eds. G. Gilmore, I. Parry & S. Ryan). p. 107. ASP.

LARSON, R. B. 1973 *MNRAS* **161**, 133.

LARSON, R. B. 1977. In *The Evolution of Galaxies and Stellar Populations* (eds. B. M. Tinsley & R. B. Larson). p. 97. Yale University Observatory.

LARSON, R. B. 1978 *MNRAS* **184**, 69.

LARSON, R. B. 1981 *MNRAS* **194**, 809.

LARSON, R. B. 1982 *MNRAS* **200**, 159.

LARSON, R. B. 1985 *MNRAS* **214**, 379.

LARSON, R. B. 1986 *MNRAS* **218**, 409.

LARSON, R. B. 1990. In *Physical Processes in Fragmentation and Star Formation* (eds. R. Capuzzo-Dolcetta, C. Chiosi & A. Di Fazio). p. 389. Kluwer.

LARSON, R. B. 1991. In *Fragmentation of Molecular Clouds and Star Formation* (eds. E. Falgarone, F. Boulanger, & G. Duvert). p. 261. Kluwer.

LARSON, R. B. 1992 *MNRAS* **256**, 641.

LARSON, R. B. 1995 *MNRAS* **272**, 213.

LEITHERER, C. 1998. In *The Stellar Initial Mass Function* (eds. G. Gilmore, I. Parry & S. Ryan). p. 61, San Francisco: ASP.

LEJEUNE, C., & BASTIEN, P. 1986 *ApJ* **309**, 167.

LIZANO, S., & SHU, F. H. 1989 *ApJ* **342**, 834.

LOEWENSTEIN, M., & MUSHOTZKY, R. F. 1996 *ApJ* **466**, 695.

LU, L., SARGENT, W., CHURCHILL, C., & VOGT, S. 1996 *ApJS* **107**, 475.

LUHMAN, K. L. & RIEKE, G. H. 1998 *ApJ* **497**, 354.

MAGNIER, E. A., BATTINELLI, P., LEWIN, W. H. G., HAIMAN, Z., VAN PARADIJS, J., HASINGER, G., PIETSCH, W., SUPPER, R., & TRUEMPER, J. 1993 *A&A* **278**, 36.

MANDELBROT, B. B. 1977. *Fractals: Form, Chance and Dimension*. Freeman.

MANDELBROT, B. B. 1983. *The Fractal Geometry of Nature*. Freeman.

MARCONI, G., TOSI, M., GREGGIO, L., & FOCARDI, P. 1995 *AJ* **109**, 173.

MASSEY, P. 1998. In *The Stellar Initial Mass Function* (eds. G. Gilmore, I. Parry & S. Ryan). p. 17. ASP.

MASSEY, P., JOHNSON, K. E., & DEGIOIA-EASTWOOD, K. 1995a *ApJ* **454**, 151.

MASSEY, P., LANG, C. C., DEGIOIA-EASTWOOD, K., & GARMANY, C. D. 1995b *ApJ* **438**, 188.

MASSEY, P. & HUNTER, D. A. 1998 *ApJ* **493**, 180.

MEGEATH, S. T. 1996 *A&A* **311**, 135.

MÉRA, D., CHABRIER, G., & BARAFEE, I. 1996 *ApJ* **459**, L87.

MEURER, G. R. 1995 *Nature* **375**, 742.

MEZGER, P. G. & SMITH, L. F. 1977. In *Star Formation* (eds. T. de Jong & A. Maeder). p. 133. Reidel.

MILLER, G. E. & SCALO, J. 1979 *ApJS* **41**, 513.

MOTTE, F., ANDRÉ, P., & NERTI, R. 1998 *A&A* **336**, 150.

MOUSCHOVIAS, T. CH. 1991. In *Physics of Star Formation and Early Stellar Evolution* (eds. C. J. Lada & N. Kylafis) p. 61. Kluwer Academic Publishers.

MURRAY, S. D. & LIN, D. N. C. 1996 *ApJ* **467**, 728.

MYERS, P. C. 1991. In *Fragmentation of Molecular Clouds and Star Formation* (eds. E. Falgarone, F. Boulanger, & G. Duvert). p. 221. Kluwer.

MYERS, P. C., LADD, E. F., & FULLER, G. A. 1991 *ApJ* **372**, L95.

MYERS, P. C. & FULLER, G. A. 1993 *ApJ* **396**, 631.

MYERS, P. C. & KHERSONSKY, V. K. 1995 *ApJ* **442**, 186.

NAKAJIMA, Y., TACHIHARA, K., HANAWA, T., & NAKANO, M. 1998 *ApJ* **497**, 721.

NAKANO, T. 1984 *Fund. Cosmic Phys.* **9**, 139.

NAKANO, T. 1998 *ApJ* **494**, 587.

NAKANO, T., HASEGAWA, T., & NORMAN, C. 1995 *ApJ* **450**, 183.

NEUFELD, D. A., LEPP, S., & MELNICK, G. J. 1995 *ApJS* **100**, 132.

NISSEN, P., GUSTAFSSON, B., EDVARDSSON, B., & GILMORE, G. 1994 *A&A* **285**, 440.

NOTA, A. 1998. In *Unsolved Problems in Stellar Evolution* (ed. M. Livio). Cambridge University Press. poster paper.

OEY, M. S. & MASSEY, P. 1995 *ApJ* **452**, 210.

PADOAN, P., NORDLUND, A., & JONES, B. J. T. 1997 *MNRAS* **288**, 145.

SAGAR, R., MIAKUTIN, V. I., PISKUNOV, A. E., & DLUZHNEVSKAIA, O. B. 1988 *MNRAS* **234**, 831.

PARKER, J. W., GARMANY, C. D., MASSEY, P., & WALBORN, N. R. 1992 *AJ* **103**, 1205.

PFENNIGER, D. & COMBES, F. 1994 *A&A* **285**, 94.

PHELPS, R. L. & JANES, K. A. 1993 *AJ* **106**, 1870.

PIOTTO, G., COOL, A. M., & KING, I. R. 1997 *AJ* **113**, 1345.

POUND, M. W., & BLITZ, L. 1995 *ApJ* **444**, 270.

PRICE, N. M., & PODSIADLOWSKI, PH. 1995 *MNRAS* **273**, 1041.

REID, I. N. 1998. In *The Stellar Initial Mass Function* (eds. G. Gilmore, I. Parry, & S. Ryan). p. 121. ASP.

REID, I. N. & GAZIS, J. E. 1997a *AJ* **113**, 2246.

REID, I. N. & GAZIS, J. E. 1997b *AJ* **114**, 1992.

RENZINI, A., CIOTTI, L., D'ERCOLE, A., & PELLEGRINI, S. 1993 *ApJ* **419**, 52.

RIEKE, G. H. 1991. In *Massive Stars in Starbursts* (eds. C. Leitherer, N. R. Walborn, T. M. Heckman, & C. A. Norman). p. 205. Cambridge University Press.

RIEKE, G. H., LEBOFSKY, M. J., THOMPSON, R. I., LOW, F. J. & TOKUNAGA, A. T. 1980 *ApJ* **238**, 24.

RIEKE, G. H., LOKEN, K., RIEKE, M. J., TAMBLYN, P. 1993 *AJ* **412**, 99.

SATYAPAL, S., ET AL. 1995 *ApJ* **448**, 611.

SATYAPAL, S., WATSON, D. M., PIPHER, J. L., FORREST, W. J., GREENHOUSE, M. A., SMITH, H. A., FISCHER, J., & WOODWARD, C. E. 1997 *ApJ* **483**, 148.

SALPETER, E. E. 1955 *ApJ* **121**, 161.

SANTIAGO, B. X., ELSON, R. W., & GILMORE, G. F. 1996 *MNRAS* **281**, 1363.

SCALO, J. M. 1985. In *Protostars and Planets II* (eds. D. C. Black, & M. S. Matthews). p. 201. University of Arizona.

SCALO, J. M. 1986 *Fundam. Cosmic Physics* **11**, 1.

SCALO, J. M. 1990. In *Physical Processes in Fragmentation and Star Formation* (eds. R. Capuzzo-Dolcetta, C. Chiosi & A. Di Fazio). p. 51. Kluwer.

SCALO, J. 1998. In *The Stellar Initial Mass Function* (eds. G. Gilmore, I. Parry, & S. Ryan). p. 201. ASP.

SCALO, J. M. & PUMPHREY, W. A. 1982 *ApJ* **258**, L29.

SCHAERER, D. 1996 *ApJ* **467**, L17.

SHU, F. H., ADAMS, F. C., & LIZANO, S. 1987 *ARAA* **25**, 23.

SILK, J. 1977 *ApJ* **214**, 718.

SILK, J. 1982 *ApJ* **256**, 514.

SILK, J. 1995 *ApJ* **438**, L41.

SILK, J., & TAKAHASHI, T. 1979 *ApJ* **229**, 242.

SIMON, M. 1997 *ApJ* **482**, 81.

SMITH, D. A., HERTER, T., HAYNES, M. P., BEICHMAN, C. A., & GAUTIER, T. N., III 1995 *ApJ* **439**, 623.

SMITH, D. A., HERTER, T., & HAYNES, M. P. 1998 *ApJ* **494**, 150.

SREENIVASAN, K. R. 1991 *Ann. Rev. Fluid Mech.* **23**, 539.

STAHLER, S. W., PALLA, F., & HO., P. T. P. 1998 In *Protostars and Planets IV* (eds. A. P. Boss & S. S. Russell), in press. Univ. Arizona Press.

STASIŃSKA,G., & LEITHERER, C. 1996 *ApJS* **107**, 427.

STUTZKI, J., BENSCH, F., HEITHAUSEN, A., OSSENKOPF, V., & ZIELINSKY, M. 1998 *A&A* **336**, 697

SUNG, H., BESSELL, M. S., & LEE, S.-W. 1998 *AJ* **115**, 734.

TATEMATSU, K., ET AL. 1993 *ApJ* **404**, 643.

TELESCO, C. M. 1988 *ARA&A* **26**, 343.

THUAN, T. X. 1991. In *Massive Stars in Starbursts* (eds. C. Leitherer, N. R. Walborn, T. M. Heckman, & C. A. Norman). p. 183. Cambridge University Press.

TOHLINE, J. E. 1980 *ApJ* **239**, 417.

TSUJIMOTO, T., YOSHII, Y., NOMOTO, K., MATTEUCCI, F., THIELEMANN, F.-K., & HASHIMOTO, M. 1997 *ApJ* **483**, 228.

VERSCHUEREN, W. 1990 *A&A* **234**, 156.

VOGELAAR, M. G. R., & WAKKER, B. P. 1994 *A&A* **291**, 557.

WANG, B., & SILK, J. 1993 *ApJ* **406**, 580.

WALTER, F. M., & BOYD, W. T. 1991 *ApJ* **370**, 318.

WRIGHT, G. S., JOSEPH, R. D., ROBERTSON, N. A., JAMES, P. A., & MEIKLE, W. P. S. 1988 *MNRAS* **233**, 1.

WYSE, R. F. G. 1997 *ApJ* **490**, L69.

WYSE, R. F. G. 1998. In *The Stellar Initial Mass Function* (eds. G. Gilmore, I. Parry, & S. Ryan), p. 89. ASP.

YOSHII, Y., & SAIO, H. 1985 *ApJ* **295**, 521.

ZIMMERMANN, T., & STUTZKI, J. 1992 *Physica A* **191**, 79.

ZIMMERMANN, T., & STUTZKI, J. 1993 *Fractals* **1**, 930.

ZINNECKER, H. 1982. In *Symposium on the Orion Nebula to honor Henry Draper* (eds. A. E. Glassgold, P. J. Huggins, & E. L. Schucking) p. 226. New York Academy of Science.

ZINNECKER, H. 1984 *MNRAS* **210**, 43.

ZINNECKER, H. 1996. In *The Interplay between Massive Star Formation, the ISM, and Galaxy Evolution* (eds. D. Kunth, et al.). p. 151. Editions Frontieres.

ZINNECKER, H., MCCAUGHREAN, M. J., & WILKING, B. A. 1993. In *Protostars and Planets III* (eds. E. H. Levy & J. I. Lunine). p. 429. University of Arizona Press.

ZUCKERMAN, B., & BECKLIN, E. E. 1992 *ApJ* **386**, 260.

Brown dwarfs: From mythical to ubiquitous

By JAMES LIEBERT

Steward Observatory, University of Arizona, Tucson, AZ 85721

Astrophysical objects below the stellar mass limit but well above the mass of Jupiter eluded discovery for nearly three decades after Kumar first proposed their existence, and for two decades after Tarter proposed the name "brown dwarfs." The first unambiguous discoveries of cluster brown dwarfs (Teide 1, PPL 15) and of planetary (51 Peg B) and brown dwarf (Gliese 229B) companions occurred about three years ago. Yet while extrasolar planets are now being discovered at a breathtaking rate, brown dwarf companions to ordinary stars are apparently rare; likewise, imaging surveys show that GL 229B is the only "methane dwarf" companion to a low mass star.

On the other hand, the deep imaging studies of the Pleiades and several imbedded young clusters show that the mass function (i.e. of single objects) extends in substantial numbers down to at least 40 Jupiter masses. The high mass/stellar density Orion Nebula Cluster may have relatively fewer low mass objects.

In the field of the solar neighborhood, the infrared sky surveys DENIS and especially 2MASS show that brown dwarfs, certified by the lithium test, exist in significant numbers. These appear to include most of the newly-defined spectroscopic class of L dwarfs; these objects are cooler than, and with different atomic and molecular absorption features than, the late M dwarfs. If the first 1% of sky analyzed is not atypical, over a thousand L dwarfs should be detected in the 2MASS survey.

---

1. A checkered history

Kumar (1963) first showed that the minimum mass for stellar objects to fully stabilize themselves by hydrogen-burning is 0.07 $M_\odot$ (or a bit higher). The calculations of Grossman, & Graboske (1974) suggested a terminus of 0.08 $M_\odot$, and the best calculations today predict that the boundary falls between the above two values (for solar composition). Tarter (1975) first proposed the term "brown dwarfs" for objects below this main sequence limit, but substantially larger than the mass of Jupiter (M_J). Her thesis advanced the hypothesis that such objects could provide the "missing mass" known dynamically to exist in galaxies and clusters of galaxies. Other suggested designations for these hypothetical objects, such as black or infrared dwarfs, were also proposed.

By 1986, the time of the first conference dedicated to this subject (Kafatos, Harrington & Maran 1986), Tarter's label was widely accepted (Tarter 1986); however, not a single unambiguous case of a substellar object of this type was known. In particular, the object featured on the cover of that particular conference proceedings turned out to be perplexingly mythical.

The next several years featured many claims of the discovery of brown dwarfs (hereafter BDs) and extrasolar planets—both in the field, in clusters, and as companions to known stars. However, these cases proved unsustainable or ambiguous. A big part of the problem was that nobody knew where the terminus really lay in an observational H-R diagram. Fainter and fainter objects were being found that appeared to have temperatures and luminosities that might be interpreted as an extension of the main sequence, both in the field and in clusters (especially the Hyades and Pleiades). The most extreme case by the late 1980s was GD 165B, found by Becklin & Zuckerman (1989) with an estimated $M_{bol} = 14.99$. Even this object could be interpreted barely as stellar, if interior models of high opacity were adopted.

In particular, D'Antona & Mazzitelli (1985, DM85) predicted that the main sequence could stretch a factor of ten lower in luminosity in comparison with earlier models, and pointed out the existence of what they called objects of "transition mass." Their calculations showed that, near 0.07 $M_\odot$, the configuration initiates proton-proton burning, for a period of time which decreases with mass. However, since the pressure support generated from the nuclear reactions never quite achieves 100% of what is needed to prevent a slow contraction of the star, the increasing densities bring an increasing degenerate electron pressure. The onset of energy transport by conduction results in the eventual decrease of the central temperature, which shuts off the nuclear reactions—but, for their 0.075 $M_\odot$ model, only after 10^{10} years!

The results of DM85 meant that the main sequence terminus itself, for solar composition at least, is quite "fuzzy." That is, the criteria for defining a BD become fuzzy as well. Should the 0.075 $M_\odot$ model of DM85 be called a brown dwarf or a star? Moreover, even objects with a limited or no hydrogen-burning phase traverse paths close to the zero age main sequence, spending substantial time in gravitational contraction to their final radii. Whether stellar or substellar, the mass assignable to a given luminosity is therefore very sensitive to the age. The small differences in T_{eff} at a given luminosity are well below the accuracies of current models in fitting observations, for a wide range of mass.

The above quandary obviously underscores the value of searches for substellar objects in the nearest clusters. The age of the cluster is known (with possible caveats), and the young brown dwarfs are more luminous anyway than older counterparts. However, for those of us searching in the field of the local disk, the burden of proof is very high—it is necessary to determine or bound the mass of the object independently of the HR Diagram. One long sought solution is to find an astrometric or spectroscopic binary with a substellar component. Indeed, Latham et al. (1989) detected a radial velocity companion to the F9 V star HD 114762 with a mass function implying $M_2 \sin i_{orb} = 0.011$ $M_\odot$. However, Cochran, Hatzes & Hancock (1991) presented strong evidence based on studying the line profiles of the star that our view of this system is close to pole-on, underscoring the possibility that the companion need not be substellar.

Fortunately, an easier solution has emerged, pioneered by Rebolo, Martin & Magazzu (1992): This is the so-called "lithium test"—the detection of Li in the spectrum for a configuration predicted to be completely convective throughout. Since Li is depleted at central temperatures of a few million degrees, its survival means that hydrogen burning (requiring a slightly higher temperature) is not taking place. However, even the Li I resonance transitions in the red spectrum generally are weak lines of an element of low abundance. Thus the test requires high resolution spectra, of very cool objects having precious little red flux.

By the mid-1990s, even the cluster searches were yet to uncover ironclad candidates, and neither the appropriate binary nor a candidate showing lithium had turned up in the field. Then, two nearly-simultaneous discoveries were announced in the same volume of *Nature*. These were (1) the first extrasolar planet, 51 Cyg B, from radial velocity searches (Mayor & Queloz 1995), and (2) the first unambiguous brown dwarf, Gliese 229B (Nakajima et al. 1995). It should be noted, however, that the discovery of the Pleiades cluster brown dwarf Teide 1 (discussed in Section 3) actually predates the above objects (Rebolo, Zapatero Osorio & Martin 1995), although its confirmation via the lithium test came later. Indeed, in the last three years a plethora of extrasolar planets and BDs have been unveiled.

The BD discoveries can be divided into those found (1) as companions to known nearby stars, (2) as members of nearby star clusters, including very young objects in

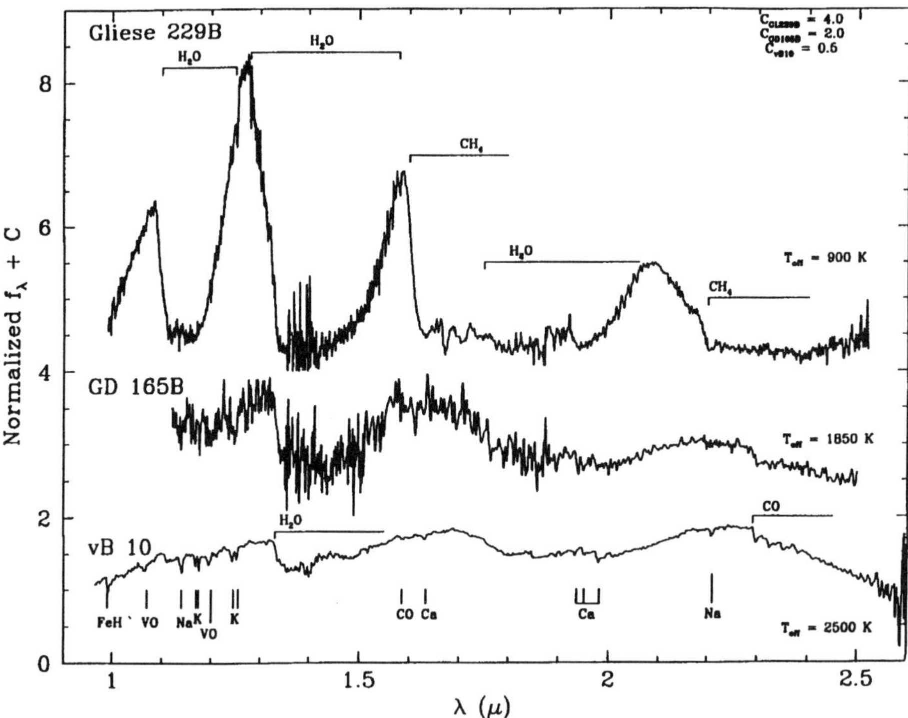

FIGURE 1. The near IR spectrum of GL 229B compared with spectra of vB10 (M8 V) and GD 165B (L2 V, see section 4) . The great strengths of the CH_4 and H_2O bands in GL 229B give it a blue J-K color. From Geballe et al. 1996, Jones et al. 1994, and Oppenheimer et al. 1998b.

giant molecular clouds, and (3) as field objects in the solar neighborhood. For reasons of space and the degree of my own involvement, a relatively greater emphasis is placed on the discovery of field objects, especially results from the Two Micron All Sky Survey (2MASS).

2. Brown dwarf companions to nearby Solar-Type stars

While other likely brown dwarf companions have been reported, GL 229B has remained unique (at the time of this writing) as the only object with T_{eff} near 1,000 K and methane in its spectrum. In Fig. 1 is shown the near-IR spectrum of this object, showing deep absorption features due to H_2O and CH_4, compared with spectra of GD 165B (an L dwarf) and the benchmark late M dwarf vB 10 (from Oppenheimer et al. 1998b). The need to explain the unusual atmospheric composition of GL 229B, the role of chemical equilibrium of gas species and various kinds of dust, mixing processes, the likely spatial and time variability ("weather")—make it a new astronomical field all to itself. Further discussion is beyond the scope of this paper, but the reader is referred to Oppenheimer et al. (1998b) and to Burrows & Sharp (1998) for, respectively, an observational and theoretical sampling.

In contrast, the inferred discoveries of extrasolar planetary companions to nearby main sequence stars by Butler & Marcy (1996), the Mayor group, and others have continued at a fast and furious pace. At first it appeared also that the radial velocity companions included a substantial fraction of BDs—defined (somewhat arbitrarily) as 0.01 $M_\odot$ and larger—generally in quite eccentric orbits. Recently, however, Mayor et al. (1998) reported that orbital inclination determinations based on HIPPARCOS astrometry show that several of the companions previously believed to be substellar have orbital inclinations high enough to lie near or above the stellar mass limit. A few known cases remaining with lower-limit masses well below the stellar limit do *not* have accurate orbital inclination determinations. The possibility exists at the time of this writing that the fairly close substellar companions amenable to detection from radial velocity surveys include brown dwarfs only rarely, if at all.

In support of this conclusion may also be the imaging survey of Oppenheimer et al. (1998a); following their early discovery of GL 229B, they have up to now reported no new substellar companions to the over 100 solar neighbors in their survey. Thus, it now appears from both types of observations that BDs orbiting solar-type stars are rare.

3. Brown dwarfs in star clusters

3.1. *The Pleiades*

The studies of star clusters began bearing real fruit—verifiable brown dwarf members—also in the last few years. This followed a series of false starts and ambiguous results, many involving the Hyades cluster and the Taurus clouds. The need to survey large areas with difficult background fields have plagued the studies of these two stellar aggregations. The Pleiades cluster is younger but more distant than the Hyades. Coincidentally, the two effects nearly offset, so that the BDs were predicted to have about the same luminosities as the Hyades counterparts. Moreover, the Pleiades is a much more compact cluster, with a less difficult background field. It is not surprising, therefore, that it has now yielded some of the most solid BD candidates. Here we sketch incompletely some major findings.

The first important Pleiades candidate, PPL 15, emerged from a deep CCD survey of Stauffer, Hamilton and Probst (1994). Basri, Marcy and Graham (1996) detected lithium for the first time in this cluster candidate, but even the apparent passing of the lithium test left an ambiguous situation. The age of the Pleiades has been estimated variously between 70 Myrs for upper main sequence interiors with no convective core overshooting, to 120 Myrs if overshooting adds more hydrogen fuel to the core. With the former value, the luminosity of PPL 15 corresponds to a substellar mass. However, the Li detection was weak enough that the authors found that it had been partially depleted. The most consistent picture at the time was provided by concluding that the age was closer to 120 Myrs, and the mass right at the hydrogen-burning mass limit. Lower luminosity BD candidates were soon found, though we shall return to the saga of PPL 15.

Deeper CCD surveys performed primarily in the Canaries by European investigators found two important brown dwarf candidates, Teide 1 and Calar 3. Both have spectral types of M8 (Martin, Rebolo & Zapatero Osorio 1996), yielding masses of 0.055 ± 0.015 $M_\odot$ (Rebolo et al. 1996), assuming the 120 Myr age. Most importantly, high resolution spectra show that these have kinematics consistent with cluster membership, and detections of apparently-undepleted lithium. In the last two years, even deeper surveys have been performed which emphasize redder CCD bandpasses like I and Z (cf. Zapatero Osorio et al. 1997) and resulting in the discoveries of the even-fainter "Roque" candidates (named for the unique rock formation at the summit of La Palma). The faintest of these appears to cross into the L spectral class discussed in Sect. 4.2 with a corresponding mass below 40 M_J (Martin et al. 1998).

At the time of this writing, there are of the order 50 photometric BD candidates from these studies. The inferred luminosity function of the Spanish/European groups suggests that the number of BDs in the cluster down to 30–40 M_J could be of the order of a few hundred, a number density comparable to that of stars. The mass contribution would of course be considerably less, though the slope of the mass function in logarithmic units appears to be flat to slightly negative (slightly more objects with decreasing mass). Moreover, a consistent lithium boundary line—below which lithium always appears, above which it does not—lies near the hydrogen-burning mass limit for the Pleiades age.

Finally, PPL 15 was shown to be the first binary brown dwarf, using *HIRES* on Keck I (Basri & Martin 1998). The excessive luminosity is explained by two components, each of mass near 0.065 $M_\odot$, contributing roughly equally to the total light. The period is 5.8 days, at a separation of a few $R_\odot$, and an orbital eccentricity near 0.5.

3.2. *Imbedded clusters*

Younger, nearby star-formation regions offer more luminous brown dwarfs, though one usually has to cope with dust extinction by working at infrared wavelengths. The following examples offer only an incomplete summary of studies of the low end of the mass function in several imbedded regions. The ρ Ophiuchis cluster is one of the nearest but most imbedded (Comeron et al. 1993; Luhman & Rieke, in preparation, see also Strom, Kepner, & Strom 1995). Others include NGC 2024 (Comeron et al. 1996), L1495E in Orion (Luhman & Rieke 1998), and IC348 (Luhman et al. 1998). There is little doubt that these studies find candidates extending well into the substellar mass range, perhaps to the 30–40 M_J reached in the Pleiades. These authors generally conclude that the log mass function—again in apparent agreement with the Pleiades—is flat, or with a small negative slope, for the clusters they have studied. On the other hand, Hillebrand (1997) in a comprehensive study of the rich Orion Nebula Cluster (ONC) finds that its mass function peaks near 0.2 $M_\odot$ and falls rapidly towards lower masses. This ONC region of higher stellar density with many massive stars may thus be a less hospitable environment for the formation of very low mass stars and brown dwarfs. However, the ONC is also generally more distant than the other clusters, so that survey completeness below the stellar mass limit may be a more difficult issue.

That the above photometric studies are finding legitimate BD members of the clusters is again demonstrated by spectra. In Fig. 2 several spectra are shown (courtesy of Kevin Luhman), which demonstrate that these are young BD members of low mass. Their spectra contrast with those of GL 229B and the older, field BD candidates discussed in the next section. The young imbedded cluster candidates are both warmer and more luminous: an M7–M9 spectrum can come from a young object of several tens of Jupiter masses. Moreover, these differ from the spectra of dwarfs. The gravity-dependent features indicate objects intermediate in luminosity between dwarfs and giants. Finally, the identification of legitimate BDs is important in order to study whether proto-planetary disks can form around such objects, analogous to what produced the Galilean moon system around Jupiter, and whether these low mass objects show T Tauri-like activity, or far-infrared excesses.

4. The first field brown dwarfs

Both the results from the clusters discussed above and the discoveries of GD 165B and GL 229B many tens of a.u. from their companion stars indicated that brown dwarfs could be found in the solar neighborhood. The problem is that such objects might represent

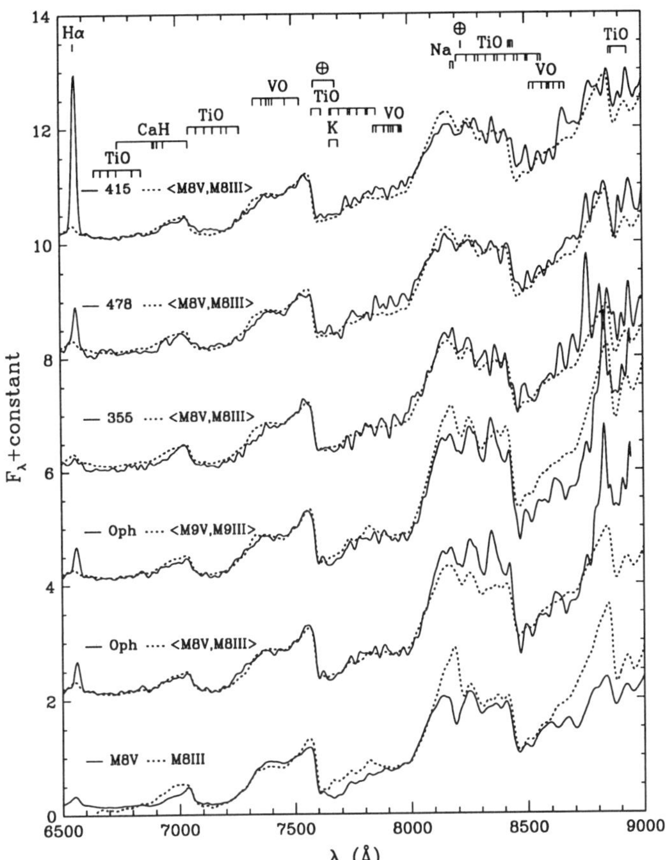

FIGURE 2. Four brown dwarfs well below the stellar mass limit, found by Luhman et al. (1998) in IC 348 and ρ Oph. The three latest sources observed in IC 348 (415 = M7.5, 478 = M7.5, 355 = M8) and ρ Oph 162349.8−242601 (M8.5) (solid lines) are plotted with averages of standard M8 and M9 dwarfs and giants (dotted lines). Features which are sensitive to surface gravity are apparent in the comparison of the M8 V and M8 III spectra. All spectra are normalized at 7500 Å.

a wide range of age in the Galactic disk, and even the distances would be initially unknown. On the other hand, the lithium test might be applied, or, for extremely cool cases, the methane test. For GD 165B, even an excellent Keck II spectrum lacks sufficient flux at 6700 Å to test for lithium. The new study by Kirkpatrick et al. (1999a) makes the case that it is a brown dwarf, although the minimum age assignable from the cooling time of the white dwarf companion suggests that it is undergoing at least a long transitory phase of nuclear-burning. (I reiterate that the definition of a "brown dwarf" is ambiguous!) However, in the last few years some cool field objects have been found which pass the lithium test, which means that the central temperatures are low enough that any hydrogen-burning phase would have been brief.

From UK and ESO Schmidt plates, Thackrah, Jones & Hawkins (1997) found an M6 dwarf which shows a distinct Li I 6707 Å detection. Depending on the age, $T_{\rm eff}$, implied Li abundance, and the uncertainties due to models, it still appears to straddle the stellar boundary. A second, much cooler object is Kelu 1, found in the Chilean proper motion survey by Ruiz, Leggett & Allard (1997). The model atmospheres generated by Allard suggest that $T_{\rm eff}$ is about 1,900 K. Its spectrum places this object among the L dwarfs discussed in Sect. 4.2. The presence of Li makes it probable that this object is substellar. A third case has been found from an old Luyten proper motion catalog, LP 944-20 (Tinney 1998). This object has been assigned spectral type M9 V by Kirkpatrick, Henry & Simons (1995), making the T_{eff} intermediate between the first two objects discussed above. Tinney (1998) argues that the mass is near 0.06 $M_{\odot}$, and the age 475–700 Myr. Again, the sequence of Li observations in the Pleiades—where ages and luminosities are known—lends credence to the model-dependent interpretations of the field objects.

4.1. *The DENIS and 2MASS Surveys*

Finding the nearest, ultracool solar neighbors is extremely valuable since these will be the brightest such objects, and most amenable to detailed followup studies. Fortunately, within the last few years, two groups have begun the first substantial sky surveys at near-infrared wavelengths. The DEep Near Infrared Sky survey (DENIS) was started in January 1996 by a consortium of European investigators for the southern hemisphere. The Two Micron All Sky Survey (2MASS) began in May 1997 at Mt. Hopkins for the northern sky, and at CTIO in February 1998 for the southern sky (Skrutskie et al. 1997).

In the paper by Delfosse et al. (1997), DENIS announced the discovery of the first brown dwarf candidates from an infrared survey; in a companion paper, Martin et al. (1997) reported the detection of lithium in one of these candidates, demonstrating that it is a BD near 60 M_J. The DENIS candidates also showed spectra obviously later than the end of the defined M sequence (M9/M9.5 V), with qualitatively different spectral features. It is in the latter citation above that a brief proposal first appears in a refereed journal that these objects be called L dwarfs, following a suggestion by Kirkpatrick (1997). We shall discuss the characteristics of these objects below.

In the first 400 square degrees of 2MASS data, plus some other fields analyzed with protocamera scans, 20 L dwarfs were found (Kirkpatrick et al. 1999b, hereafter K99). A total of 25 were known at the time of the preparation of this paper, including the three DENIS objects, Kelu 1, and GD 165B. The 2MASS selection criterion which has a high degree of success in identifying very late M and L dwarfs is that J–K > 1.3, while R–K > 5.5, where "R" is the photographic red magnitude from the Palomar Observatory Sky Survey (POSS 1 or 2).

4.2. *The L dwarfs*

A complete classification system has been developed in K99; some principal features are summarized here. In Fig. 3 spectrophotometry is shown of one late M dwarf, one middle and one very late 2MASS L dwarf. These cover "far red" wavelengths (6200–10000 Å), and were obtained with the Keck II low resolution spectrograph (LRIS).

The strongest features at these wavelengths in late M dwarfs are the prolific band systems of titanium and vanadium oxides, just as water and carbon monoxide dominate at longer wavelengths. So strong are these that nearby "pseudo-continuum" peaks tower high above the wavelengths where the absorptions are strongest, though at no wavelength is the true continuum reached. By the latest M types, however, the TiO strengths have peaked out and begun to weaken. At the slightly cooler T_{eff} where VO begins to weaken as well, this is defined as the beginning of the L sequence (L0 V). With progressively increasing L type (and decreasing T_{eff}), the TiO and VO bands weaken (in the L3 V dwarf of Fig. 3) and then disappear entirely (the L8 V object).

Our understanding of the physics of the above phenomena is as follows: The M dwarf temperature scale is not known to better than 10% or so, but it is believed that below about 2500 K, first the TiO and then VO molecules in the atmospheres precipitate out as dust grains (Tsuji et al. 1996ab; Allard 1997; Burrows and Sharp 1998). By removing

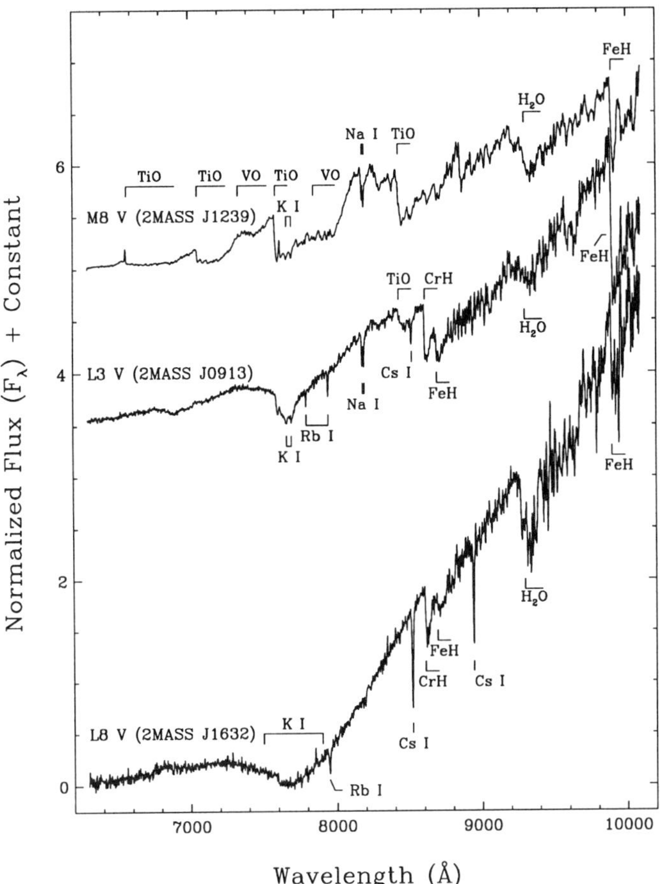

Wavelength (Å)

FIGURE 3. Red spectra of an M8 V dwarf (top) an L3 V dwarf (middle) and very late L8 V dwarf (bottom). Note the dominating strength and width of the K I resonance doublet in the L8 V spectrum, and other resonance lines of rarer alkalis, along with the disappearance of the TiO and VO bands.

these principal sources of opacity from the atmospheres, dwarfs a few hundred degrees cooler than late M stars have more transparent atmospheres at these wavelengths. Among the remaining absorbers are hydride bands with limited wavelength coverage. Otherwise, there is little continuum opacity, due to the extremely low electron density (for H^- and H_2^-). Thus, the atomic resonance lines of the neutral alkalis become quite strong for an abundant species like K I, and easily detected for rare species like Rb I, Cs I, and if undepleted, Li I. That is, the crucial spectral identifier (Li I) of high mass brown dwarfs appears strongly enough at an undepleted abundance to be detectable on low resolution spectra, rather than requiring the high resolution needed for M dwarfs—see Fig. 4, discussed below. At late L types, these trends carry to extreme. As is evident for the L8 V dwarf in Fig. 3, the K I doublet with great width and strength becomes the dominant feature of the red spectrum. (The Na I resonance doublet would be even stronger, if our spectra extended to short enough wavelengths.)

Late L dwarf optical spectra are in one respect analogous to those of cool white dwarfs, which may also show strong, pressure-broadened lines. On the other hand, in the near infrared, the L dwarfs show very strong CO and H_2O, but these are the same features seen in M dwarfs (see GD 165B in Fig. 1). The effective temperatures of the coolest

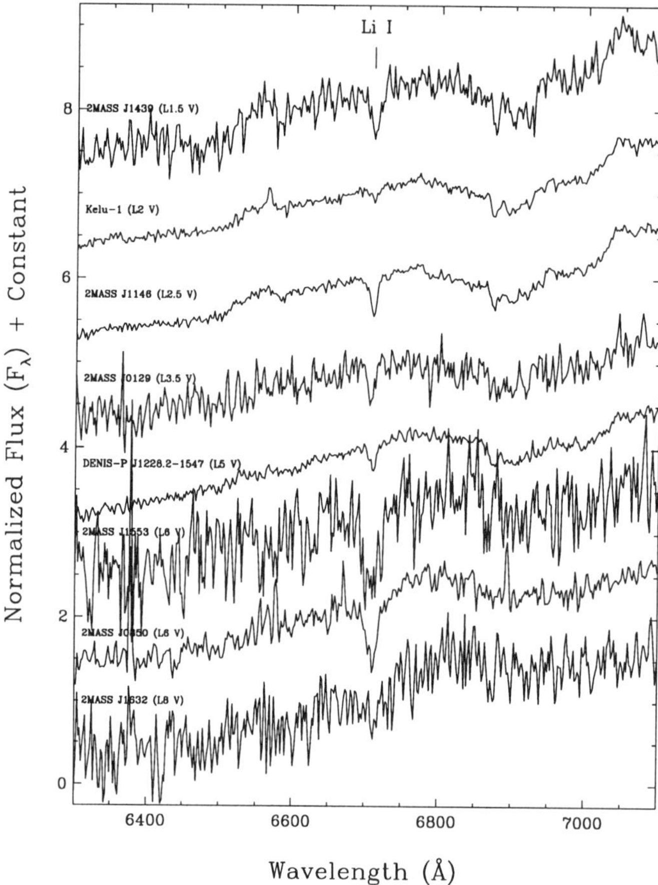

Wavelength (Å)

FIGURE 4. Spectra of a sequence of L dwarfs, increasing in type from top to bottom, centered on the Li I 6707Å resonance doublet. The line is generally detected, but highly variable in strength. The Hα line is also included, but appears convincingly in emission only for Kelu 1.

L dwarfs are not known, but in any case they are not cool enough for the CO molecule to give way to CH_4, as it is predicted to do near 1500 K (Burrows & Sharp 1998). The only known "methane dwarf" remains GL 229B.

The onset of methane in the near-infrared spectrum will reverse the trend in J–K color with decreasing temperature (Burrows et al. 1997). GL 229B has a very blue J–K = −0.1. Thus the selection criteria must be modified to search for objects with blue or neutral J–K color, increasing enormously the numbers of apparent point sources which must be screened. At the time of this writing, it must be concluded that we have not yet searched effectively for field "methane dwarfs."

The yield of the L dwarfs is substantial enough to conclude that they are present in significant numbers in the field around the Sun. It is elementary to do the math: Some 15 were found in the first 400 square degrees (1%) of sky, which means that the entire sky may yield at least 1,000 objects to the 2MASS survey depth. (This neglects the likelihood that the current color selection techniques do not find 100% of the L dwarfs, not to mention methane dwarfs.) It is too early for a first estimate of the space density of substellar objects, since the distances and luminosities are not yet known (not to mention the ages and masses).

We can show that many are substellar, however. In Fig. 4, a handful of our objects are illustrated in spectra centered on the Li I 6707 Å resonance doublet. The candidates observed with Keck II fall into three categories: (1) those with Li I detections at equivalent widths of several Angstroms, due to the high atmospheric transparency discussed earlier; (2) those with good enough spectra to show that Li I is depleted and undetected; and (3) those too faint for a decent spectrum to be achieved at these wavelengths, even with a 10 meter telescope. The statistics suggest that most of the L dwarfs are indeed brown dwarfs, but the reader is referred to K99 for justification of this conclusion.

This work was supported by NASA's Jet Propulsion Lab through contract number 961040NSF, a core science grant to the Two Micron All Sky Survey science team. I also acknowledge support from the National Science Foundation through grant AST 92–17961.

REFERENCES

ALLARD, F. 1998. In *Very Low Mass Stars and Brown Dwarfs in Clusters and Associations*.

BASRI, G., MARCY, G., & GRAHAM, J. R. 1996 *ApJ* **458**, 600.

BASRI, G. & MARTIN, E. 1998. In *Very Low Mass Stars and Brown Dwarfs in Clusters and Associations*.

BECKLIN, E. E. & ZUCKERMAN, B. 1988 *Nature* **336**, 656.

BURROWS, A. & SHARP, C. M. 1998 *ApJ*, submitted.

BURROWS, A., MARLEY, M., HUBBARD, W. B., LUNINE, J. I., GUILLOT, T., SAUMON, D., FREEDMAN, R., SUDARSKY, D., & SHARP, C. 1997 *ApJ* **491**, 856.

BUTLER, R. P. & MARCY, G. W. 1996 *ApJ* **464**, L153.

COCHRAN, W. D., HATZES, A. P. & HANCOCK, T. J. 1991 *ApJ* **380**, L35.

COMERON, F., RIEKE, G. H., BURROWS, A., & RIEKE, M. J. 1993 *ApJ* **416**, 185.

COMERON, F., RIEKE, G. H., & RIEKE, M. J. 1996 *ApJ* **473**, 294.

D'ANTONA, F. & MAZZITELLI, I. 1985 *ApJ*, **296**, 502. (DM85)

DELFOSSE, X., TINNEY, C. G., FORVEILLE, T., EPCHTEIN, N., BERTIN, E., BORSENBERGER, J., COPET, E., DE BATZ, B., FOUQUE, P., KIMESWENGER, S., LE BERTRE, T., LACOMBE, F., ROUAN, D. & TIPHENE, D. 1997 *A&A* **327**, L25.

GEBALLE, T. R., KULKARNI, S. R., WOODWARD, C. E., & SLOAN, G. C. 1996 *ApJ* **467**, L101.

GROSSMAN, A. S., HAYS, D., & GRABOSKE, H. C. 1974 *A&A* **30**, 95.

HILLEBRAND, L. A. 1997 *AJ* **113** 1733.

JONES, H. R. A., LONGMORE, A. J., JAMESON, R. F. & MOUNTAIN, C. M. 1994 *MNRAS* **267**, 413.

KAFATOS, M. C., HARRINGTON, R. S., & MARAN, S. P. 1986 *Astrophysics of Brown Dwarfs*. Cambridge Univ. Press.

KIRKPATRICK, J. D. 1997. In *Very Low Mass Stars and Brown Dwarfs* (eds. Rebolo, R., Martin, E. L. & Zapatero Osorio, M. R.) ASP Conf. Series, v. 134. ASP Press.

KIRKPATRICK, J. D., ALLARD, F., BIDA, T., ZUCKERMAN, B., BECKLIN, E. E., CHABRIER, G., & BARAFFE, I. 1999a *ApJ*, submitted.

KIRKPATRICK, J. D., REID, I. N., LIEBERT, J., CUTRI, R. M., NELSON, B., BEICHMAN, C. A., DAHN, C. C., MONET, D. G., SKRUTSKIE, M. F., & GIZIS, J. 1999b *Apj*, submitted (K99).

KUMAR, S. 1963 *ApJ* **137**, 1121.

LATHAM, D. W., MAZEH, T., STEFANIK, R., MAYOR, M. & BURKI, G. 1989 *Nature* **339**, 38.

LUHMAN, K. L., & RIEKE, G. H. 1998 *ApJ* **497**, 354 (L1495E).

LUHMAN, K. L., RIEKE, G. H., LADA, C. J., & LADA, E. A. 1998 *ApJ* **508**, 347 (IC348).

MARTIN, E. L., BASRI, G., DELFOSSE, X., & FORVEILLE, T. 1997 *A&A* **327**, L29.

MARTIN, E. L., BASRI, G., ZAPATERO OSORIO, M. R., REBOLO, R., & GARCIA LOPEZ, R. J. 1998. In *Very Low Mass Stars and Brown Dwarfs in Clusters and Associations*.

MARTIN, E. L., REBOLO, R., & ZAPATERO OSORIO, M. R. 1996 *ApJ* **469**, 706.

MAYOR, M. & QUELOZ, D. 1995, *Nature* **378**, 355.

MAYOR, M., ARENOU, F., HALBWACHS, J., QUELOZ, D., & UDRY, S. 1998 In *Extrasolar Planets: Formation, Detection and Modelling*.

NAKAJIMA, T., OPPENHEIMER, B. R., KULKARNI, S.R., GOLIMBOWSKI, D. A., MATTHEWS, K., & DURRANCE, S. T. 1995 *Nature* **378**, 463.

OPPENHEIMER, B. R., KULKARNI, S. R., GOLIMOWSKI, D. A. & MATTHEWS, K. 1998a. In *Very Low Mass Stars and Brown Dwarfs in Clusters and Associations*.

OPPENHEIMER, B. R., KULKARNI, S. R., MATTHEWS, K. & VAN KERKWIJK, M. H. 1998b *ApJ*, submitted.

REBOLO, R., MARTIN, E. L., BASRI, G., MARCY, G.W., & ZAPATERO OSORIO, M. R. 1996 *ApJ* **469**, L53.

REBOLO, R., MARTIN, E. L. & MAGAZZU, A. 1992 *ApJ* **389**, L83.

REBOLO, R., ZAPATERO OSORIO, M. R., & MARTIN, E. L. 1995 *Nature* **377**, 129.

RUIZ, M. T., LEGGETT, S. K., & ALLARD, F. 1997 *ApJ* **491**, L107.

SKRUTSKIE, M. F. ET AL. 1997. In *The Impact of Large Scale Near-IR Sky Surveys* (eds. F. Garzon et al.). p. 25. Kluwer.

STAUFFER, J. R., HAMILTON, D., & PROBST, R. 1994 *AJ* **108**, 155.

STROM, K. M., KEPNER, J. & STROM, S. E. 1995, *ApJ* **438**, 813.

TARTER, J. C. 1975 Ph.D. thesis, University of California, Berkeley.

TARTER, J. C. 1986. In *Astrophysics of Brown Dwarfs* (eds. M. C. Kafatos, R. S. Harrington, and S. P. Maran) p. 121. Cambridge Univ. Press.

THACKRAH, A., JONES, H., & HAWKINS, M. 1997 *MNRAS* **284**, 507.

TINNEY, C. G. 1998 *MNRAS* **296**, L42.

TSUJI, T., OHNAKA, K., & AOKI, W. 1996 *A&A* **305**, L1.

TSUJI, T., OHNAKA, K., AOKI, W., & NAKAJIMA, T. 1996 *A&A* **308**, L29.

ZAPATERO OSORIO, M. R., REBOLO, R., MARTIN, E. L., BASRI, G., MAGAZZU, A., HODGKIN, S. T., JAMESON, R. F., COSSBURN, M. R. 1997 *ApJ* **491**, L81.

Applications of accurate masses, radii, and luminosities

By J. ANDERSEN

Astronomical Observatory; Niels Bohr Institute for Astronomy, Physics & Geophysics;
University of Copenhagen, Denmark (*ja@astro.ku.dk*)

The determination, status, and interpretation of accurate data on the masses, radii, and luminosities of normal stars are briefly reviewed. With modern techniques, errors in M and R of well-studied eclipsing binary stars can now be reduced to such a level (1–2%) as to introduce negligible uncertainty in the comparison with models. The error budget in critical model tests is then dominated by the transformation from observed colours to effective temperature when the metal abundance is known. Unfortunately, observational data on metal abundances are still missing for most systems. Still, the present body of data is complete and accurate enough to demonstrate significant discrepancies with some of the currently competing stellar models. Accurate stellar masses and radii can thus help to narrow down the choice of solutions to some of the unsolved problems in stellar evolution, and prospects for rapid progress are excellent.

1. Introduction

A stellar evolution model is specified by the mass and chemical composition of the star. These, together with the physics built into the model, uniquely predict all observable properties of the star as functions of age. The main observable parameters of ordinary stars are luminosities and colours, from which the intrinsic properties radius, effective temperature, and chemical composition can be derived if good distance indicator and colours calibrations are available. Such observations form the basis for the classical models tests performed by comparison of computed isochrones with colour-magnitude diagrams (CMDs) of star clusters. However, while cluster CMDs offer a number of detailed diagnostic morphological features, they do not allow the direct test of a specific model that is possible when the initial stellar mass is independently known.

2. The role of binary star data

It is elementary textbook material that direct empirical mass determinations are only possible for stars in binary systems (and for the Sun). What is not elementary is how to determine individual masses and radii for stars in binary systems such that these data may help address some of the current unsolved problems in stellar evolution. For this, observational errors must have negligible effect compared to those of the effective temperatures and metallicities of the stars: Since the latter are significant at a level of some 5%, errors in M and R should be kept below 1% or at most 2%. Sections 3 and 4 of this paper review these issues, while Section 5 summarises the available data. Although most of the actual data in the classic review by Popper (1980) have now been superseded by more modern analyses, the general discussion in that paper contains much general wisdom that remains relevant today.

Even if known with infinite precision, the mass and radius of a star yield very limited useful information on stellar evolution theory if other significant parameters, such as effective temperature and chemical composition remain unconstrained by observation, just as for parallaxes (Lebreton, this volume). General agreement with theory within broad limits may be aesthetically pleasing, but does not advance our understanding: Only when the data and their uncertainties are sufficiently complete and precise that *some* of the competing theories can be ruled out are we making progress in understanding stellar evolution. Section 6 discusses how to do this, while the concluding Section 7 provides a forward look of the probable future development of the subject.

3. Progress in techniques

Until accurate *absolute* interferometric orbits, parallaxes, and angular diameters of both components of visual binary systems may become available, the determination of masses of *individual* stars in binary systems must rely on spectroscopic orbits of both components. Similarly, accurate radii can currently only be determined in eclipsing binary systems. Because the derived masses depend on the cube of the individual velocity amplitudes, the latter must have (random *and* systematic) errors of no more than $\sim 0.5\%$, not a trivial task. To remain within the total error budget of 1–2% required for data to be used in testing stellar models, the accompanying light curve and/or visual orbit solutions are subject to similar or even stricter standards.

Nonetheless, complete determinations at this level of accuracy are now possible fairly routinely, as described by Andersen (1991, 1997) and briefly updated in the following. A fundamental prerequisite is a basic observational material (spectra, light curves, etc.) that is complete and of adequate quality (see Andersen (1991) for details and examples). Further, the data must be analysed in a way that ensures complete consistency between all available information of every kind, drawing on other kinds of data as needed; for particularly tricky cases, see e.g. Andersen et al. (1985, 1991).

Modern, accurate spectroscopic orbits are derived from high-resolution digital spectra from which radial velocities are derived numerically. Like traditional line-by-line measurements, simple one-dimensional cross-correlation techniques applied to double-lined spectra are susceptible to systematic errors due to line blending effects. If not accounted for, these will propagate as systematic mass errors far in excess of those allowable in stellar model tests. Two recent developments in the analysis of double-lined spectra provide accurate and testable modelling of these effects:

First, the presence of two stars in the spectrum of the binary is explicitly modeled in the two-dimensional cross-correlation algorithm TODCOR by Zucker & Mazeh (1994). As demonstrated by Latham et al. (1996), TODCOR greatly reduces systematic errors from line blending relative to simple cross-correlation with a single template spectrum. However, as shown by Torres et al. (1997a), the derived orbital solution should always be tested by constructing a set of synthetic double-lined spectra with known radial velocities, matching the real data as closely as possible and analysed in exactly the same way as the real data. Any residual systematic errors in the analysis can be evaluated by this procedure and corrected for if necessary.

In the independent, so-called "disentangling" method of Simon & Sturm (1993), a set of double-lined binary spectra is subjected to a single-value decomposition analysis. The result is not only a set of accurate orbital parameters, but also an optimum reconstructions of the mean spectrum of each of the two binary components. These spectra can then be analysed by standard spectroscopic techniques for determination of effective temperatures and metallicity, key parameters in the discussion of masses and radii in terms of stellar models. The method appears to be of great promise but still needs to be stringently tested at the highest level of accuracy; the preliminary results of Hynes & Maxted (1998) are encouraging, however.

Photometric models of binary stars are used to derive the orbital inclination and individual stellar radii and surface fluxes from observed light curves and have reached a high degree of physical realism and numerical accuracy. Yet, the problem is highly non-linear, and in some frequently-encountered types of eclipsing binary (partial eclipses of similar stars), a wide range of parameters may produce near-identical light curves. Additional spectroscopic or other information must then be included for the analysis to yield truly reliable parameters. Formal errors from standard light-curve analysis programs alone are rarely realistic indicators of the true uncertainty of the parameters, and the importance of comprehensive consistency checks including a wide variety of data cannot be over-emphasized (see Andersen et al. (1980) for a general discussion and Andersen et al. (1985, 1991) for illustrative examples).

A technique which has recently begun to compete in this field is the application of optical interferometry to the visual orbits of double-lined spectroscopic binaries. As demonstrated by Hummel et al. (1994) and Torres et al. (1997b,c), the combination of an accurate interferometric determination of the projected relative orbit with accurate radial-velocity curves for both components is beginning to yield masses of competitive accuracy. At the same time, a very accurate parallax for the system and, therefore, individual luminosities for the two stars are determined in a fundamental manner. Assuming that precise effective temperatures can be derived, individual radii may then be computed which, while not derived in as fundamental a manner as those in eclipsing systems, are beginning to rival the latter in precision.

4. Observed colours vs. effective temperatures

Using the methods outlined above, fundamental determinations of stellar masses and radii can be made, reliably and almost routinely, to an accuracy of 0.5–1%. The situation is far less satisfactory with regard to the third fundamental parameter in comparisons with stellar models, effective temperature.

The determination of effective temperatures of real stars relies on observed colours and theoretical colour-temperature relations as derived from stellar atmosphere models (e.g. Edvardsson et al. (1993)) and/or from empirical calibrations (e.g. Alonso et al. (1996)). Conversely, colours can be derived from the theoretical surface parameters of the models using basically the same transformations. Present limitations of stellar-atmosphere theory (line blanketing, convection, surface inhomogeneities, etc.) result in uncertainties in these transformations which are *the* main current source of ambiguity in the comparison of stellar models with observations. For well-known reasons, these problems are at their worst when broad-band colours (such as UBV are used as the link between theory and observations.

It must be emphasized that this is a fundamental issue that affects the interpretation of all kinds of data, not only on binary stars but also, e.g. of clusters. A recent illustrative example is the open cluster study by Nordström et al. (1997). The thorough review of the problem given there need not be repeated here; but it must be borne in mind that any discussion in the following that involves the effective temperatures of real stars or, conversely, the colours of stellar models, is subject to the significant limitations of current effective temperature-colour transformations.

5. Available data

The available data of sufficient precision for critical studies of stellar evolution (errors below 2% in M and R) are listed and referenced in detail in the review by Andersen (1991); recent additions and corrections (including an error in Table 1 of that paper) are listed by Andersen (1997) and Pols et al. (1997). Andersen (1991) also reviewed the ways in

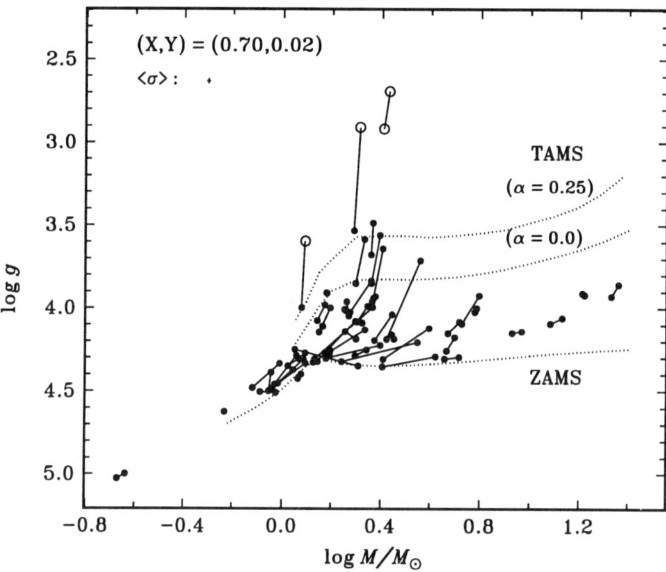

FIGURE 1. log$M$–log$g$ diagram for the current sample of well-observed binary components. Lines connect stars in a given binary system; open circles denote subgiant and giant stars.

which the data can be analysed in increasing detail as first only masses and radii, then effective temperatures and finally metal abundances can be included in the discussion. Unfortunately, the final key parameter in the comparison with models—a spectroscopic metal abundance—is still only available for the same three systems as in 1991. For basic observational reasons, most of the data pertain to (near) main-sequence stars in the range 1–20 M$_\odot$. No evidence exists that any of these systems have experienced significant mass exchange or mass loss during their evolution.

Only a couple of the most informative diagrams of the data will be shown here. The basic parameters derived most directly from the data are M and R. They are, for our purposes, best displayed as the combination log$M$–log$g$ (Fig. 1): In this diagram, the main-sequence band is almost horizontal, highlighting subtle evolutionary differences in radius within the main sequence; log$g$ of the best-observed systems pinpoint these stars to within 2% of the width of the main-sequence band, a sensitivity matched by few if any other kinds of stellar data. In the absence of mass loss, evolutionary tracks are vertical in Fig. 1, and the lines defining the upper boundary of the main-sequence band (TAMS) by standard and overshooting models (Claret (1995); $\alpha = 0$ and $\alpha = 0.25$, respectively) hint that significant information is contained already in these basic data.

Another popular diagram combining both M, R, and effective temperature is "the" mass-luminosity relation (Fig. 2), luminosity being computed from the absolute stellar radii and effective temperatures derived from observed colours as discussed above. In contrast to Fig. 1, the data form an apparently very tight and well-defined sequence that invites a fit describing "the" $M - L$ relation of the sample; and the relation is indeed often referred to as unique in the literature. We note that spectroscopic-interferometric binaries (see above) can provide additional points of high accuracy in the $M - L$ diagram.

The relation of Fig. 2 is, however, in reality not unique at all by the standards of the observational errors. Fig. 3 shows the residuals of the individual data points of Fig. 2 from a polynomial fit to the overall relation (main-sequence stars only). Broad agreement of models with a rough $M - L$ relation from binary data was regarded as

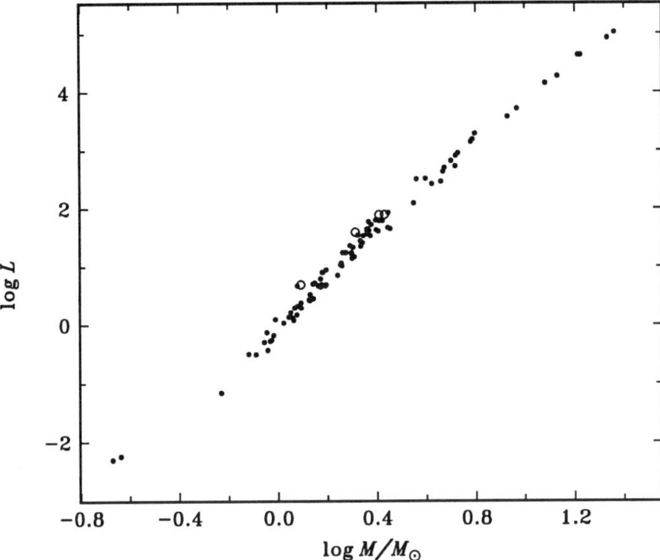

FIGURE 2. Mass-luminosity relation for the stars in Fig. 1.

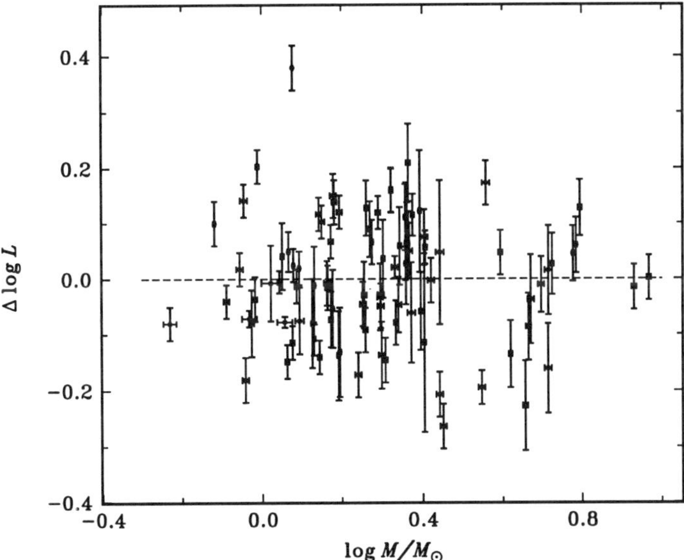

FIGURE 3. Residuals of individual main-sequence stars from the average mass-luminosity relation of Fig. 2.

a major success for early stellar evolution theory, but Fig. 3 emphasizes that the best modern luminosities data for binary stars deviate far more from a mean relation than can be ascribed to observational errors alone. The deviations are, of course, now readily explained by stellar models in terms of differences in age and metallicity, but is it not always appreciated how clearly these effects are in fact demonstrated by the data.

6. Interpretation of the data

Two fundamental assumptions are made when binary data are applied to test models for single stars: First, that no mass exchange or other interaction has modified the evolution of the stars significantly; and second, that the stars in a given system were formed simultaneously from the same material, differing only in mass. Accordingly, it is assumed that single-star models for the appropriate initial masses and chemical composition should reproduce all observed parameters for *both* stars for a single value of the age. The most stringent test of the models is therefore achieved in systems where age is the only parameter unconstrained by observation.

Unfortunately, still only three binary systems out of the fifty or so with accurate masses and radii have accurate spectroscopic metallicity determinations. More such determinations are under way (J. V. Clausen, priv. comm.), but in the interim, some comparisons must be made using samples that are heterogeneous with respect to age and metallicity. Although such procedures lead to less stringent conclusions, some results can be obtained, even with the use of just mass and radius.

The lines connecting two stars in the same binary in Fig. 1 are, in effect, observed isochrone segments that should be matched by the predictions of theoretical models for the observed metal abundance. A striking example was provided by the young B-type binary GG Lup (Andersen et al. (1993)): In the $\log M$–$\log g$ diagram, the least-evolved star (at $\log M/\mathrm{M}_\odot = 0.4$ in Fig. 1) is found to be on or even below the horizontal ZAMS of models for solar or higher metallicity, while the significant slope of the observed isochrone as defined by the two stars shows that at least the primary is already evolved well above the ZAMS. The firm (if model-dependent) conclusion is that the metallicity of GG Lup is well below that of the much older Sun, an observational fact of some significance for models of the chemical evolution of our Galaxy.

A more detailed comparison of models in the $\log M$–$\log g$ diagram was made recently by Torres et al. (1997a); in the particular main-sequence binary studied, the difference in mass (only 3%) was too small to clearly discriminate between different models in the $\log M$–$\log g$ diagram, but the potential of the method was clearly illustrated.

A prominent overall feature of Fig. 1 supports a robust conclusion: The number of stars that have evolved out of the main-sequence band as defined by the TAMS of standard models (convective cores computed using the Schwarzschild criterion, with no extra mixing) is much larger than predicted. Overshooting models perform much better in this regard as well as in fitting individual systems, as demonstrated already by Andersen et al. (1990) and later rediscovered by Pols et al. (1997).

In view of the vigorous debate on the physical nature of the mixing phenomenon that causes larger-than-classical convective cores and is commonly referred to as "overshooting," it is worth emphasizing that a diagram such as Fig. 1 does not in itself validate any particular convection model. What the figure does demonstrate is that the evolution in radius of models with enhanced mixing is consistent with that of real stars while that of standard models definitely is not. Testing physical recipes in more detail than this requires more data and a more stringent type of analysis.

The most critical tests of stellar models using binary data are made by detailed simulation of individual well-observed binaries with specially computed evolutionary tracks. No metallicity determinations for eclipsing binaries having been added recently, the most striking examples of such analyses probably still remain those of Andersen et al. (1988) and Andersen et al. (1991). Both systems analysed in these studies (AI Phe and TZ For) consist of stars with closely similar masses, yet in very different evolutionary stages and thus particularly suitable for highlighting the effects of differential stellar evolution within these binary systems.

In AI Phe, standard models calibrated to fit the radius and luminosity of the Sun at the solar age were shown to also fit both components of the binary with extraordinary precision: With mixing length and helium abundance determined by the solar fit and the masses and metal abundance by observation, age is the only free parameter allowed in fitting the position of the least-evolved star in the HR diagram. With the age fixed by this fit, there are then *no* free parameters allowed in the computation of a model for the more massive star. Yet, in AI Phe the fit was such that models differing by as much as 1% in age from the secondary would not fit the observed primary within the errors.

In contrast, standard models were found to be inadequate to fit both stars in the higher-mass system TZ For at a single age (Andersen et al. (1991)). Moreover, both stars were in unlikely, short-lived stages of evolution and the primary should have been through a He-flash episode, when its radius would have exceeded its Roche lobe and mass exchange would have radically changed the evolution of the components. On the other hand, models with enhanced mixing ("overshooting") from the convective core were able to match all observational facts of the system at a single age—not just the masses, radii, and luminosities, but also the unusual contrast of a circular orbit and synchronized primary star with a strongly super-synchronously rotating secondary (Claret & Giménez (1995) and Claret & Cunha (1997)). The requirement to simultaneously fit such a wide variety of qualitatively different, accurate data constitutes a very strong test.

Nonetheless, two caveats must be recalled: First, the practice of "normalizing" models by applying constant shifts in temperature and luminosity to the theoretical values in order to achieve consistency with the present Sun is basically meaningless when applied to overshooting models: Inasmuch as the Sun has no convective core, it can provide no test of the physics governing the extent of such cores. Second, even when models do fit all the data, this is again no proof that their physical basis is correct in any detail; all we can say is that the output of such models fits the real stars to a certain accuracy. However, if the models *disagree* significantly with the properties of even one real star, they must be rejected as inadequate in some respect.

An apparently subtle, but important point for the conclusions of the analysis concerns the treatment of temperature errors. Absolute values of the effective temperatures are derived from observed colours and contain the—usually dominant—contribution of the systematic errors of the calibration used, in addition to the observational error of the colour indices. However, the relative depth of primary and secondary eclipse in the light curves and the colour changes during eclipses contain direct and accurate information on the temperature *ratio* between the two components of a binary. Similarly, if eclipses are total, the luminosity *ratio* between the stars is determined far more accurately than suggested by propagated radius and absolute temperature errors.

As a result, the *slope* of the observed isochrone in the $\log T_e$–$\log g$ or HR diagrams is often defined by the two binary components with much greater accuracy than suggested by summary application of a data compilation. If such "details" are ignored, key information inherent in the data may be missed. E.g. Pols et al. (1997) find their fits with standard and overshooting models to be of nearly equal quality for a system like S Cen, whereas the accurate observed isochrone slope in the $\log T_e$–$\log R$ diagram is in fact quite incompatible with the former models. Conversely, significant discrepancies are sometimes blamed on unspecified systematic errors in the data or effects of mass exchange in the systems. While these possibilities must always be kept in mind, such conjectures should be supported by an analysis of the actual data and not merely be based on disagreement with some favourite set of models.

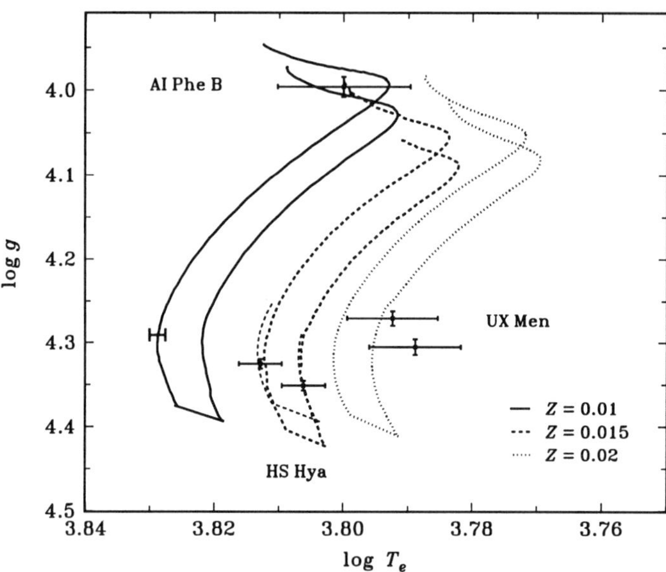

FIGURE 4. Evolution of AI Phe, HS Hya, and UX Men in the $\log T_e$–$\log g$ diagram. Thick lines: Models by Claret (1995); thin line: models for HS Hya by VandenBerg & Dowler.

A recent illustration of the increasing ability of accurate data to challenge model predictions was offered by the coincidence that the primary masses in the three systems AI Phe, HS Hya, and UX Men are identical within a total range of 1%, and the secondaries as well (Torres et al. (1997a)). Thus, the three systems differ only in metal abundance and age, HS Hya being the youngest and AI Phe the oldest. Fig. 4 shows the three sets of model pairs for the observed identical masses but different metal abundances, $Z = 0.010$, 0.015, and 0.020 for AI Phe, HS Hya, and UX Men, respectively. As is seen, the Claret (1995) models do an excellent job of fitting the observed differences in effective temperature as caused by the different metallicities of the three systems.

Yet, not all is still quite well, as shown by the second set of tracks for HS Hya at $Z = 0.015$ by VandenBerg & Dowler (priv. comm.). The small, but highly significant difference in ZAMS $\log g$ (only 0.02 dex) leads not only to very different age estimates from the two model series, but also to doubts as to how precisely model predictions can be trusted, even near the ZAMS. The unevolved HS Hya does not allow to discriminate between the two theoretical isochrone slopes, but systems with larger mass ratios and/or higher ages should provide more definitive information on this point.

Another discrepancy was been pointed out recently by Popper (1997). In his analysis of new data for binaries with masses slightly below solar, he found that the slopes of the observed isochrones in the $\log M$–$\log R$ and $\log M$–$\log L$ planes were shallower than predicted by models. In other words, the lower-mass components seemed to be more evolved than the primaries, an unlikely situation that suggests a failure of current models to correctly describe the evolution in radius of stars near 0.8 $M_\odot$.

While this mass range is not well covered by the best modern models, there seems to be no similar problem in fitting open cluster sequences in the same range (VandenBerg, priv. comm.). At the same time, the available binary data near 0.8 $M_\odot$ come largely from systems with short periods, fast rotations, and considerable spot activity, hence complicated light curves requiring more elaborate models and additional free (spot) parameters for their analysis. It would appear prudent to await additional data and/or more refined analyses before concluding that the models are seriously inadequate.

To conclude this section, we comment briefly on the combination of (open) cluster and binary data. A well-defined cluster sequence in a well-calibrated CMD contains information in the shape of the isochrone which is a far more detailed and subtle diagnostic than afforded by just two points as sampled by a binary system; on the other hand, the masses of cluster stars are generally unknown *a priori*. Clearly, an accurately studied double-lined eclipsing binary in a cluster with a well-observed CMD would allow both types of data to be combined, but few if any such clusters exist in which both the cluster and the binary have been analysed with sufficient accuracy.

However, if a well-studied binary and a cluster are known to have (near-) identical chemical composition and comparable ages, it may still be possible to combine both types of constraint on the models. Such a case was recently studied by Nordström et al. (1997), who fit a variety of standard and overshooting models to a new CMD of the open cluster NGC 3680. Two of the overshooting models provided fits of essentially equal quality in the CMD, but their predicted turnoff masses differed by 7%, an apparently small amount. Yet, by including the binary system TZ For in the analysis, of the same metal abundance but 25% younger than NGC 3680 (Andersen et al. (1991)), one set of models could be eliminated as predicting a significantly higher turnoff mass than observed, while the other set remained consistent with both types of data simultaneously.

7. Still-unsolved problems and outlook

This volume is devoted to an impressive array of still-unsolved problems in stellar evolution. Accurate data on masses and radii can be used, as outlined in this chapter, to address some of the more fundamental of these. Much progress is expected on both the observational and theoretical fronts in the near future, notably in the following respects.

First, more high-quality data will result from the new generation of large telescopes and efficient instruments, combined with advanced analysis methods as mentioned in Sect. 3. In particular, a small effort will suffice to provide the metallicity data needed to realise the full potential of the existing accurate mass and radius determinations, e.g. in discussions of the proper modelling of small convective cores—which again determines the age scale for most galactic disk stars. New data will also help to fill in the now sparsely-populated regions of Fig. 1, *viz.* the high- and low-mass ends of the main sequence. More data are also needed for giant stars, where optical interferometry of longer-period binaries will be most valuable; it is noteworthy that while impressive model fits to cluster CMDs are being obtained for the main sequences, few if any models are yet able to match the precise locations and detailed morphology of the observed giant branches.

Second, important progress will result from advances in stellar atmosphere theory, with more comprehensive opacity data and better physical descriptions of the convective motions and spatial inhomogeneities observed in the Sun and other real stars. More reliable transformations between theoretical and observed stellar parameters will then improve the discussion of stellar masses and radii in terms of stellar models, but the atmosphere models will also yield better outer boundary conditions for the interior models themselves, particularly important for very hot, very cool, or very evolved stars.

And third, when the ground has thus been reliably paved, precise stellar masses and radii will be integrated in the battery of tools—along with cluster sequences, elemental and isotopic surface abundances, apsidal motion and asteroseismology, to name a few— that will make it harder to hold full-scale conferences on our topic in the next century.

The inspiring collaboration over many years of, in particular, Birgitta Nordström, Jens Viggo Clausen, Dan Popper, and Don VandenBerg is gratefully acknowledged. I thank the organiser of the meeting and Editor of this volume, Mario Livio, for inviting me to a most interesting meeting, and for much patience with this contribution. Research on stellar masses and radii in eclipsing binaries at Copenhagen University is supported through Danish and ESO observing time at La Silla, Chile, and by continuing grants from the Danish Natural Science Research Council and the Carlsberg Foundation.

REFERENCES

ALONSO, A., ARRIBAS, S. & MARTINEZ-ROGER, C. 1996 *A&A* **313**, 873.

ANDERSEN, J. 1991 *A&AR* **3**, 91.

ANDERSEN, J. 1997. In *Fundamental Stellar Properties: The Interaction Between Observation and Theory (IAU Symposium No. 189)* (eds. T. Bedding, A. Booth & J. Davis). p 99. Kluwer.

ANDERSEN, J., CLAUSEN, J. V. & GIMÉNEZ, A. 1993 *A&A* **277**, 439.

ANDERSEN, J., CLAUSEN, J. V., GUSTAFSSON, B., NORDSTRÖM, B. & VANDENBERG, D. A. 1988 *A&A* **196**, 128.

ANDERSEN, J., CLAUSEN, J. V. & NORDSTRÖM, B. 1980. In *Close Binary Stars: Observation and Interpretation (IAU Symp. No. 88)* (eds. M. J. Plavec, D. M. Popper & R. K. Ulrich). p. 81. Reidel.

ANDERSEN J., CLAUSEN J. V., NORDSTRÖM, B. & POPPER, D. M. 1985 *A&A* **151**, 228.

ANDERSEN, J., CLAUSEN, J. V., NORDSTRÖM, B., TOMKIN, J. & MAYOR, M. 1991 *A&A* **246**, 99.

ANDERSEN, J., NORDSTRÖM, B. & CLAUSEN, J. V. 1990 *ApJ* **363**, L33.

CLARET, A. 1995 *A&AS* **109**, 441.

CLARET, A. & CUNHA, N. C. S. 1995 *A&A* **318**, 187.

CLARET, A. & GIMÉNEZ, A. 1995 *A&A* **296**, 180.

EDVARDSSON, B., ANDERSEN, J., GUSTAFSSON, B., LAMBERT, D. L., NISSEN, P. E. & TOMKIN, J. 1993 *A&A* **275**, 101.

HUMMEL, C. A., ARMSTRONG, J. T., QUIRRENBACH, A., BUSCHER, D. F., MOZURKEVICH, D., ELIAS, N. M., WILSON, R. E. 1994 *AJ* **104**, 1859.

HYNES, R. I. & MAXTED, P. F. L. 1998 *A&A* **331**, 167.

Latham, D. W., Nordström, B., Andersen, J., Torres, G., Stefanik, R. P., Thaller, M. & Bester, M. 1996 A&A **314**, 864.

NORDSTRÖM, B., ANDERSEN, J., ANDERSEN, M. I. 1997 *A&A* **322**, 460.

POLS, O. R., TOUT, C. A., SCHRÖDER, K.-P., EGGLETON, P. & MANNERS, J. 1997 *MNRAS* **289**, 869.

POPPER, D. M. 1980 *ARA&A* **18**, 115.

POPPER, D. M. 1997 *AJ* **114**, 1195.

SIMON, K. P. & STURM, E. 1993 *A&A* **281**, 286.

TORRES, G., STEFANIK, R. P., ANDERSEN, J., NORDSTRÖM, B., LATHAM, D. W., CLAUSEN, J. V. 1997a *AJ* **114**, 2764.

TORRES, G., STEFANIK, R. P., LATHAM, D. W. 1997b *ApJ* **474**, 256.

TORRES, G., STEFANIK, R. P., LATHAM, D. W. 1997c *ApJ* **479**, 268.

ZUCKER, S. & MAZEH, T. 1994 *ApJ* **420**, 806.

Science results for stellar structure and evolution from Hipparcos

By YVELINE LEBRETON

DASGAL, CNRS URA 335, Observatoire de Paris, Place J. Janssen, 92195 Meudon, France;
e-mail address: Yveline.Lebreton@obspm.fr

The ESA Hipparcos satellite observed stars during 37 months from 1989 to 1993. First astrophysical results were presented in May 1997 during the ESA Hipparcos Venice '97 Symposium and just after the Hipparcos and Tycho Catalogues were released to the Astrophysical Community (see Perryman 1997). The Hipparcos astrometry mission has provided accurate positions, trigonometric parallaxes and proper motions for 118,218 stars. Photometric information can also be found in the Catalogue resulting from a combination of satellite and ground-based photometry: magnitudes, colour indices, information on the variability. Hipparcos has also shown its ability to detect double and multiple stars.

This paper reviews important selected results for stellar structure and evolution obtained with both the Hipparcos data and ground-based data derived from high resolution spectroscopy on large telescopes. Some of the results are still preliminary and incomplete but they show what can be learned using distances derived from parallaxes accurate to about 1 milliarcsecond (mas) on the average. After Hipparcos, the parallaxes are no longer the major sources of uncertainties for stellar studies and improvements in the determinations of abundances, effective temperatures and gravities of stars are highly required, implying progress in the theoretical description of the stellar atmospheres.

Very accurate Hertzsprung-Russell diagrams of nearby stars in the thin and thick disks and in the halo have been obtained and used to discuss the position of the stars as a function of their helium content and metallicity. In particular Hipparcos has given a more precise location of the main sequence of the local halo. This has permitted to estimate the age of the local halo and to derive the distances of several globular clusters by main-sequence fitting of the subdwarf sequence. Hipparcos has observed all the open clusters up to distances of about 300 parsecs. Individual distances have been obtained for the nearest Hyades which has allowed to estimate the initial helium content and age of the cluster. The other clusters are still under study and active work is in progress concerning the Pleiades which have been found to be fainter than expected. Hipparcos has also observed about twenty white dwarfs which has allowed to place more precisely these stars in the mass-radius plane and to discuss the mass-radius relation predicted by the theory of stellar degeneracy.

---

1. The ESA Hipparcos Space Astrometry Mission

The Hipparcos satellite has observed 118,218 stars during the period 1989–93. The contents of the Hipparcos Catalogue (ESA 1997) were presented by Perryman (1997) at the ESA Hipparcos Venice '97 Symposium, in May 1997 and the Catalogue was published soon after.

Stars of all types were observed down to a limiting magnitude $V \simeq 12.4$. The completeness of the Catalogue varies in the range $V = .3$–9.0 depending on the Galactic latitude and spectral type of the star.

Hipparcos has provided very precise astrometric parameters: stellar positions, proper motions and trigonometric parallaxes. Hipparcos has also given information on the duplicity or multiplicity of stars: 15 percent of the entries (17,917) in the Hipparcos catalogue have shown manifestations of multiplicity. Among those, 13,211 entries are resolved systems with astrometric data on each component; the rest consists in astrometric systems, for which only the photocentres could be observed. An interesting point is that more than one third of the stars classified as multiple were not previously known to be non-single. The results for double stars can be found in the Double and Multiple System Annex of the Catalogue (see Lindengren 1997).

The median precision on the parallax is typically 1 mas and is even better for the brighter stars. Precisions of the order of 1 mas on the positions and 1 mas/yr on the proper motions have been reached. With Hipparcos, 17,900 single stars and 1,600 double or multiple stars have parallaxes known to better than 10 percent and 49,500 stars have parallaxes known to better than 20 percent (see Mignard 1997).

The Hipparcos Catalogue also includes detailed and *homogeneous* photometric information for each star observed. The satellite data combined with ground-based observations have provided the Johnson V magnitude with a typical accuracy of 0.01 mag. The broad-band Hipparcos (Hp) magnitude is provided with a median precision of 0.015 mag for Hp < 9 mag. It corresponds to the specific passband of the instrument spanning the wavelength interval ~ 350–800 nanometers (see Fig. 3 in van Leeuwen 1997). Two colours B_T and V_T magnitudes were derived from the observations of Tycho (star mapper), corresponding to passbands close to the Johnson B and V bands. The precision is of the order of 0.012 mag for stars with $V_T < 9$. The combination of satellite data with ground-based observations yielded the Johnson B–V and Cousins V–I colour indices.

The mean number of photometric observations per star is 110 over the observational period 1989–93. This allows detailed variability classification and characterization of stars (see van Leeuwen 1997). The number of entries in the Catalogue which have shown to be variable or possibly variable are 11,597. This includes 2,712 periodic variables among which 970 new periodic variables were discovered (2 Cepheids, 9 δ Scuti and SX Phe stars, 36 eclipsing binaries).

Detailed descriptions of the Hipparcos and Tycho Catalogues and first astrophysical studies based on Hipparcos data are presented in the proceedings of the Hipparcos Venice Symposium (ESA SP-402, 1997). Astrometric results as well as results on stellar physics, galactic physics and on the cosmic distance and age scale have been obtained. In this paper selected results concerning stellar internal structure and evolution are presented.

2. Nearby disk and halo stars

Among the stars observed by Hipparcos the subsample constituted of the nearby stars is particularly interesting: the corresponding parallaxes are the most accurate and for most of these stars photometric and spectroscopic observations have been carried on from the ground with the best reachable accuracies. This allows to construct very precise empirical Hertzsprung-Russell (HR) diagrams which can be interpreted by means of stellar evolution models. High quality observations are valuable tools to discuss and validate the physics entering the stellar models calculations or to infer some of the stellar characteristics which are not accessible through observations.

2.1. *The fine structure of the HR Diagram of Stars in the Solar Neighbourhood*

Lebreton et al. (1997a) examined a sample of 114 late-type nearby stars, closer than 25 parsecs, of spectral types F, G and K, which were observed by Hipparcos. Among these stars, they selected a subsample of 32 stars with the aim to get a highly accurate and homogeneous HR diagram of stars of the solar neighbourhood.

The selected stars have parallaxes determined with an accuracy better than 5 percent and are not suspected to be unresolved binaries.

The bolometric fluxes of the stars were derived by Alonso et al. (1995) directly by integrating UBVRIJHK photometry which eliminates the problem of the value of the

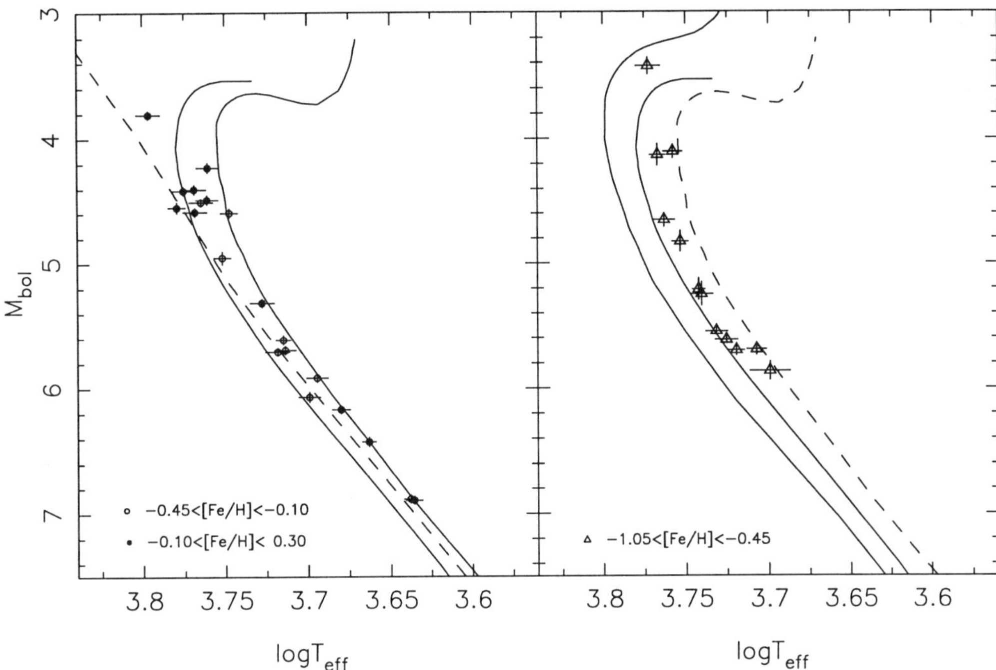

FIGURE 1. The Hipparcos HR diagram of the 32 best known nearby stars split in two domains of metallicity. On the left (Fig. 1a) are plotted stars of solar metallicity and close to it with [Fe/H] values in the interval $[-0.10, +0.30]$ (filled circles) and $[-0.45, -0.10]$ (open circles). On the right (Fig. 1b) are plotted the moderately deficient stars with [Fe/H] in the range $[-1.05, -0.45]$ (open triangles). The accuracy on the parallax is better than 5 percent and bolometric fluxes are available for each star. Individual error bars are given for each object by the horizontal and vertical sizes of the cross. Theoretical isochrones have been added to the figures: a 10 Gyr isochrone with $Y = .236$, $[Fe/H] == -1.0$ and $[\alpha/Fe] == +0.4$ (left continuous line in Fig. 1b), a 10 Gyr isochrone with $Y = .256$, $[Fe/H] == -0.5$ and $[\alpha/Fe] == +0.4$ (left continuous line in Fig. 1a and right continuous line in Fig. 1b), a 8 Gyr isochrone with $Y = .32$, $[Fe/H] == +0.3$ and $[\alpha/Fe] == 0.0$ (right continuous line in Fig. 1a and right-dashed line in Fig. 1b), and a solar ZAMS (dashed-line in Fig. 1a).

bolometric correction. The average absolute error on the bolometric magnitude M_{bol} resulting from both the parallax and bolometric flux uncertainties is of typically 0.045 mag.

The effective temperatures T_{eff} of the stars were obtained by Alonso et al. (1996) from the bolometric fluxes and using a grid of theoretical model line-blanketed flux distributions (Kurucz 1991). The mean error on T_{eff} is around 1.5 percent for temperatures greater than 4200 K (Alonso et al. 1996).

Results of detailed spectroscopic analysis are available for each star. They appear in the recent [Fe/H] catalogue of Cayrel de Strobel et al. (1997). The [Fe/H] ratio, i.e. the logarithm of the number abundances of iron to hydrogen relative to the solar value, ranges from about -1.0 to $+0.3$ dex corresponding to Population I and thick disk population. The mean error on [Fe/H] is of the order of 0.1 dex.

Moreover observations of the abundances of the α-elements (O, Mg, Si, Ca and Ti) have also shown that metal-deficient stars with $[Fe/H] < -0.5$ generally exhibit an α-elements enrichment relative to the Sun (Wheeler et al. 1989). Gratton et al. (1997a) have derived an α-elements enrichment $[\alpha/Fe] = +0.26 \pm 0.08$ dex from high resolution spectra of a hundred metal deficient field dwarfs with $[Fe/H] < -0.5$.

Fig. 1a and Fig. 1b present the resulting positions of the 32 selected stars in the $(M_{bol}, \log T_{eff})$ diagram with their individual error bars. The sample has been split into two subsamples: Fig. 1a corresponds to stars of solar metallicity or close to it $(-0.45 \leq [Fe/H] \leq +0.30)$ while Fig. 1b represents the moderately metal deficient stars $(-1.0 \leq [Fe/H] \leq -0.45)$. Theoretical isochrones associated to the limits of the metallicity range considered have been superimposed on the diagrams. The description of the stellar models and of their input physics is given in Lebreton et al. (1998a). In particular the helium content of the models, Y, varies as a function of the metallicity Z according to the usual relation $\Delta Y/\Delta Z = (Y-Y_p)/Z$ where Y_p is the primordial helium abundance and $\Delta Y/\Delta Z$ is the helium to metal enrichment ratio. The $\Delta Y/\Delta Z$ -value is derived from the calibration of the solar model: the solar model has to reach at solar age the observed luminosity and radius; this fixes the initial helium content of the Sun (here $Y = 0.266$) once the observed solar metallicity ($Z = 0.0175$) and the input physics of the solar model are specified. Moreover the models of metallicity lower than $[Fe/H] = -0.5$ are calculated with an input mixture and opacities corresponding to an α-elements enrichment relative to the Sun $[\alpha/Fe] = +0.4$ dex. The mixing-length to pressure scale-height ratio α entering the description of convection is derived from the solar calibration which yields $\alpha = 1.65$.

As shown in Fig. 1a, stars of solar luminosity and close to it lie in the regions defined by the theoretical models corresponding to their associated metallicity range. On the other hand Fig. 1b shows that the moderately metal deficient stars are outside the theoretical band corresponding to their metallicity range. In fact the stars have a tendency to lie on a theoretical isochrone corresponding to a metallicity which is higher than the metallicity derived from the detailed analysis of high resolution spectra (see Lebreton et al. 1998a for more details). Such a problem was already pointed out by Axer et al. (1995) when they studied the evolutionary stage of subdwarfs; they found that the evolutionary models corresponding to observed metallicities predicted stars systematically hotter than those observed.

This trend is well illustrated by the study of the star μ Cas A. This star is one of the 32 stars presented above and is also the A-component of a well-known visual binary system. This gives an additional very important information on the star, its mass $M = 0.757 \pm 0.059$ $M_\odot$, which was derived from the astrometric and relative orbits by Drummond et al. (1995). The $[Fe/H]$ ratio is -0.86 ± 0.06 (Axer et al. 1994) and the observed $[\alpha/Fe]$ ratio is $+0.30 \pm 0.08$ (Fuhrmann et al. 1995). The metallicity of the system is representative of the thick disk whereas its kinematics indicate that it belongs to the halo.

Lebreton et al. (1998a) made a careful modeling of μ Cas A by means of stellar models. Fig. 2 shows the observed position of μ Cas A in the HR diagram together with its error bar and the position of a theoretical isochrone of 12 Gyr calculated with the observed $[Fe/H]$ and $[\alpha/Fe]$ abundances and with an initial helium abundance $Y = 0.245$, close to the primordial value $Y_p = 0.243 \pm 0.003$ (Izotov et al. 1997). On the isochrone, the point corresponding to the magnitude of μ Cas A has a mass $M = 0.727$ $M_\odot$ which is well inside the observational range. However the effective temperature of that point is 235 ± 82 K hotter than the observed point (82 K being the uncertainty on the observed T_{eff}).

The uncertainties on both the observed position of μ Cas A and on the position of the associated theoretical isochrone have been examined by Lebreton et al. (1998a).

The first point is that the knowledge of the mass of the star (and of its luminosity) puts rather strong constraints on the initial helium abundance of the star. When the helium content entered in the stellar models is decreased the mass corresponding to a given

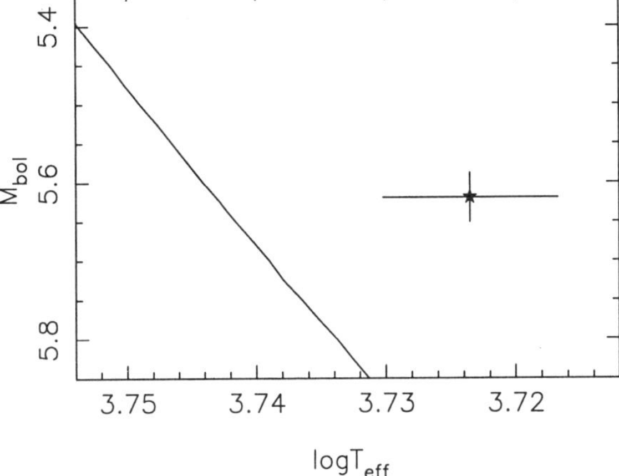

FIGURE 2. The observed position of μ Cas A in the HR diagram together with its error bars. The continuous line represents the associated theoretical isochrone of 12 Gyr calculated with $Y = 0.245$, [Fe/H] $= -0.86$, [α/Fe] $= +0.4$.

luminosity on the isochrone is increased. The helium abundance chosen must however lead to a mass inside the error bar. In the case of μ Cas A the mass is not so accurate but it indicates that the helium abundance must be close to the primordial value to get a mass well inside the observed range.

The location of the theoretical isochrone crucially depends on the adopted value of [Fe/H]. As discussed by Nissen et al. (1997), uncertainties on $T_{\rm eff}$ and weaknesses in the atmosphere models, mainly resulting from an inadequate treatment of convection, severely affect the [Fe/H] determination. Moreover the [Fe/H] value of Axer et al. (1994) is based on atmosphere models in local thermodynamic equilibrium (LTE). Thvenin & Idiart (1998) have examined the effects of departures from LTE on the [Fe/H] determination based on Fe I and Fe II lines. They found that when metallicity decreases, due to the lack of absorbers, the UV flux is increased producing an over-ionization of strongly ionized elements. They suggest that for a dwarf star like μ Cas A for which they got a LTE reference value [Fe/H] $= -0.70$ dex, a correction of $+0.15$ dex on [Fe/H] has to be applied to account for non-LTE effects. On the other hand they find that no appreciable correction is required in the case of dwarfs of solar metallicity.

The discrepancies shown in Fig. 1b would therefore be lessened if the [Fe/H] value of the metal-deficient stars was to be increased to account for departures from LTE. For instance, increasing the [Fe/H] value of μ Cas by $+0.15$ dex leads to a shift to the right of the theoretical isochrone by about 80 K, that is about one third of the discrepancy between models and observations. Furthermore this would not alter the agreement found in Fig. 1a for stars of solar metallicity which are not affected by this effect. To be able to conclude on that point it is important to quantify properly the corrections to apply to take non-LTE effects into account.

The models calculated by Lebreton et al. (1998a) are standard stellar models which do not take into account the microscopic diffusion processes. In low mass stars microscopic diffusion by gravitational settling makes helium and heavy elements sink towards the center. This modifies the surface abundances as well as the inner abundance profiles. In the Sun microscopic diffusion is invoked to explain the low helium abundance in the convection zone inferred from helioseismological measurements (Pérez Hernández & Christensen-Dalsgaard 1994; Bahcall et al. 1995). In metal deficient stars the density at the bottom of the convection zone is smaller than in stars of solar metallicity which implies a greater efficiency of gravitational settling. Since the mass of the convection zone is also smaller in metal deficient stars, diffusion induces higher depletions of the surface abundances. Furthermore metal deficient stars are old stars in which diffusion had plenty of time to proceed. However it is also well-known that microscopic diffusion is inhibited as soon as turbulent processes, such as turbulence induced by rotation, take place (see e.g. Zahn 1992).

It is possible to estimate the maximum effect expected from helium diffusion at the surface of a star like μ Cas A from the models by Proffitt & VandenBerg (1991). These models take into account gravitational settling of helium during the whole evolution of the star, giving at 12 Gyr an effective temperature decrease of about 80 K with respect to standard models. This again represents about one third of the discrepancy between models and observations. Furthermore when diffusion is taken into account the surface metallicity decreases with time; that means that to get the metallicity now observed at the surface of the star it is necessary to start the calculations with a higher initial metallicity. This would add to the effective temperature decrease of the models.

For μ Cas A, a global additional uncertainty $\sigma_{T_{eff}} \simeq 100$ K results from the uncertainties coming from the magnitude ($\sigma_{T_{eff}} \simeq \pm 20$ K), the age ($\sigma_{T_{eff}} \simeq 20$ K), the mixing-length ratio ($\sigma_{T_{eff}} \simeq 40$ K) and the $[\alpha/Fe]$ enrichment ($\sigma_{T_{eff}} \simeq 20$ K).

Combining all sources of errors would probably put models and observations in agreement for the particular case of μ Cas A. Concerning the whole sample of stars considered, detailed investigations are now required. Microscopic diffusion has to be examined in all its aspects. Microscopic diffusion brings helium and heavy elements down to the center of the stars and modifies their evolutionary course. On the other hand due to microscopic diffusion the surface abundances are modified and do not necessarily reflect the initial abundances, therefore the initial abundances entered as an input of the stellar models are different from the surface abundances observed today. On the observational side, the [Fe/H] determinations should be properly corrected for non-LTE effects. Last but not least, much progress would be expected from the study of several binary stars of known mass, if the masses were accurate enough to really constrain the helium content of the stars.

2.2. *The Halo subdwarf sequence*

Cayrel et al. (1997) examined a sample of 122 population II halo subdwarfs and subgiants observed by Hipparcos. The concerned stars have metallicities [Fe/H] lower than -1.0 dex. Among these stars, a subsample composed of the best-known stars has been extracted, with the same criteria than those adopted for disk stars (see subsection 2.1 above). In addition stars were corrected for reddening and those which had an E(B–V) larger than 0.05 were excluded.

Before Hipparcos only 5 stars had parallaxes determined with a relative accuracy better than 10 percent, after Hipparcos they are 16 which represents a considerable progress. The 30 halo stars which have parallaxes accurate to better than 25 percent are plotted in Fig. 3. Subgiants and subdwarfs are present, drawing an isochrone-like shape with a turn-off region.

One of the aims of Cayrel et al. (1997) was to estimate the age of the local halo by comparing the observed Population II sequence to theoretical isochrones. For that purpose they kept the stars having a reasonably accurate parallax ($\sigma_\pi/\pi < 0.125$) and in a narrow range of metallicity, near the most frequent metallicity in the halo ([Fe/H] $= -1.5 \pm 0.3$). The position in the HR diagram of the 13 remaining stars is shown in Fig. 4a together with a theoretical isochrone of 14 Gyr derived from stellar models having

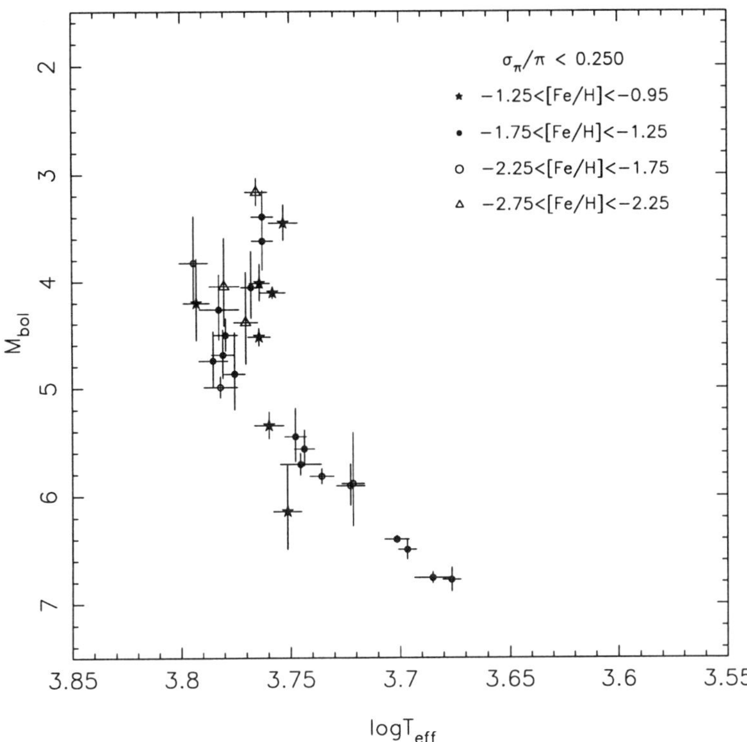

FIGURE 3. The Hipparcos HR diagram of the 30 *Hipparcos* stars with a relative error less than 25 percent on the distance. A bunch of subgiants emerges with an isochrone-like shape.

[Fe/H] $= -1.5$ and $[\alpha/Fe] = +0.4$ and calculated with the same input physics that the models presented in subsection 2.1 above. Fig. 4a shows that halo stars, like disk stars, are colder than the theoretical isochrone corresponding to their metallicity. A shift to the right of about 130 K of the theoretical isochrone would be required to get agreement between the theoretical and the observed sequence.

Nissen et al. (1997) have studied a sample of halo stars to determine their surface gravities. They used theoretical isochrones of 14 Gyr calculated by Bergbusch & VandenBerg (1992) to estimate the masses of the stars. They also had to shift the isochrones by $\Delta \log$ T$_{eff}$ amounts in the interval $[-0.02, -0.01]$ to make them agree with the observed sequence.

Pont et al. (1997a, 1997b) worked with a large sample of halo stars with the aim of estimating the age of the globular cluster M92 (see subsection 2.4 below) and also had to shift the theoretical isochrones of VandenBerg et al. (1998) by Δ(B–V) $\simeq 0.012$ mag to make them coincide with observations. Pont et al. (1997b) and Cayrel et al. 1997 estimated the age of the local halo to be in the range 12–16 Gyr.

Similar remarks than in subsection 2.1 can be made. To understand discrepancies between theory and observations at low metallicities, it is necessary to ascertain the [Fe/H] determination and to examine in detail the effects of microscopic diffusion on the theoretical models. Fig 4b illustrates the latter point. It shows two isochrones calculated by Proffitt & VandenBerg (1991) superimposed on the observed halo sequence. The isochrone on the right takes into account the microscopic diffusion of helium while the isochrone on the left does not. The isochrone with microscopic diffusion is redder and has a different shape with a lower luminosity at the turn-off. Consequently the fit with

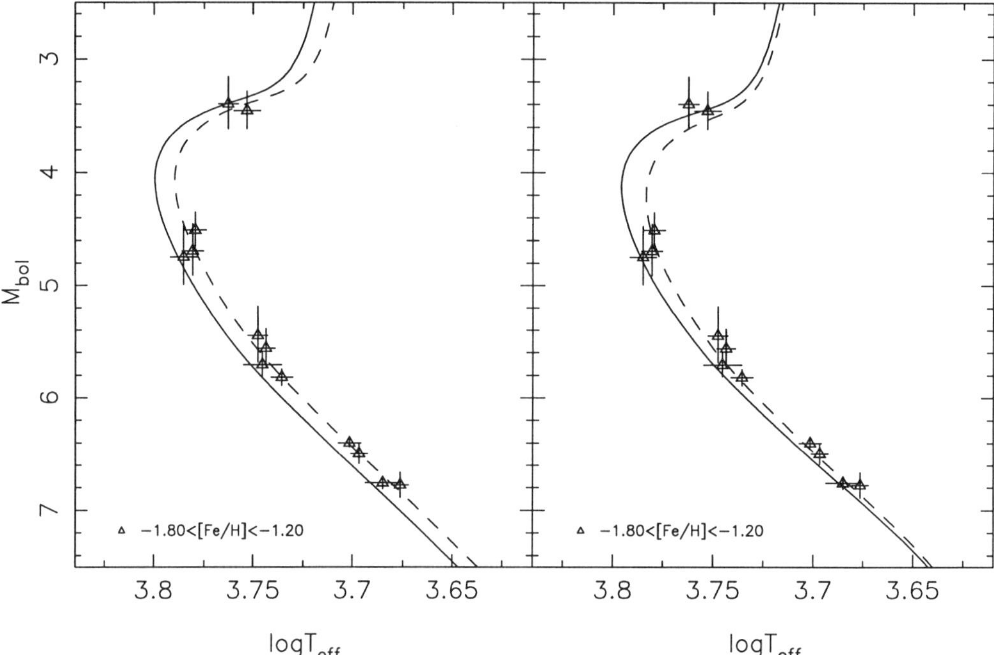

FIGURE 4. The Hipparcos HR diagram of the 13 halo stars having parallaxes determined to better than 12.5 percent and [Fe/H]-values in the interval $[-1.80, -1.20]$. The average absolute error on the bolometric magnitude M_{bol} resulting from both the parallax and bolometric flux uncertainties is of typically 0.14 mag. On the left figure is plotted an isochrone of 14 Gyr, [Fe/H] = -1.5 and [α/Fe] = $+0.4$, Y $\simeq 0.24$ (continuous line) and the same isochrone arbitrarily shifted by log T_{eff} = 0.01. On the right are plotted two isochrones of 12 Gyr, [Fe/H] = -1.3 and [O/Fe] = 0.55, without (continuous line) and with (dashed-line) microscopic diffusion of helium (Proffitt & VandenBerg 1991)

the observed sequence is better and is achieved for an age smaller by 0.5–1.5 Gyr than the age obtained without diffusion. Recent models by Castellani et al. (1997) confirm this general trend and also show that, when the sedimentation of metals is taken into account, including its effects on the matter opacity, then the shift of the isochrone is smaller than the shift obtained with helium diffusion only.

To properly estimate the age of the local halo more stars with precise parallaxes are highly required. Subgiants are about a hundred times rarer than subdwarfs in the halo and only two subgiants had parallaxes measured by Hipparcos with a relative accuracy better than 12.5 per cent (no subgiant with $\sigma_\pi/\pi < 0.05$). After Hipparcos the position of the subgiant branch is still not very well determined which limits the accuracy on the age determination of the halo stars.

2.3. *The ZAMS Position as a function of metallicity*

Combining the data for disk and halo stars provides a sample of stars spanning the whole galactic metallicity range. The non-evolved stars of this sample can be used to discuss the position of the zero age main sequence (ZAMS) as a function of metallicity and its implications for the helium abundances which cannot be obtained from spectroscopic analysis.

Fig. 5 shows the position in the HR diagram of the stars studied by Lebreton et al.(1997a) and by Cayrel et al. (1997). Only non-evolved stars were kept; according to theoretical models, stars fainter than $M_{bol} \simeq 5.5$ are still very close to their ZAMS location.

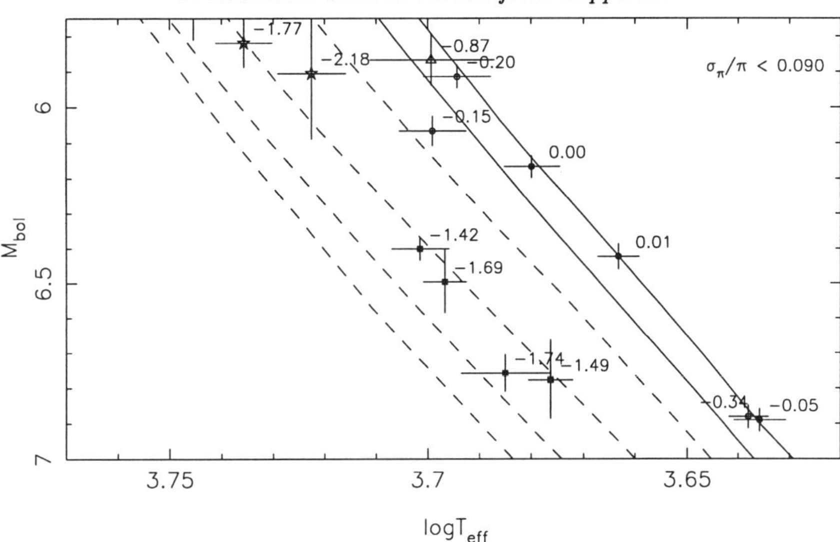

FIGURE 5. HR diagram position of the *Hipparcos* non-evolved stars with $\sigma_\pi/\pi < 0.10$. Each star is labeled with its [Fe/H] value. Theoretical isochrones are plotted with, from left to right, [Fe/H] $= -2.0, -1.5, -1.0, -0.5, 0.0, 0.3$.

Theoretical isochrones for various metallicities have been superimposed on Fig. 5. They have been calculated with values of helium derived from a solar $\Delta Y/\Delta Z$ enrichment (see subsection 2.1 above).

The observational and theoretical main sequence widths measured in Fig. 5 are in reasonable agreement. However, as discussed in details in the previous subsections, the stars are not located on the isochrone corresponding to their metallicity. For stars of solar metallicity, a small decrease of the helium abundance below the solar value would be sufficient to correct the position while it has been shown before that a large helium decrease would not acceptable for metal-deficient stars, with helium abundances already close to the primordial value.

It can be noticed that moderately metal deficient stars with accurate positions are not numerous in the non-evolved part of the HR diagram of Fig. 5. There is a gap from [Fe/H] $\simeq -1.4$ to [Fe/H] $\simeq -0.3$. It is therefore difficult to establish empirically the dependence of the ZAMS location with metallicity (see also Chaboyer et al. 1998).

2.4. *Distances and ages of globular clusters*

The globular clusters stars were beyond the possibilities of Hipparcos. However the distances of globular clusters have been estimated by several groups applying for the first time the main-sequence fitting technique using the subdwarf main-sequence defined with the subdwarfs observed by Hipparcos (Reid 1997, Gratton et al. 1997b, Pont et al. 1997a 1997b, Chaboyer et al. 1998).

The main-sequence fitting technique is the following: the non-evolved part of the subdwarf absolute main-sequence, defining the ZAMS, is compared to the non-evolved part of the globular cluster sequence; the measure of the magnitude difference between the two lines directly gives the distance d_{GC} of the globular cluster through the relation $(m–M) = 5\log d_{GC}$.

The technique is rather simple but it has to be applied with caution (Chaboyer et al. 1998). First, as discussed before in subsection 2.2, there are only few stars with parallaxes accurate enough to determine the position of the halo stars sequence with an acceptable precision. Moreover one has to ensure that the subdwarf sample used is free of Lutz-Kelker type biases. Chaboyer et al. (1998) only retained the stars with parallax error smaller enough ($\sigma_\pi/\pi < 0.10$) for biases to be negligible according to Brown et al. (1997). Pont et al. (1997b) and Gratton et al. (1997b) carefully dealt with the selection effects.

Furthermore, as shown in subsection 2.3 above and pointed out by Chaboyer et al. (1998), due to the small number of subdwarfs in each interval of metallicity it is not possible to establish properly the variation of the observed ZAMS position with metallicity. To estimate the distance of a given globular cluster it is therefore safer to choose halo stars in a very narrow metallicity interval around the metallicity of the cluster and to ensure that α elements excesses are similar. Gratton et al. (1997b) and Pont et al. (1997b) avoided this problem by applying theoretical "colour" corrections to the subdwarfs data to account for metallicity differences between the globular clusters and subdwarfs. In all cases it is crucial that the abundances of both the globular cluster stars and halo subdwarfs are accurate and on a consistent scale. For that purpose Gratton et al. (1997b) determined abundances from high resolution spectra yielding $\sigma_{[Fe/H]} \simeq 0.07$–$0.10$ dex while Pont et al. (1997b) derived new abundances using the CORAVEL radial velocity spectrometer with $\sigma_{[Fe/H]} \simeq 0.15$–$0.20$ dex.

In order to get accurate globular clusters distances it is important to examine unresolved known or suspected binaries carefully, for they are located above the ZAMS and can introduce errors on the definition of the ZAMS position. In the different papers quoted above unresolved binaries, known or suspected, were either excluded (Chaboyer et al. 1998, Gratton et al. 1997b) or an average correction of 0.375 mag has been applied to them (Pont et al. 1997b).

Similarly, stars which have evolved off the ZAMS have to be excluded. They introduce errors in distances derived from main-sequence fitting, since there is no certitude that globular clusters and halo dwarfs have exactly the same age. For instance, according to Chaboyer et al. (1998), a difference of 2 Gyr between a calibrating subdwarf at $M_V = 5$ and a globular cluster would lead to a systematic error of 0.14 mag in the distance modulus. Therefore, one has to ensures that the non-evolved part of both the halo stars sequence and the globular cluster sequence are used. From theoretical models it is estimated that stars fainter than $M_V \simeq 5.5$ are essentially unevolved.

The globular cluster sequence has to be determined from good photometry well below the main sequence turn-off. Furthermore it is clear that the interstellar reddening to apply to correct the sequences has to be well-determined.

The number of globular clusters studied by the different authors quoted above varies because of the different criteria and techniques chosen to select the subdwarfs samples. Anyway they agree in the general conclusion that globular clusters distances derived from main sequence fitting are larger than previously found. In particular, Gratton et al. (1997b) found a distance scale greater by about 0.2 mag on the average.

Furthermore, as discussed by Chaboyer et al. (1998), there are five independent means to obtain the globular cluster distance scale: distances can be obtained from astrometric data, from white dwarf sequence fitting, from subdwarf main-sequence fitting, from calibration of the mean magnitudes of RR Lyrae stars in the LMC and from theoretical models of horizontal branch stars. They now all indicate that globular clusters are farther than previously believed, implying a reduction in age estimates by typically 2–3 Gyr. Chaboyer's et al. (1998) best estimate of the age of the oldest globular clusters is 11.5 ± 1.3 Gyr which has to be compared to the "old" age interval of 10–18 Gyr quoted by VandenBerg et al. (1996) in their review. Also, Chaboyer et al. (1998) pointed out that with the new distance scale, the fit to calculated isochrones near the cluster turn-off is worse suggesting problems with stellar radii due to difficulties with stellar atmospheres and convection modeling.

For a recent review on the ages of globular clusters and cosmological implications, see the paper by A. Renzini (these Proceedings).

3. Open clusters

Hipparcos has observed all the open clusters closer than 300 parsecs and 8 rich clusters located between 300 and 500 parsecs. This provides valuable material for distance scaling of the Universe as well as for studies of kinematical and chemical evolution of the Galaxy.

The absolute locations of the cluster sequences in the HR diagram can now be obtained directly from Hipparcos distances and do not rely on chemical composition considerations. Since an open cluster is constituted of stars spanning a broad range of masses and which can be considered as coeval and of the same initial chemical composition, the study of several clusters, each of a different age, allows to test stellar evolution theory for various stellar characteristics.

3.1. *The Hyades*

The Hyades is the closest open cluster and is used as the zero-point of the galactic and extragalactic distance scale. Hipparcos has measured individual distances, with a mean accuracy of 5 percent, and proper motions providing a consistent picture of the Hyades distance, structure and dynamics (Perryman et al. 1998).

It is worth to compare the Hyades distance modulus derived from Hubble Space Telescope observations of seven Hyades stars by van Altena et al. (1997), M–m = 3.42 ± 0.09 mag, to the distance modulus obtained from Hipparcos observations of 134 stars within 10 parsecs of the cluster center, M–m = 3.33 ± 0.01 mag. The two results are in good agreement and we have gained a large factor on the parallax accuracy.

The location of the lower main sequence of the Hyades in the HR diagram has been determined using 40 Hipparcos stars, previously subjected to detailed spectroscopic analysis (see Perryman et al. 1998 and Lebreton et al. 1997a). The internal errors on T_{eff} are in the range 50–75 K and the mean [Fe/H] is 0.14 ± 0.05. The comparison of the lower part of this sequence, constituted of the non-evolved stars, with theoretical ZAMS corresponding to the observed metallicity yields an estimate of the initial helium abundance of the cluster Y = 0.26 ± 0.02, the metallicity in mass fraction being Z = 0.024 ± 0.04 (Fig. 6). Metallicity is the dominant source of uncertainty on Y.

Perryman et al. (1998) have estimated the age of the Hyades by comparing the whole cluster sequence in the (M_V, B–V) plane to theoretical isochrones calculated with the chemical composition derived from the ZAMS analysis. For that purpose the whole HR diagram of the cluster (131 stars) was considered and 69 stars were kept after elimination of known or suspected binaries, of variable stars and of rapid rotators. The V and B–V values were taken from the Hipparcos catalogue ($\sigma_{(B-V)} < 0.05$ mag). The transformations from the theoretical (M_{bol}, log T_{eff}) plane to the observational plane have been examined carefully. The optimum fit was achieved with an isochrone of 625 ± 50 Myr obtained from models including convective core overshooting (Fig. 7). The quoted uncertainty only results from the visual fitting of the isochrones. Errors on age result from several factors: unrecognized binaries, rotating stars, colour calibrations and bolometric corrections, weaknesses of the theoretical models in particular through the parameterization of overshooting (Lebreton et al. 1995). An actual uncertainty on age of at least 100 Myr is estimated (Lebreton et al. 1997b).

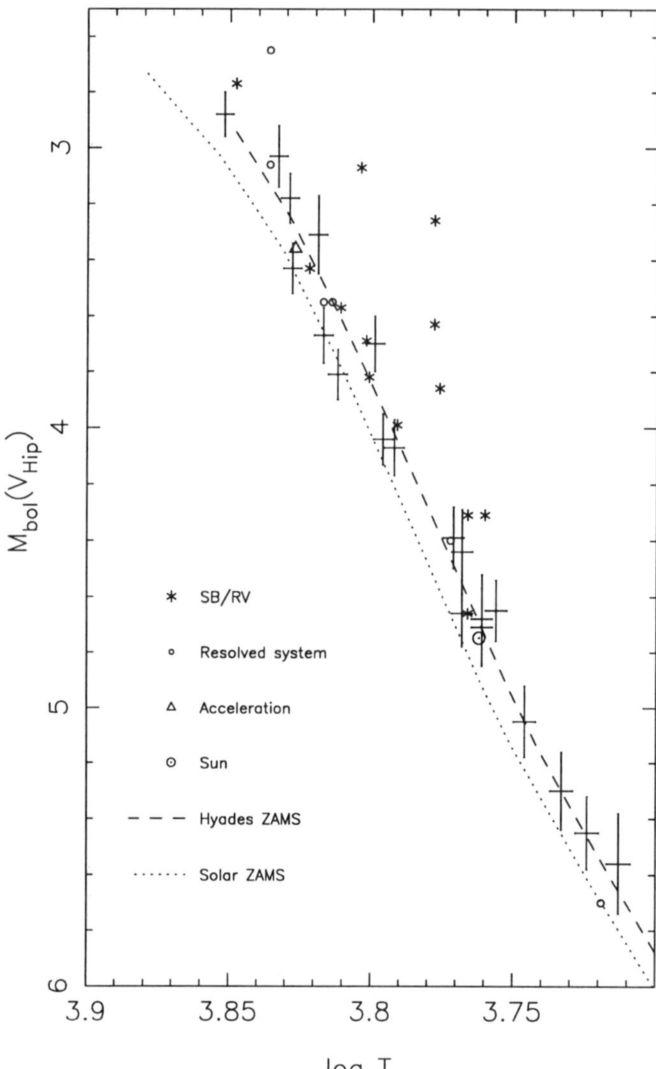

FIGURE 6. Hipparcos HR diagram for 40 selected stars in the Hyades. The 19 stars with error bars are not suspected to be double or variable. Theoretical ZAMS loci are given for the Hyades (dashed line, Y = 0.26 Z = 0.024) and solar (dotted line, Y = 0.266 Z = 0.0175) chemical compositions.

Additional information on the Hyades have recently been provided in three papers by Torres et al. (1997 a,b,c) who have determined the masses of stars in four binary systems in the Hyades (51 Tauri, Finsen 342, θ^1 and θ^2). The masses of the components of 51 Tauri were also derived from Hipparcos data by Söderhjelm et al. (1997) giving results in agreement with those of Torres et al. (1997a). Moreover very accurate masses for the components of the well-known double-line spectroscopic binary VB22 were obtained by Peterson & Solensky (1988). Nine of the stars are on the main-sequence. On Fig. 8 the corresponding empirical mass-luminosity (M-L) relation is compared to the theoretical relation derived from the isochrone of 625 Myr calculated with Y = 0.26 and Z = 0.024, showing an excellent agreement. The lower part of the M-L relation is defined by the masses of the two components of VB22 which are very accurate. This allows to give more

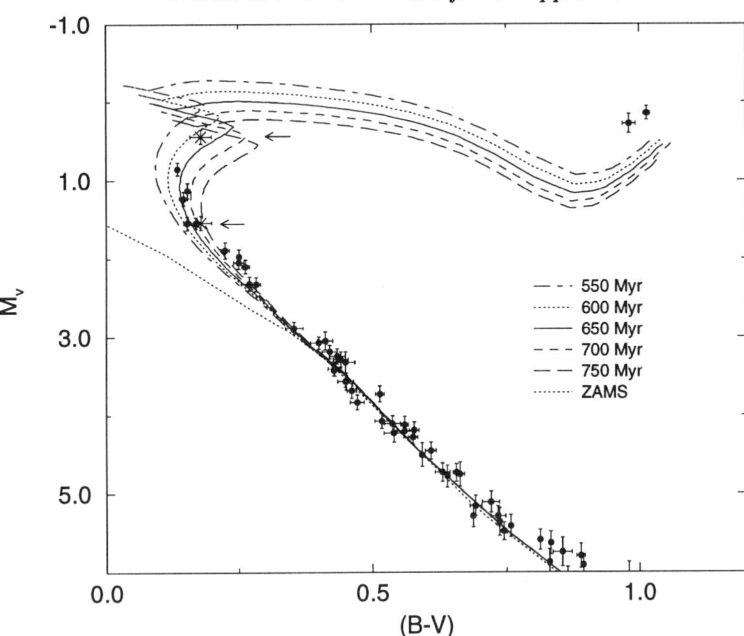

FIGURE 7. Hipparcos colour-magnitude diagram of the Hyades. The locus of the ZAMS and of theoretical isochrones calculated with overshooting are indicated. Arrows indicate the position of the components of the binary θ^2 Tau used by Perryman et al. (1998) for the age determination.

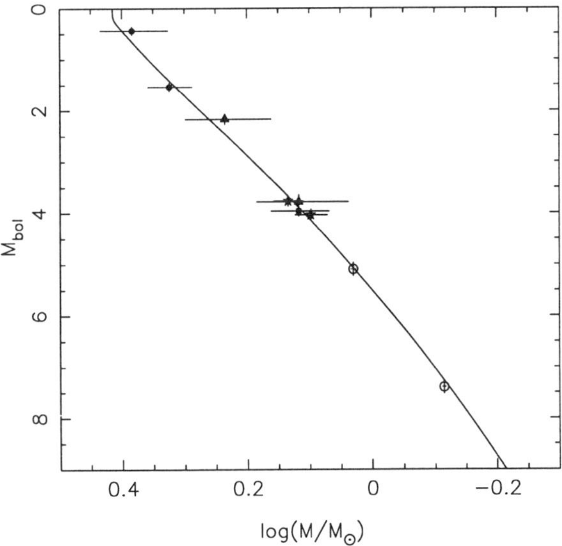

FIGURE 8. The Hyades empirical and theoretical mass-luminosity relations

constraints on the helium abundance ($Y = 0.26 \pm 0.02$) derived from ZAMS calibration. Preliminary work seems to suggest that the helium abundance of the Hyades is preferentially between $Y = 0.24$–0.26 with [Fe/H] values in the range 0.09–0.14 (Lebreton et al. 1998b).

3.2. *The Pleiades and other open clusters*

Mermilliod et al. (1997) and Robichon et al. (1997) computed the membership of stars in seventeen open cluster fields and derived the mean parallax and proper motion of the clusters from intermediate Hipparcos data. They obtained accuracies from 0.2 to 0.5 mas for parallaxes and 0.1 to 0.5 mas per year for proper motions.

As described by Robichon et al. (1997), the membership selection was made assuming that all stars in the cluster have the same space velocity and lie into a 10 parsec radius sphere roughly corresponding to the tidal radius. The selection was done iteratively with the requirement that the star parallax and proper motion were consistent with the mean parallax and mean proper motion of the cluster at a 3σ level. In a further step, the double stars which duplicity could bias the mean parallax and proper motion were excluded as well as stars departing from the cluster sequence in a colour-magnitude (C-M) diagram obtained from photometric data from the "Base des Amas" (Mermilliod 1988).

In clusters, stars are located within a few degrees in the sky and have often been observed in the same field of view of the satellite which induces correlations between their measurements. In order to obtain an optimal mean parallax, Robichon et al. (1997) did not make a straight average of the parallaxes of the cluster stars but calculated it from Hipparcos intermediate data (i.e. parallax and proper motion of the cluster center and position of each cluster member) assuming that all cluster stars have the same parallax.

Mermilliod et al. (1997) and Robichon et al. (1997) compared the sequences of the various clusters in C-M diagrams derived from different photometric systems and found puzzling results. On one hand, there are examples of clusters with different metallicities which define the same main sequence, for instance the Coma Ber sequence is similar to that of Praesepe (or Hyades). On the other hand, some clusters have a sequence abnormally faint with respect to the previous one. In particular the Pleiades sequence, obtained using the distance derived from Hipparcos (M$-$m $= 5.36 \pm 0.06$ mag), lies ~ 0.4 mag below the Praesepe or Hyades sequence. This difference cannot be explained in terms of [Fe/H] variations assuming that the helium abundance varies from the solar one according to a $\Delta Y/\Delta Z = (\Delta Y/\Delta Z)_\odot$ relation. Using the [Fe/H] value derived from spectroscopic analysis by Boesgaard & Friel (1990), [Fe/H] $= -0.034 \pm 0.024$, the value of helium required to reconcile theory and observations would be at least Y ~ 0.31 which is hardly supported by observations (Mermilliod et al. 1997).

The distances of the 5 closest open clusters (Hyades, Pleiades, alpha Per, Praesepe and Coma Ber) obtained from Hipparcos trigonometric parallaxes can be compared to those recently derived from main sequence fitting by Pinsonneault et al. (1998). In order to obtain the distance modulii, these authors have compared theoretical isochrones, translated into the C-M plane by means of colour calibrations, to observational data both in the (M_V , B$-$V) and (M_V , V$-$I) planes. The colour indices are sensitive to metallicity, especially B$-$V (Alonso et al. (1996)) and Pinsonneault et al. (1998) derive the value of the metallicity which gives the same distance modulus in the two planes. Pinsonneault et al. (1998) found distance modulii in good agreement with Hipparcos results except for the Pleiades and Coma Ber but they suggest that for the latter the problem could be due to the colours. In the case of the Pleiades the discrepancy with Hipparcos amounts to 0.24 mag, while the [Fe/H] value derived from main-sequence fitting in the two colours planes are in agreement with the spectroscopic determination. Pinsonneault et al. (1998) examined different possible origins of the discrepancy (erroneous metallicity, age related effects, reddening, helium variations) and concluded that none of them is likely to be responsible for the problem of the Pleiades.

According to Soderblom et al. (1998), the Pleiades cannot be a unique case in the Galaxy and there should exist some young nearby solar-type stars also looking anomalously faint. These authors looked for active single stars among a sample of field stars with very safe parallaxes and selected 50 young objects, but none of them was found to lie 0.3 mag below the solar ZAMS. Then they examined the subluminous stars observed by Hipparcos with $\sigma_\pi/\pi < 0.05$ and $\sigma_{B-V} < 0.025$. They selected randomly 6 stars among the stars lying about 0.3 mag below the ZAMS and made new spectroscopic observations of them. They found that those stars are metal deficient with respect to the Sun. Moreover, they found that all the subluminous stars that they had selected have kinematics typical of stars of thick disk or halo population.

From these studies Soderblom et al. (1998) and Pinsonneault et al. (1998) concluded that it is abnormal that no star similar to the Pleiades is found in the field. They claimed that the distance obtained from multi-colour main-sequence fitting, which is determined to a precision of about 0.05 mag, is correct and concluded that the distance obtained from the analysis of Hipparcos data is wrong at the 1 mas level, greater than the mean random error. They invoked the existing correlations between right ascension and parallax (ρ_α^π) for stars within a small angular region. Pinsonneault et al. (1998) showed that the brightest stars, highly concentrated near the cluster center, and with the highest ρ_α^π are those which bring the average parallax to a value greater than expected from main sequence fitting.

Although errors in Hipparcos data cannot be precluded, various checks have been done concerning the parallax zero-point and the systematic error was found to be negligible on the average (Arénou et al. 1997). Further tests are now in progress using the best ground-based parallaxes available for distant objects, stars and clusters and looking for parallax biases for stars with high ρ_α^π. Up to now there is no indication of additional systematic error in Hipparcos data and no explanation for a particular problem in the Pleiades (Robichon et al. 1998).

However one should also keep in mind that metallicity determinations are still subject to uncertainties. As pointed out by Mermilliod et al. 1997, photometric and spectroscopic approaches may produce differing results. In fact a recent photometric determination of the metallicity of the Pleiades indicates that they could be slightly metal-deficient with respect to the Sun, with [Fe/H] $= -0.12 \pm 0.02$ (Grenon, private communication). This result appears to be confirmed by metallicity determinations using the CORAVEL spectrometer data which yield [Fe/H] $= -0.09 \pm 0.03$ (Mermilliod, private communication). These [Fe/H] values therefore differ from spectroscopic values. We point out that with such lower values of [Fe/H], a lower helium abundance $Y \sim 0.28$ would be sufficient to explain the faintness of the Pleiades found by Hipparcos.

Furthermore the RI photometry data used for main sequence fitting by Pinsonneault et al. 1998 are from different and rather old sources and different systems and transformations were required to put all data on the same scale. It would therefore be worth to verify the quality and precision of these data by making new measurements of stars in clusters.

Also, as pointed out by Soderblom et al. 1998, the detection and measurement of visual binary orbits in the Pleiades, within the capabilities of experiments on board of the Hubble Space Telescope, could provide an independent valuable estimate of the distance.

4. The white dwarf mass-radius relation

The white dwarf mass-radius (M-R) relation was first described by Chandraseckar (1931) and is a result of the theory of stellar electron degeneracy. The M-R relation is widely used as a basic underlying assumption in most studies of white dwarf properties but before the release of Hipparcos data there was no convincing empirical evidence for this relation (Schmidt 1996).

In order to test the mass-radius relation it is necessary to have precise determinations of masses and radii of white dwarfs. The radius is obtained from the comparison of the absolute energy flux leaving the stellar surface to the flux reaching the Earth, provided the distance of the star is known. The absolute flux is derived from the value of effective temperature and surface gravity obtained by comparison of the observed spectrum to spectra predicted by theoretical atmosphere models (Schmidt 1996). The only way to determine the masses of single field white dwarfs is to draw them from the surface gravity $(g = GM/R^2)$ once the radius is known. Masses of white dwarfs in visual binary systems can be obtained from the orbital parameters through the third Kepler's law. The third method to determine the mass of a white dwarf is to measure the redshift of the star's spectral lines caused by the strong gravitation field at the star's surface. This requires to distinguish the line shift due to gravitational velocity $v_{grs} = GM/Rc$ (c is the speed of the light) from that due to the Doppler effect. This is possible for white dwarfs members of visual binary systems or of common proper motion (CPM) pairs where the system velocity can be determined from the companion or for white dwarfs belonging to stellar clusters of known velocity.

All methods require high resolution spectra and good parallaxes. Before the start of Hipparcos project more than 15 years ago the parallax was the most uncertain parameter in obtaining the white dwarf mass-radius relation for the accuracy on ground-based parallaxes was not better than 10 mas. During the last ten years great progress has been made with the use of CCD detectors on dedicated telescopes leading to improvements on the parallax determination of a factor of about 2. In the meantime Hipparcos has measured the parallax of about 20 white dwarfs. Although the white dwarfs are generally close to the faint magnitude limit of Hipparcos, the mean accuracy of Hipparcos white dwarfs parallaxes is $\sigma_\pi \simeq 3.6$ mas (Vauclair et al. 1997). In parallel, improvements in spectroscopic observations and analysis have lead to more accurate atmospheric parameters, $T_{\rm eff}$ and $\log g$.

Among the white dwarfs (WD) observed by Hipparcos there are 4 WD which are members of visual binary systems. Prior to Hipparcos, Sirius B was the only star which had observed mass and radius compatible with a theoretical mass-radius relation, the three others (Procyon B, 40 Eridani B and Stein 2051) were 1.5 σ below the position expected from theory (see Fig. 1 in Provencal et al. 1998). As shown in Fig. 1 by Shipman et al. (1997) the Hipparcos parallax has increased the mass of 40 Eri B by 14 percent and the star is now placed on the zero-temperature M-R relation for carbon cores of Hamada & Salpeter (1961). The important point is that the position of 40 Eri B is now compatible with single star evolution. The case of Procyon B remains puzzling for the star still lies well below the theoretical predictions for carbon cores (Provencal et al. 1997) and close to a M-R relation corresponding to white dwarfs with iron or iron-enriched cores. Provencal et al. (1998) also examined 7 white dwarfs members of CPM pairs which have Hipparcos distances and gravitational redshift measurements. They showed that 2 of these objects also lie on theoretical M-R relation corresponding to iron cores. If this situation was to be confirmed after improvements on effective temperatures determinations then the explanation would be a challenge for stellar evolution theory (see D. Winget, these Proceedings).

Vauclair et al. (1997) and Provencal et al. (1998) have studied the whole sample of 20 WD observed by Hipparcos adding 11 field WD to the most accurate sample of WD in binaries and CPM systems. These authors find that the M-R relation is now more firmly supported on observational grounds since most points are within 1 σ of recent models of evolved white dwarfs with hydrogen surface layers calculated by Wood (1995). However it is still difficult to confirm the theoretical shape of the relation empirically mainly because the observed white dwarfs are concentrated in a small interval around 0.6 $M_\odot$ and there are too few objects to verify the M-R relation in the regions of high and low mass. Also the error bars are still too large to distinguish between zero temperature and evolutionary models, or "thin" or "thick" hydrogen envelopes.

In conclusion parallel improvements in distance measurements with Hipparcos and in high resolution spectroscopy have allowed to better assess the theoretical mass-radius relation for white dwarfs and have shown evidence for difficulties for a few objects which do not appear to have carbon cores. Further progress is expected from further parallax improvements but also from better orbital parameters, T_{eff}, v_{grs} and magnitudes combined with a better understanding of atmosphere structure, in particular of convection.

5. Conclusions and prospects

Hipparcos has greatly enlarged the available stellar samples with accurate and homogeneous astrometric and photometric data. As discussed in this paper, this has allowed to confirm several basis of the stellar internal structure theory and to infer more precise characteristics of individual stars and clusters.

Because the uncertainties on distances have largely been reduced, other uncertainties are now very acute, preventing important further progress in the fine characterization of stellar structure. Progress in atmosphere modeling is highly demanded, for it has implications on colour calibrations and on determinations of effective temperatures, gravities, abundances and bolometric corrections. On the other hand, theoretical stellar models still suffer from unperfect understanding of physical processes at work in stellar interiors. This includes turbulent processes, such as convection and overshooting, or the effects of rotation or of microscopic diffusion on evolution.

A further step forward is expected from future astrometric missions. The ESA candidate mission GAIA will be dedicated to the observation of about one billion objects down to V $\simeq$ 20 mag with a typical parallax accuracy of 10 microarcsecond at V = 15 mag and will also provide multi-colour, multi-epoch photometry for each object. Stars will be observed in various distant regions of the Galaxy, including halo, bulge, thin and thick disk, spiral arms. If accepted, the mission is aimed to be launched in 2009 (Perryman et al. 1997).

In addition to parallax improvements, the future measurements of stellar oscillations frequencies will provide very strong constraints for the interior of stars. Future asteroseismological experiments are under study. A first step ahead is expected from the spatial mission COROT, still in project and expected to be launched in 2002, which is designed to detect oscillations in a few solar-type stars and δ Scuti stars (Baglin et al. 1998; see also the review by T. Brown, these Proceedings).

I enjoyed very much working with Ana Gómez, Michael Perryman, João Fernandes, Giusa and Roger Cayrel, Annie Baglin and Marie-Noël Perrin on the various subjects presented here. I am very grateful to Noël Robichon, Jean-Claude Mermilliod, Frédéric

Arénou, Catherine Turon, Corinne Charbonnel, Marie-Jo Goupil and Frédéric Thévenin for many fruitful discussions. I would like to express my sincere thanks to Mario Livio for his warm welcome at ST ScI.

REFERENCES

ALONSO, A., ARRIBAS, S., & MARTÍNEZ-ROGER, C. 1995 *A&A* **297**, 197.

ALONSO, A., ARRIBAS, S., & MARTÍNEZ-ROGER, C. 1996 *A&AS* **117**, 227.

VAN ALTENA, W. L., LEE, J. T., HOFFLEIT ET AL. 1997 *ApJL*.

ARÉNOU, F., MIGNARD, F., & PALASI, J. 1997 *The Hipparcos and Tycho Catalogues* ESA SP-1200, vol. 3, chapter 20.

AXER, M., FUHRMANN, K., & GEHREN, T. 1994 *A&A* **291**, 895.

AXER M., FUHRMANN, K., & GEHREN, T. 1995 *A&A* **300**, 751.

BAHCALL, J. N., PINSONNEAULT, M. H., & WASSERBURG, G. 1995 *Rev. Mod. Phys.* **67**, 781.

BERGBUSCH, P. A. & VANDENBERG, D. A. 1992 *ApJS* **81**, 163.

BOESGAARD, A. M. & FRIEL, E. D. 1990 *ApJ* **351**, 467.

BROWN, A. G. A., ARÉNOU, F., VAN LEEUWEN, F. ET AL. 1997. In *Hipparcos Venice '97*, ESA SP-402, 63.

CASTELLANI V., CIACIO, F., DEGL'INNOCENTI, S. ET AL. 1997 *ApJ* **322**, 801.

CAYREL, R., LEBRETON, Y., PERRIN M.-N. ET AL. 1997. In *Hipparcos Venice '97*, ESA SP-402, 219.

CAYREL DE STROBEL, G., SOUBIRAN, C., FRIEL, E. D. ET AL. 1997 *A&AS* **124**, 1.

CHABOYER, B., DEMARQUE, P., KERNAN, P. J. ET AL. 1998 *ApJ* **494**, 96.

CHANDRASECKAR, S. 1931 *MNRAS* **91**, 456.

DRUMMOND, J. D., CHRISTOU, J. D., & FUGATE, R. Q. 1995 *ApJ* **450**, 380.

FUHRMANN, K., AXER M., K., & GEHREN, T. 1995 *A&A* **301**, 492.

GRATTON, R. G., CARRETTA, E., CLEMENTINI, G., & SNEDEN, C. 1997. In *Hipparcos Venice '97*, ESA SP-402, 339.

GRATTON, R. G., FUSI PECCI, F., CARRETTA, E. ET AL. 1997 *ApJ* **491**, 749.

HAMADA, T., SALPETER, E. E. 1961 *ApJ* **134**, 683.

IZOTOV, Y. I., THUAN, T. X., & LIPOVETSKY, V. A. 1997 *ApJS* **108**, 1.

KURUCZ, R. L. 1991. In *Stellar atmospheres: Beyond Classical Models* (eds. L. Crivarelli, I. Hubeny, D. G. Hummer) NATO ASI Series C, Vol. 341.

LEBRETON, Y., FERNANDES, J., BAGLIN, A. ET AL. 1998 *A&A*, in preparation.

LEBRETON, Y., GÓMEZ, A. E., MERMILLIOD, J. C. ET AL. 1997. In *Hipparcos Venice '97*, ESA SP-402, 231.

LEBRETON, Y., MICHEL, E., GOUPIL, M. J. ET AL. 1995. In *Astronomical and Astrophysical Objectives of Sub-Milliarcsecond Optical Astrometry* (eds. Høg, E., Seidelman, P. K.) IAU Symp. 166. p. 135. Kluwer.

LEBRETON, Y., PERRIN, M.-N., CAYREL, R. ET AL. 1998 *A&A*, submitted.

LEBRETON, Y., PERRIN, M.-N., FERNANDES, J. ET AL. 1997. In *Hipparcos Venice '97*, ESA SP-402, 379.

LINDEGREN, L. 1997. In *Hipparcos Venice '97*, ESA SP-402, 13.

MERMILLIOD, J.-C. 1988 *Bull. Inf. Centres des Données stellaires* **35**, 77.

MERMILLIOD, J.-C., TURON, C., ROBICHON, N. ET AL. 1997 In *Hipparcos Venice '97*, ESA SP-402, 643.

MIGNARD, F. 1997. In *Hipparcos Venice '97*, ESA SP-402, 5.

NISSEN, P. E., HØG, E., & SCHUSTER, W. J. 1997. In *Hipparcos Venice '97*, ESA SP-402, 225.

PETERSON, D. M. & SOLENSKY, R. 1988 *ApJ* **333**, 256.

PÉREZ HERNÁNDEZ, F. & CHRISTENSEN-DALSGAARD, J. 1994 *MNRAS* **267**, 111.

PERRYMAN, M. A. C. 1997. In *Hipparcos Venice '97*, ESA SP-402, 1.

PERRYMAN, M. A. C., LINDEGREN, & L., TURON, C. 1997. In *Hipparcos Venice '97*, ESA SP-402, 743.

PERRYMAN, M., BROWN, A. G. A., LEBRETON, Y. ET AL. 1998 *A&A* **331**, 81.

PONT, F., CHARBONNEL, C., LEBRETON, Y. ET AL. 1997. In *Hipparcos Venice '97*, ESA SP-402.

PONT, F., MAYOR, M., TURON, C., & VANDENBERG, D. 1997 *A&A* **329**, 87.

PINSONNEAULT, M. H., STAUFFER, J., SODERBLOM, D. R. ET AL. 1998 *ApJ* **504**, 170.

PROFFITT, C. R. & VANDENBERG, D. A. 1991 *ApJS* **77**, 473.

PROVENCAL, J. L., SHIPMAN, H. L., WESEMAEL, F. ET AL. 1997 *ApJ* **480**, 777.

PROVENCAL, J. L., SHIPMAN, H. L., HØG E., & THEJLL, P. 1998 *ApJ* **494**, 759.

REID, I. N. 1997 *AJ* **114**, 161.

ROBICHON, N., TURON, C., ARÉNOU ET AL. 1997. In *Hipparcos Venice '97*, ESA SP-402, 567.

ROBICHON, N., ARÉNOU, F., MERMILLIOD, J.-C., TURON, C. 1998 *A&A*, in preparation.

SCHMIDT, H. 1996 *A&A* **311**, 852.

SHIPMAN, H. L., PROVENCAL, J. L., HØG E., & THEJLL, P. 1997 *ApJ* **488**, L46.

SODERBLOM, D. R., KING, J. R., HANSON, R. B. ET AL. 1998 *ApJ* **504**, 192.

SÖDERHJELM, S., LINDENGREN, L., & PERRYMAN, M. A. C. 1997. In *Hipparcos Venice '97*, ESA SP-402, 251.

THÉVENIN, F. & IDIART, T. 1998 *A&A*, in preparation.

TORRES, G., STEFANIK, R. P., & LATHAM, D. W. 1997 *ApJ* **474**, 256.

TORRES, G., STEFANIK, R. P., & LATHAM, D. W. 1997 *ApJ* **479**, 268.

TORRES, G., STEFANIK, R. P., & LATHAM, D. W. 1997 *ApJ* **485**, 167.

VAN LEEUWEN, V. 1997. In *Hipparcos Venice '97*, ESA SP-402, 19.

VANDENBERG, D. A., BOLTE, P. B., & STETSON, P. B. 1996 *ARA&A* **34**, 461.

VANDENBERG, D. A., SWENSON, F.AJ., & ROGERS, F. J. 1998, in preparation.

VAUCLAIR, G., SCHMIDT, H., KOESTER, D., & ALLARD, N. 1997. In *Hipparcos Venice '97*, ESA SP-402, 371.

WHEELER, J. C., SNEDEN, C., & TRURAN, J. W. 1989 *ARA&A* **27**, 279.

WOOD, M. A. 1995. In *Proc. 9th European Workshop on White Dwarfs* (eds. D. Koester & K. Werner). p. 41. Springer.

ZAHN, J. P. 1992 *A&A* **265**, 115.

Solar neutrinos

By JOHN N. BAHCALL

Institute for Advanced Study, Princeton, NJ 08540

1. Introduction

Five experiments have observed solar neutrinos, three of which are radiochemical experiments: Homestake (chlorine detector) (Cleveland et al. 1998), GALLEX (gallium detector) (Hampel et al. 1996), and SAGE (gallium detectors) (Gavrin et al. 1997). These radiochemical detectors register all neutrinos above a fixed threshold energy (0.8 MeV for chlorine, 0.2 MeV for gallium), with no further information about energy. Three other experiments, Kamiokande (Fukuda et al. 1996), SuperKamiokande (Suzuki 1998), and SNO (which, as of this writing, is just turning on), measure the energies in real time of electrons produced in neutrino interactions in water. Kamiokande and SuperKamiokande use neutrino-electron scattering in ordinary (but very pure) water and SNO uses neutrino absorption and neutrino disassociation of deuterium in heavy water. These water detectors are designed to measure electrons, from neutrino interactions, with more than 5 MeV of energy.

I will discuss the results of each of this experiments after I first answer a series of background questions. Why study solar neutrinos? What does the combined standard model (solar plus electroweak) predict for solar neutrinos? Why are the calculations of neutrino fluxes robust? What are the three solar neutrino problems? I will conclude by summarizing what have we learned in the first 35 years of solar neutrino research and what we hope to learn in the next decade.

Many of the results given here are adapted from the recent Bahcall-Pinsonneault (1998) (BP98) solar model calculations (for a general review of the subject see Bahcall 1989). If you want to obtain some of the numerical data that are discussed in this talk, you can copy them from my Web site: http://www.sns.ias.edu/~jnb .

2. Why study solar neutrinos?

Astronomers study solar neutrinos for different reasons than physicists. For astronomers, solar neutrino observations offer an opportunity to test directly the theories of stellar evolution and of nuclear energy generation. With neutrinos one can look into the interior of a main sequence star and observe the nuclear fusion reactions that are ultimately responsible for starlight via the general reaction

$$4\,^1\mathrm{H} \longrightarrow {}^4\mathrm{He} + 2e^+ + 2\nu_e \ . \tag{2.1}$$

The optical depth of the sun for a typical neutrino produced by nuclear fusion is $\sim 10^{-9}$, about 20 orders of magnitude smaller than the optical depth for a typical optical photon. As we shall see, solar neutrino experiments constitute quantitative, well-defined tests of the theory of stellar evolution. Since stellar evolution theory is used widely in interpreting astronomical observations, direct tests of this theory are of importance to astronomers.

Table 1 shows the principal nuclear reactions that accomplish equation (2.1) via the proton-proton chain of reactions. In what follows, we shall refer often to the reactions listed in this table.

Solar neutrinos are of interest to physicists because they can be used to perform unique particle physics experiments. Many physicists believe that solar neutrino experiments

Reaction Number	Reaction	Neutrino Energy (MeV)
1	$p + p \rightarrow {}^2\mathrm{H} + e^+ + \nu_e$	0.0 to 0.4
2	$p + e^- + p \rightarrow {}^2\mathrm{H} + \nu_e$	1.4
3	${}^2\mathrm{H} + p \rightarrow {}^3\mathrm{He} + \gamma$	
4	${}^3\mathrm{He} + {}^3\mathrm{He} \rightarrow {}^4\mathrm{He} + 2p$	
	or	
5	${}^3\mathrm{He} + {}^4\mathrm{He} \rightarrow {}^7\mathrm{Be} + \gamma$	
	then	
6	$e^- + {}^7\mathrm{Be} \rightarrow {}^7\mathrm{Li} + \nu_e$	0.86, 0.38
7	${}^7\mathrm{Li} + p \rightarrow {}^4\mathrm{He} + {}^4\mathrm{He}$	
	or	
8	$p + {}^7\mathrm{Be} \rightarrow {}^8\mathrm{B} + \gamma$	
9	${}^8\mathrm{B} \rightarrow {}^8\mathrm{Be} + e^+ + \nu_e$	0 to 15

TABLE 1. The Principal Reactions of the $p\text{--}p$ Chain

may have already provided evidence for a non-zero neutrino mass and that electron flavor (or the number of electron-type neutrinos) may not be conserved.

For some of the theoretically most interesting neutrino parameters, solar neutrino experiments are more sensitive tests for neutrino transformations in flight than experiments that can be carried out with laboratory sources. The reasons for this exquisite sensitivity are: 1) the great distance between the accelerator (the solar interior) and the detector (on earth); 2) the relatively low energy (MeV) of solar neutrinos; and 3) the enormous path length of matter ($\sim 10^{11}\mathrm{gm}\ \mathrm{cm}^{-2}$) that neutrinos must pass through on their way out of the sun.

One can quantify the sensitivity of solar neutrinos relative to laboratory experiments by considering the proper time that would elapse for a finite-mass neutrino in flight between the point of production and the point of detection. The elapsed proper time is a measure of the opportunity that a neutrino has to transform its state and is proportional to the ratio, R, of path length divided by energy:

$$\text{Proper Time} \ \propto \ R = \frac{\text{Path Length}}{\text{Energy}}. \tag{2.2}$$

Future accelerator experiments with multi-GeV neutrinos are expected to reach a sensitivity of order $R = 10^2$ km GeV^{-1}. Reactor experiments are planned that will reach a level of sensitivity of $R = 10^5$ km GeV^{-1} for neutrinos with MeV energies. Solar neutrino experiments, because of the enormous distance between the source (the center of the sun) and the detector (on earth) and the relatively low energies (1 MeV to 10 MeV) of solar neutrinos involve much larger values of neutrino proper time,

$$R(\text{solar}) = \frac{10^8}{10^{-3}} \left(\frac{\text{km}}{\text{GeV}} \right) \sim 10^{11} \left(\frac{\text{km}}{\text{GeV}} \right). \tag{2.3}$$

Because of the long proper time that is available to a neutrino to transform its state, solar neutrino experiments are sensitive to very small neutrino masses that can cause neutrino oscillations (changes between different neutrino states) in vacuum. Quantitatively, vacuum neutrino oscillations are sensitive to masses as small as

$$m_\nu(\text{solar level of sensitivity}) \sim 10^{-6}\mathrm{eV} \ \text{to} \ 10^{-5}\mathrm{eV} \ , \tag{2.4}$$

provided the electron-type neutrino that is created by beta decay contains appreciable portions of at least two different neutrino mass states. Technically, this amounts to saying that the mixing angle between different neutrino states is relatively large. Neutrino mixing is analogous to photon polarization in that photons can be in different polarization states (e.g. linear or circular) and there are mixing angles which describe the relations between the different states of the photon.

Laboratory experiments have achieved a sensitivity to electron-type neutrino masses of order 1 eV. Observations of neutrinos produced by cosmic ray interactions in the earth's atmosphere have suggested the existence of neutrino oscillations involving neutrino mass differences of order ~ 0.07 eV.

Resonant neutrino oscillations, which may be induced by neutrino interactions with electrons in the sun (the famous Mikheyev-Smirnov-Wolfenstein [MSW] effect), can occur even if the electron neutrino produced in the sun is not well mixed with any of the other neutrinos (i.e. even if the mixing angles between e and μ neutrinos and between e and τ neutrinos are tiny). Standard solar models indicate that the sun has a high central density, $\rho(\text{central}) \sim 1.5 \times 10^2$ g cm^{-3}, which allows electron-type neutrinos to be resonantly converted to the more difficult to detect μ or τ neutrinos by the MSW effect, provided the difference in neutrino mass satisfies a resonance condition involving the local electron density and the neutrino energy.

Given the solar parameters from standard models, the planned and operating solar neutrino experiments are sensitive to neutrino masses in the range

$$10^{-4} \text{ eV} \lesssim m_\nu \lesssim 10^{-2} \text{ eV}, \tag{2.5}$$

via matter-induced resonant oscillations (MSW effect).

The range of neutrino masses given by equations (2.4) and (2.5) is included in the range of neutrino masses that are suggested by attractive particle-physics generalizations of the standard electroweak model.

Both vacuum neutrino oscillations and matter-enhanced neutrino oscillations can change electron-type neutrinos to the more difficult to detect μ or τ neutrinos. In addition, the likelihood that a neutrino will have its type changed may depend upon its energy, affecting the shape of the energy spectrum of the surviving electron-type neutrinos. The shape of the energy spectrum for solar neutrinos produced by any particular nuclear process must be the same as the shape measured in the laboratory unless some new physics (e.g. neutrino oscillations) is occurring. Currently operating solar neutrino experiments will measure the shape of the energy spectrum.

3. What does the combined standard model tell us about solar neutrinos?

In this section, I will describe the combined standard model (standard solar model and standard electroweak theory) that is used to decide if solar neutrino experiments have revealed something unexpected. Then I will present the calculated solar neutrino spectrum as predicted by the standard model.

3.1. *The Combined Standard Model*

In order to interpret solar neutrino experiments, one must have a quantitative solar model. Unlike many other areas of astronomy in which one can make important discoveries by identifying new classes of objects (such as quasars, or X-ray sources, or γ-ray sources), solar neutrino research requires a reliable theoretical model for comparison with the observations in order to determine whether one has found something surprising. Our physical intuition is not yet sufficiently advanced to know if we should be surprised by, for example, 10^{-2}, by 10^0, or by 10^2 neutrino-induced events per day in a chlorine tank the size of an Olympic swimming pool.

I will use the most conservative model for comparison with experiments, the combined standard model, unless explicitly stated otherwise. The combined standard model is the standard model of solar structure and evolution and the standard electroweak model of particle physics.

A solar model is required in order to predict the number of neutrinos created in a given energy range per unit time. On a fundamental level, a solar model is required in order to predict the rate of nuclear fusion by the p–p chain (shown in Table 1 and discussed below) and the rate of fusion by the CNO reactions [originally favored by H. Bethe in his epochal 1939 study of nuclear fusion reactions (Bethe 1939)].

In our discussion, I will assume a result common to all modern solar models, namely, that the CNO reactions contribute only a very small fraction of the luminosity of the sun. Although the dominance of the p–p chain is often taken for granted in theoretical analyses of solar neutrino experiments, it is not an *a priori* obvious result.

A precise solar model is required to calculate accurately which nuclear reaction occurs more often at the two principal branching points of the p–p fusion chain. Referring to Table 1, the branching points occur between reactions 4 and 5 and between reactions 6 and 8. If the p–p chain is terminated by reaction 4, only low-energy (< 0.4 MeV) p–p neutrinos are produced, but if the termination occurs via reaction 5, then higher-energy ^{7}Be and ^{8}B neutrinos are created.

Many authors have claimed that the p–p neutrino flux is essentially determined by the solar luminosity. This is wrong. Depending upon whether reaction 4 or reaction 5 is dominant, the flux of p–p neutrinos can vary by a factor two. If reaction 4 is dominant, two p–p neutrinos are produced each times four protons are burned to produce an alpha particle (see Eq. 2.1). If reaction 5 is dominant, only one p–p neutrino plus a ^{7}Be neutrino (or rarely a ^{8}B neutrino) is produced for each alpha particle that is formed. A solar model is required to decide which reaction, 4 or 5, is dominant.

The ratio of the rates for reaction 6 and reaction 8 determines how often ^{7}Be neutrinos (two lines: 0.86 MeV and 0.38 MeV) are produced rather than the rare, but more easily detected ^{8}B neutrinos (maximum energy ~ 14 MeV) are produced. The predicted rates in solar neutrino experiments depend sensitively upon the relative frequencies of these crucial reactions. Fortunately, the theoretical uncertainties in the predicted neutrino fluxes are not very large. For the important fluxes, the uncertainties vary from $\sim 1\%$ to $\sim 20\%$, depending upon the neutrino source in question. The rates for individual detectors are determined by the energy spectrum, by the neutrino type of the incoming solar neutrinos, and by the interaction cross sections of the different detectors.

A particle physics model is required to predict what happens to the neutrinos after they are created. I will use the simplest version of the standard electroweak model, according to which nothing happens to the neutrinos after they are created in the interior of the sun. In this theory, neutrinos are massless and neutrino flavor (electron type) is conserved. The standard electroweak model has had many successes in precision laboratory tests; modifications of this theory will be accepted only if incontrovertible experimental evidence forces a change.

3.2. *The Solar Neutrino Spectrum*

Figure 1 shows the calculated neutrino spectrum for the most important neutrino sources from the sun. I will discuss briefly the p–p (and *pep*) neutrinos, the ^{7}Be neutrinos, and

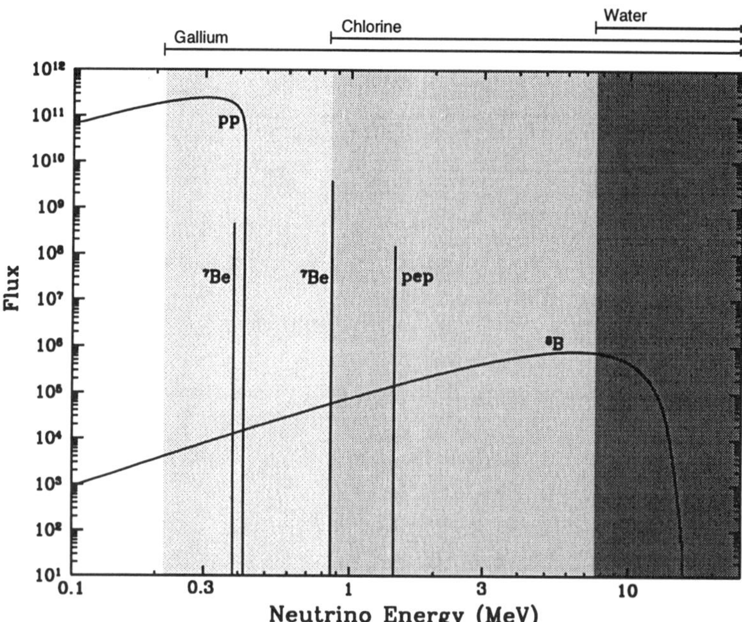

FIGURE 1. Solar Neutrino Spectrum. This figure shows the energy spectrum of neutrinos from the p–p chain that is predicted by the standard solar model. The neutrino fluxes from continuum sources (p–p and ^{8}B) are given in the units of number per cm^2 per second per MeV at one astronomical unit. The line fluxes (pep and ^{7}Be) are given in number per cm^2 per second. The arrows at the top of the figure indicate the energy thresholds for the ongoing neutrino experiments. The higher-energy ^{7}Be line is just above threshold in the chlorine experiment. For simplicity, CNO neutrinos are omitted.

the ^{8}B neutrinos. I will concentrate on the reliability of the predictions and will indicate the role of each of these neutrinos in the ongoing experiments.

The dominant source of solar neutrinos is the first reaction listed in Table 1, the basic p–p reaction ($p + p \longrightarrow D + e^+ + \nu_e$), which creates neutrinos with energies less than 0.4 MeV. Most of the nuclear energy that emerges as sunlight begins with this reaction. The theoretical uncertainty in the p–p neutrino flux is about 1%. Among the solar neutrino experiments that are currently operating or that are being constructed, only the GALLEX and SAGE gallium experiments have energy thresholds low enough to detect the p–p neutrinos. About 0.2% of the p–p fusions are believed to occur via the *pep* reaction, the second reaction in Table 1. This reaction produces a neutrino with a fixed energy, a 'neutrino line' that contributes a small part of the calculated event rate in the chlorine and gallium experiments.

The next most important source of neutrinos is from the ^{7}Be neutrino line at 0.86 MeV, which is produced by reaction 6 of Table 1. About 15% of the solar luminosity is produced by reactions which go through this channel; the uncertainty in the neutrino flux is $\sim 9\%$. The ^{7}Be neutrinos contribute significantly, according to standard model calculations, to the chlorine and the gallium experiments, but are too low in energy to be detected in the Kamiokande experiment. In an experiment under development called BOREXINO, ^{7}Be neutrinos will be detected by the unique signature they produce in scintillation light caused by neutrino-electron scattering.

The ^{8}B neutrino flux, produced by reaction 9 of Table 1, is tiny, $\sim 10^{-4}$ of the flux of p–p neutrinos. However, the ^{8}B neutrinos are crucial for solar neutrino physics and astronomy. Because of their high energy (~ 10 MeV, which takes advantage of a superallowed transition from the nuclear ground-state of chlorine to an excited state of argon), ^{8}B neutrinos dominate the predicted capture rate for the chlorine experiment. They are also the only significant source of neutrinos above the energy threshold in the water Cherenkov experiments, Kamiokande, SuperKamiokande, and SNO. Unfortunately, the theoretical uncertainty in the predicted ^{8}B neutrino flux is relatively large, $\sim 19\%$.

4. Why are the predicted neutrino fluxes robust?

The predicted event rates in the different solar neutrino experiments have been remarkably stable over the past 30 years. Our published estimate in 1968 (Bahcall, Bahcall, and Shaviv 1968), which accompanied the first report by Davis and his collaborators of measurements with the chlorine experiment, was 7.5 ± 1.0 Solar Neutrino Units (SNU); the most recent and detailed calculation yielded in 1998 a predicted rate of $7.7^{+1.2}_{-1.0}\pm$ SNU (Bahcall, Basu, and Pinsonneault 1998). A SNU is a convenient unit to describe the measured rates of solar neutrino experiments: 10^{-36} interactions per target atom per second. The theoretical errors are intended to be as close as possible to effective 1σ errors; they are obtained by carrying out detailed calculations using 1σ uncertainties on all the measured input data and, for the theoretical errors (which are generally less important), by taking the extreme range of theoretical calculations to be 3σ uncertainties.

In the intervening three decades since the first experimental report on solar neutrinos, my colleagues and I have calculated many different models with steady improvements in measured and calculated input data and in the physics used to describe the solar interior. The best-estimate predictions have bounced around every few years, as different improvements were included, but the answers have always remained with ± 2 SNU of the 1968 best-estimate.

There are three reasons that the theoretical calculations of the neutrino fluxes are robust: 1) the availability of precision measurements and precision calculations of input data; 2) the connection between neutrino fluxes and the measured solar luminosity; and 3) the measurement of the helioseismological frequencies of the solar pressure-mode (p-mode) eigenfrequencies.

Over the past three decades, many hundreds of researchers have performed precision measurements of crucial input data including nuclear reaction cross sections and the abundances of the chemical elements on the solar surface. Many other researchers have calculated accurate opacities, equations of state, and weak interaction cross sections. By now, these input data are relatively precise and their uncertainties are quantifiable.

The solar neutrino fluxes and the solar luminosity both depend upon the rates of the nuclear fusion reactions in the solar interior. Since we know experimentally the solar luminosity (to an accuracy of $\sim 0.4\%$), the calculated neutrino fluxes are strongly constrained by the fact that the standard solar models must yield precisely the measured solar luminosity.

Could the solar model calculations be wrong by enough to explain the discrepancies between predictions and measurements for solar neutrino experiments? Helioseismology, which confirms predictions of the standard solar model to high precision, suggests that the answer is probably "No."

Thousands of p-mode helioseismological frequencies have been measured to an accuracy of 1 part in 10^4. The standard solar models discussed here reproduce these p-mode frequencies to a rms accuracy of better than 1 part in $1,000$.

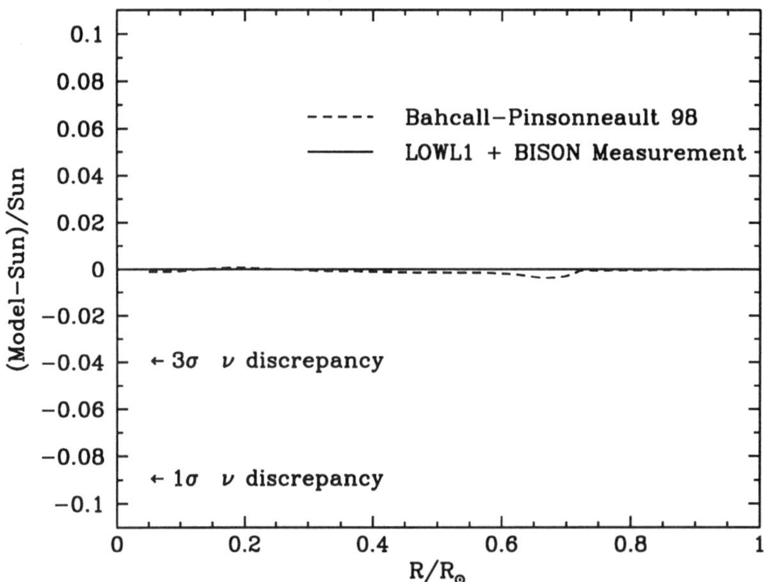

FIGURE 2. Predicted versus Measured Sound Speeds. This figure shows the excellent agreement between the calculated (solar model BP98, Model) and the measured (Sun) sound speeds, a fractional difference of 0.001 rms for all speeds measured between $0.05R_\odot$ and $0.95R_\odot$. The vertical scale is chosen so as to emphasize that the fractional error is much smaller than generic changes in the model, 0.03 to 0.08, that might significantly affect the solar neutrino predictions.

Figure 2 shows the fractional differences between the most accurate available sound speeds measured by helioseismology (Basu et al. 1997) and sound speeds calculated with our best solar model with no free parameters (Bahcall, Basu, and Pinsonneault 1998). The horizontal line corresponds to the hypothetical case in which the model predictions exactly match the observed values. The rms fractional difference between the calculated and the measured sound speeds is 1.1×10^{-3} for the entire region over which the sound speeds are measured, $0.05R_\odot < R < 0.95R_\odot$. In the solar core, $0.05R_\odot < R < 0.25R_\odot$ (in which about 95% of the solar energy and neutrino flux is produced in a standard model), the rms fractional difference between measured and calculated sound speeds is 0.7×10^{-3}.

Helioseismological measurements also determine two other parameters that help characterize the outer part of the sun (far from the inner region in which neutrinos are produced): the depth of the solar convective zone (CZ), the region in the outer part of the sun that is fully convective, and the present-day surface abundance by mass of helium (Y_{surf}). The measured values, $R_{\mathrm{CZ}} = (0.713 \pm 0.001)$ R$_\odot$ (Basu and Antia 1995), and $Y_{\mathrm{surf}} = 0.249 \pm 0.003$ (Basu and Antia 1997), are in satisfactory agreement with the values predicted by the solar model BP98, namely, $R_{\mathrm{CZ}} = 0.714$ R$_\odot$, and $Y_{\mathrm{surf}} = 0.243$. However, we shall see below that precision measurements of the sound speed near the transition between the radiative interior (in which energy is transported by radiation) and the outer convective zone (in which energy is transported by convection) reveal small discrepancies between the model predictions and the observations in this region.

If solar physics were responsible for the solar neutrino problems, how large would one expect the discrepancies to be between solar model predictions and helioseismological observations? The characteristic size of the discrepancies can be estimated using the results of the neutrino experiments and scaling laws for neutrino fluxes and sound speeds.

All recently published solar models predict essentially the same fluxes from the fundamental pp and pep reactions (amounting to 72.4 SNU in gallium experiments, cf. Table 1), which are closely related to the solar luminosity. Comparing the measured gallium rates (reported at Neutrino 98) and the standard predicted rate for the gallium experiments, the ^{7}Be flux must be reduced by a factor N if the disagreement is not to exceed n standard deviations, where N and n satisfy $72.4 + (34.4)/N = 72.2 + n\sigma$. For a 1σ (3σ) disagreement, $N = 6.1(2.05)$. Sound speeds scale like the square root of the local temperature divided by the mean molecular weight and the ^{7}Be neutrino flux scales approximately as the 10th power of the temperature (Bahcall and Ulmer 1996). Assuming that the temperature changes are dominant, agreement to within 1σ would require fractional changes of order 0.09 in sound speeds (3σ could be reached with 0.04 changes), if all model changes were in the temperature.† This argument is conservative because it ignores the contributions from the ^{8}B and CNO neutrinos which contribute to the observed counting rate (cf. Table 1) and which, if included, would require an even larger reduction of the ^{7}Be flux.

I have chosen the vertical scale in Fig. 1 to be appropriate for fractional differences between measured and predicted sound speeds that are of order 0.04 to 0.09 and that might therefore affect solar neutrino calculations. Fig. 1 shows that the characteristic agreement between solar model predictions and helioseismological measurements is more than a factor of 30 better than would be expected if there were a solar model explanation of the solar neutrino problems.

The calculated solar neutrino fluxes are, after 30 years of intense study, known to reasonable accuracy because of the many precise measurements and calculations of input data, because of the strong constraint imposed on the models by the measured total solar luminosity, and because of the important tests of solar structure that are provided by helioseismological measurements.

5. What are the three solar neutrino problems?

I will compare in this section the predictions of the combined standard model with the results of the operating solar neutrino experiments. We will see that this comparison leads to three different discrepancies between the calculations and the observations, which I will refer to as the three solar neutrino problems.

Figure 3 shows the measured and the calculated event rates in the five solar neutrino experiments (Homestake, GALLEX, SAGE, Kamiokande, and SuperKamiokande) for which quantitative results are available at the time of this writing. The figure reveals three discrepancies between the experimental results and the expectations based upon the combined standard model. As we shall see, only the first of these discrepancies depends at all sensitively upon predictions of the standard solar model.

5.1. *Calculated versus observed chlorine rate*

The first solar neutrino experiment to be performed was the chlorine radiochemical experiment, which detects electron-type neutrinos that are more energetic than 0.81 MeV.

---

† I have used in this calculation the GALLEX and SAGE measured rates reported by Kirsten and Gavrin at Neutrino 98. The experimental rates used in BP98 were not as precise and therefore resulted in slightly less stringent constraints than those imposed here. In BP98, we found that agreement to within 1σ with the then available experimental numbers would require fractional changes of order 0.08 in sound speeds (3σ could be reached with 0.03 changes).

Total Rates: Standard Model vs. Experiment

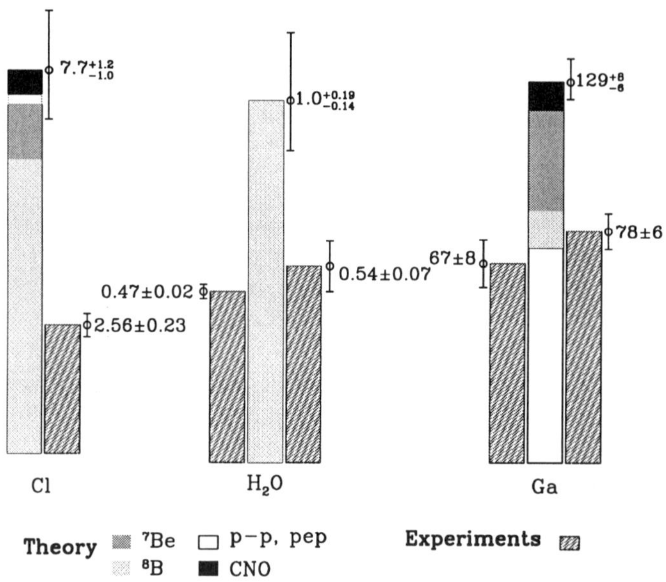

FIGURE 3. Comparison of measured rates and standard-model predictions for four solar neutrino experiments.

After more than 30 years of the operation of this experiment, the measured event rate is 2.56 ± 0.23 SNU, which is a factor 3 less than is predicted by the most detailed theoretical calculations. Most of the predicted rate in the chlorine experiment is from the rare, high-energy ^{8}B neutrinos, although the ^{7}Be neutrinos are also expected to contribute significantly. According to standard model calculations, the *pep* neutrinos and the CNO neutrinos (for simplicity not discussed here) are expected to contribute less than 1 SNU to the total event rate.

This discrepancy between the calculations and the observations for the chlorine experiment was, for more than two decades, the only solar neutrino problem. I shall refer to the chlorine disagreement as the "first" solar neutrino problem. (It used to be called "the" solar neutrino problem.)

5.2. *Incompatibility of chlorine and water experiments*

The second solar neutrino problem results from a comparison of the measured event rates in the chlorine experiment and in the Japanese pure-water experiments, Kamiokande and SuperKamiokande. The water experiments detect higher-energy neutrinos, those with energies above 7.5 MeV (Kamiokande) or 6.5 MeV (SuperKamiokande, so far), by neutrino-electron scattering: $\nu + e \longrightarrow \nu' + e'$. According to the standard solar model (see also Table 1), ^{8}B beta decay is the only important source of these higher-energy neutrinos.

The Kamiokande and SuperKamiokande experiments show that the observed neutrinos come from the sun. The electrons that are scattered by the incoming neutrinos recoil predominantly in the direction of the sun-earth vector; the relativistic electrons are observed by the Cherenkov radiation they produce in the water detector.

In addition, the water experiments measure the energies of individual scattered electrons and therefore provide information about the energy spectrum of the incident solar neutrinos. The currently most popular parameters for neutrino oscillations predict only relatively small deviations from the spectrum expected with no oscillations. Indeed, SuperKamiokande does observe a deviation that can be explained by neutrino oscillations but as of the time of this writing the statistical significance is not very high.

The event rate in the Kamiokande experiment is determined by the same high-energy ^{8}B neutrinos that are expected, on the basis of the combined standard model, to dominate the event rate in the chlorine experiment. I have shown elsewhere that solar physics changes the shape of the ^{8}B neutrino spectrum by less than 1 part in 10^5. Therefore, we can calculate the rate in the chlorine experiment that is produced by the ^{8}B neutrinos observed in the SuperKamiokande experiment (6.5 MeV threshold energy). This partial (^{8}B) rate in the chlorine experiment is 2.78 ± 0.10 SNU, which exceeds by about one standard deviation the total observed chlorine rate from all neutrino sources, of 2.56 ± 0.23 SNU.

Comparing the rates of the Kamiokande and the chlorine experiments, one finds that the net contribution to the chlorine experiment from the *pep*, ^{7}Be, and CNO neutrino sources is negative: -0.22 ± 0.10 SNU. The standard model calculated rate from *pep*, ^{7}Be, and CNO neutrinos is 1.9 SNU.

The apparent incompatibility of the chlorine and the Kamiokande experiments is the "second" solar neutrino problem. The inference that is often made from this comparison is that the energy spectrum of ^{8}B neutrinos is changed from the standard shape by physics not included in the simplest version of the standard electroweak model.

5.3. *Gallium experiments: No room for ^{7}Be neutrinos*

The results of the gallium experiments, GALLEX and SAGE, constitute the third solar neutrino problem. The average observed rate in these two experiments is 73 ± 5 SNU, which is essentially fully accounted for in the standard model by the theoretical rate of 72 SNU that is calculated to come from the basic *p–p* and *pep* neutrinos (with only a 1% uncertainty in the standard solar model *p–p* flux). The ^{8}B neutrinos, which are observed above 6.5 MeV in the SuperKamiokande experiment, must also contribute to the gallium event rate. Using the standard shape for the spectrum of ^{8}B neutrinos and normalizing to the rate observed in Kamiokande, ^{8}B contributes another 6 SNU, unless something happens to the lower-energy neutrinos after they are created in the sun. (The predicted contribution is 12 SNU on the basis of the standard model.) Given the measured rates in the gallium experiments, there is no room for the additional 34 ± 4 SNU that is expected from ^{7}Be neutrinos.

The seeming exclusion of everything but *p–p* neutrinos in the gallium experiments is the "third" solar neutrino problem. This problem is essentially independent of the previously-discussed solar neutrino problems, since it depends upon the *p–p* neutrinos that are not observed in the other experiments and whose calculated flux is approximately model-independent (if the general scheme of the *p–p* chain shown in Table 1 is correct).

The missing ^{7}Be neutrinos cannot be explained away by any change in solar physics. The ^{8}B neutrinos that are observed in the Kamiokande experiment are produced in competition with the missing ^{7}Be neutrinos; the competition is between reaction 6 and reaction 8 in Table 1. Solar model explanations that reduce the predicted ^{7}Be flux reduce much more (too much) the predictions for the observed ^{8}B flux.

I conclude that either: 1) at least three of the four operating solar neutrino experiments (the two gallium experiments plus either chlorine or Kamiokande) have yielded misleading results, or 2) physics beyond the standard electroweak model is required to change the

Property	Predicted		Observed
Direction	From the Sun		o.k.
Rates	Measurable		$\sim$ Predicted Rates (within factor of few)
Neutrino Energy	0–15 MeV		< 15 MeV
Time Dependence	Constant (except seasonal)		o.k.
p–p not CNO	If CNO :	Cl : 28 SNU Ga : 610 SNU H_2O : 0.0	2.6 SNU 73 SNU 0.44 Standard Model
Central Temperature	16×10^6 K		$T(^8B)/T_{\mathrm{model}} \gtrsim 0.98$

TABLE 2. Predictions versus observations: 1964 vs. 1995

neutrino energy spectrum (or flavor content) after the neutrinos are produced in the center of the sun.

6. What have we learned?

No solar-model solution has been found that explains the results of the five existing solar neutrino experiments. Many particle-physics solutions have been proposed that can explain the existing data. In this section, I will summarize the main astronomical lessons that have been learned from the first 30 years of solar neutrino research.

The chlorine solar neutrino experiment was proposed in 1964 as a practical test of solar model calculations (Bahcall 1964; Davis 1964). The only motivation presented in the theoretical and experimental papers was to use neutrinos "to see into the interior of a star and thus verify directly the hypothesis of nuclear energy generation in stars."

What have we learned by direct experiments about nuclear energy generation in stars? How does our 1964 understanding compare with the results of the solar neutrino experiments? Table 2 summarizes the six principal predictions that were made (or which were implicit in the theory) in 1964 and compares those predictions with the results of the four ongoing solar neutrino experiments.

The neutrinos were predicted to originate in the solar interior; the direction of origin of the neutrinos has been verified by detecting neutrino-electron scattering (as was also suggested in 1964) in the Kamiokande and SuperKamiokande experiments.

The rates of the four operating experiments are in semi-quantitative agreement with the predictions; the ratios of the observed to the predicted rates are 0.3 (chlorine), 0.5 (neutrino-electron scattering in water), and 0.6 (gallium, average). This agreement is better than any of us dared hope for in 1964, especially since the dominant neutrino flux (from ^8B beta-decay) for the first two experiments depends upon the central temperature of the sun as approximately the 24th power of the central temperature.

The energy range of the dominant neutrinos was predicted to be from 0 MeV to 14 MeV, which is consistent with the observations from the Kamiokande and Super-Kamiokande experiments.

Standard models predict that the neutrino fluxes are constant in time except for a small seasonal variation. (The Kelvin-Helmholtz cooling time for the solar interior is

$\sim 10^7$ years.) The consensus view of the experimentalists is that there are no statistically significant variations in the available data. Small deviations from the constant flux prediction are predicted by some of the popular neutrino oscillation scenarios, but improved statistics are required to test these predictions.

Standard solar models predict that the sun shines almost entirely via the p–p chain of nuclear fusion reactions, rather than the CNO reactions originally emphasized by Bethe. If CNO reactions were dominant, the event rates in solar neutrino experiments could be calculated precisely. These "all-CNO" rates, as shown in Table 2, differ from the observed rates by more than an order of magnitude. Finally, if we crudely characterize the rate of the ^{8}B neutrino emission by its approximate dependence upon the central temperature of the solar model, then the central temperature of the solar model agrees with the value obtained from the experimental rates to an accuracy of $\sim 2\%$ or better.

The pioneering solar neutrino experiments have shown directly that the sun shines by nuclear fusion reactions, thus achieving the original goal proposed in 1964. Quantitative improvements in the tests shown in Table 2 will occur with the next generation of experiments, which should also refine our knowledge of the physical properties of neutrinos. But the most important qualitative result has been established: *neutrinos have been observed from the interior of the sun in approximately the number and with the energies expected.*

In the more than three-decade long struggle to improve solar models in order to calculate more accurate solar neutrino fluxes, we have obtained a greater understanding of solar structure. The theoretical models have gradually been refined as improved input data, more accurate physical descriptions, and more precise numerical techniques have been employed. Perhaps most importantly, the complementary field of helioseismology has been developed and now provides precise data that determine the sound velocity over most of the solar interior; these beautiful measurements are used to test and to refine the standard solar model. Further improvements in the solar model are desirable and important, but the quantitative agreement, typically better than 1 part in 1,000, between the calculated eigenfrequencies of pressure modes and the measured (helioseismological) frequencies provides strong evidence for the basic correctness of the standard solar model.

7. What next?

In this section, I will summarize the goals of solar neutrino research, first in physics and then in astronomy, during the next decade or two.

7.1. *Physics goals*

The fundamental goal of physics research with solar neutrinos is to measure the energy spectrum and flavor content as a function of time of the solar neutrino flux. We want to know how many neutrinos reach the earth with a given energy and with a given flavor (i.e. neutrino type: e, μ, or τ), all as a function of time. Because of some exotic particle physics possibilities, we also want to know if the solar neutrino flux contains any anti-neutrinos.

The standard model predicts that the energy spectrum of neutrinos from any given neutrino source, e.g. from ^{8}B beta-decay, will be the same to high accuracy as the energy spectrum inferred from terrestrial laboratory measurements. In the standard electroweak theory, only massless electron-type neutrinos are created in nuclear beta decay or nuclear fusion reactions. Standard electroweak theory predicts that the solar neutrinos produced by nuclear fusion reactions are all ν_e, not ν_μ or ν_τ. (According to MSW and vacuum oscillation theories, neutrinos created in nuclear beta-decay or nuclear fusion reactions are linear combinations of different neutrino types and at least one neutrino type has a non-zero mass.) Finally, the total amount of thermal energy in the solar interior implies that the neutrino fluxes will be constant in time (for time scales less than 10^7 years) except for the seasonal dependences caused by the earth's orbital eccentricity. Any departure from these expectations will be a signal of physics beyond the standard electroweak model.

Physicists want to use solar neutrino experiments to measure, or to set stringent limits on, the elementary properties of neutrinos, especially their masses and mixing angles. It seems likely that we will make important progress toward this goal in the next decade.

7.2. *Astronomy goals*

The fundamental goal of solar neutrino astronomy is to determine the rates of different nuclear fusion reactions in the solar interior. Neutrino fluxes created by the different nuclear sources are the signatures of the fusion reactions. We must know what happens to the neutrinos after they are created in order to infer the created neutrino energy spectrum from the measured neutrino energy spectrum.

Progress in solar neutrino astronomy is held hostage to progress in particle physics. As discussed in the previous subsection, it seems likely that we will learn enough about the particle physics in the next decade to permit accurate inferences about the rates of neutrino creation in the sun from the observed rates of neutrino arrival at the earth. The discussion in this subsection presumes that the required progress in understanding the properties of the neutrino will be achieved.

Completing Hydrogen Fusion.—Table 1 shows that the two principal ways of completing nuclear fusion in the sun are reactions 4 and 5, the so-called ^{3}He–^{3}He and ^{3}He–^{4}He reactions. Because of the slightly smaller reduced mass that exists for the ^{3}He–^{3}He reaction, Coulomb barrier penetration favors this reaction over the ^{3}He–^{4}He reaction at lower temperatures. According to the standard solar model, the ^{3}He–^{4}He reaction is dominant in the innermost region of the sun (where it is 1.5 times faster than the ^{3}He–^{3}He reaction), but overall occurs in only $\sim 15\%$ of the fusion terminations that are described by equation (2.1). That is, in the most detailed solar models, the ^{3}He–^{3}He reaction is on average more than 6 times faster in completing the nuclear fusion of protons into α-particles than the competing ^{3}He–^{4}He reaction.

Is this prediction of the standard solar model correct? A determination of the p–p and ^{7}Be neutrino fluxes (corrected for what non-standard particle physics has done to them after they were created in the sun) can answer this important question. The average ratio of the total number of ^{3}He–^{4}He reactions per unit time in the sun to the total number of ^{3}He– the total number of ^{3}He–^{3}He reactions per unit time in the sun is

$$\frac{<\,^3\mathrm{He}-^3\mathrm{He}\,>}{<\,^3\mathrm{He}-^4\mathrm{He}\,>} = \frac{2\phi(^7\mathrm{Be})}{[\phi(p\text{-}p) - \phi(^7\mathrm{Be})]} , \tag{7.6}$$

where $\phi(p\text{-}p)$ and $\phi(^7\mathrm{Be})$ are, respectively, the fluxes from the p–p and ^{7}Be neutrinos.

Equation (7.6) is the most precisely-testable prediction that I know of that follows directly from the theory of stellar energy generation. The known theoretical uncertainties in the calculation of the average solar ratio of ^{3}He–^{4}He to ^{3}He–^{3}He reactions is 9%.

The ^{8}B Neutrino Flux.—The flux of neutrinos from ^{8}B beta-decay in the sun (see reaction 9 of Table 1) is, in principle, the simplest solar neutrino flux to measure. The higher energies of the ^{8}B neutrinos make them easiest to detect. For this reason, the Kamiokande, SuperKamiokande, and SNO neutrino experiments will all concentrate on the ^{8}B neutrinos.

However, one must determine the total flux of ^{8}B neutrinos, including the more difficult to detect μ or τ neutrinos that may have been produced by neutrino oscillations from the originally-created electron-type neutrinos. The total number of neutrinos of all types will be measured directly in the SNO experiment via the neutral-current disintegration of deuterium and, less directly, via electron-neutrino scattering in SuperKamiokande. (This statement presumes there are no sterile neutrinos, i.e. neutrinos that do not interact with matter.)

The magnitude of the ^{8}B flux (all neutrino flavors), which is a sensitive probe of the temperature of the solar interior, varies approximately as T^{24}_{central}. Therefore, it is important to determine experimentally the total ^{8}B solar neutrino flux.

The Temperature Profile of the Solar Interior.—A precision test of the theory of stellar structure and stellar evolution can be performed by measuring the average difference in energy between the neutrino line produced by ^{7}Be electron capture in the solar interior and the corresponding neutrino line produced in a terrestrial laboratory. This energy shift is calculated to be 1.29 keV. The energy shift is approximately equal to the average temperature of the solar core, computed by integrating the temperature over the interior of a standard solar model with a weighting factor equal to the locally-produced ^{7}Be neutrino emission. The total range of values for the shift, calculated for a number of modern solar models (going back to 1982), is 0.06 keV (Bahcall 1994).

A measurement of the energy shift is equivalent to a measurement of the central temperature distribution of the sun.

The calculated energy profile of the ^{7}Be line contains, analogous to line-broadening in classical (photon) astronomy, information about the distribution of solar interior temperatures. The theoretical shape of the ^{7}Be neutrino line is asymmetric: on the low-energy side, the line shape is Gaussian with a half-width at half-maximum of 0.6 keV, and on the high-energy side, the line shape is exponential, with a half-width at half-maximum of 1.1 keV.

The calculated shape of the ^{7}Be neutrino line is not affected significantly by vacuum neutrino oscillations, the MSW effect, or other frequently discussed weak-interaction solutions to the solar neutrino problems. This is a key result: it implies that the astronomical information contained in the line shift and in the line profile is not dependent upon further progress in neutrino physics.

Detectors are available that have the resolution to measure the line shift. Unfortunately, their current sizes are too small to permit a full-scale solar neutrino experiment. However, proposals have been made in the literature for developing detectors that are sufficiently large to be able to measure well the average shift in energy of the solar neutrino line.

More Complete Models of the Sun.—The accuracy of the physical description that is currently achieved with one-dimensional (spherically symmetric) models of the sun that include diffusion is sufficient to permit excellent quantitative agreement with the measured p-mode oscillation frequencies. Numerical experiments and theoretical arguments also suggest that further improvements are unlikely to affect significantly the calculated neutrino fluxes.

Nevertheless, current models of the sun are incomplete. They are spherically symmetric and do not take account of the two-dimensional (or three-dimensional) nature of solar structure. They do not contain a self-consistent dynamical treatment of the effects of rotation, of magnetic fields, of mass loss, or of other possible effects that may violate the currently-used approximations of spherical symmetry and quasi-static evolution. We know observationally that the sun (at least near its surface) contains magnetic fields, that it is losing mass, and that it departs from spherical symmetry by ~ 1 part in 10^5.

There are both analytic and calculational challenges in including these complicated processes in a more complete physical description in the next generation of solar models. New self-consistent methods of calculating solar models (and stellar models) must be developed, and then the appropriate numerical techniques must be worked out, tested, and applied.

The goal of developing a more complete solar model is a challenge for the next decade and beyond. Fortunately, it is a challenge that could lead to important progress since computing power is much greater than it was in the past and there is an abundance of precision data with which to make detailed comparisons.

8. Summary

The first 35 years of solar neutrino research have verified experimentally that the fundamental predictions of nuclear energy generation in stars. The next 10 or 20 years of research will, I think, concentrate on using solar neutrinos to learn more about weak interaction physics. As the weak interaction questions are being resolved, it will be possible to carry out progressively more accurate tests of the theory of nuclear energy generation and of stellar structure.

In retrospect, the history of solar neutrino research seems ironic. It began with an effort to use neutrinos, whose properties were assumed to be well known, to study the interior of the nearest star. The project was an unconventional application of microscopic physics that was designed to carry out a unique investigation of a massive, macroscopic body, the sun. It now appears likely that a large community of chemists, physicists, astrophysicists, astronomers, and engineers working together may have stumbled across physics beyond the standard electroweak model.

We may have been incredibly lucky.

REFERENCES

BAHCALL, J. N. 1964 *Phys. Rev. Lett.* **12**, 300.

BAHCALL, J. N. 1989 *Neutrino Astrophysics* Cambridge University Press.

BAHCALL, J. N. 1994 *Phys. Rev. D* 49, **No. 8**, 3923.

BAHCALL, J. N., BAHCALL, N. A. & SHAVIV 1968 *Phys.Rev. Lett.*, **20**, 1209.

BAHCALL, J. N., BASU, S. & PINSONNEAULT, M. H. 1998 *Phys. Lett. B* **433**, 1.

BAHCALL, J. N. & ULMER, A. 1996 *Phys. Rev. D* **53**, 4202.

BASU, S. & ANTIA, H. M. 1995 *M.N.R.A.S.* **276**, 1402.

BASU, S. & ANTIA, H. M. 1997 *M.N.R.A.S.* **287** 189.

BASU, S., ET AL. 1997 *M.N.R.A.S.* **292**, 234.

BETHE, H. A. 1939 *Phys. Rev.* **55**, 434.

CLEVELAND, B. T., ET AL. (CHLORINE COLLABORATION) 1998 *ApJ* **496**, 505.

DAVIS, R., JR. 1964 *Phys. Rev. Lett.* **12**, 303.

FUKUDA, Y., ET AL. (KAMIOKANDE COLLABORATION) 1996 *Phys. Rev. Lett.* **77**, 1683.

GAVRIN, V., ET AL. (SAGE COLLABORATION) 1997, In *Neutrino 96, Proceedings of the XVII International Conference on Neutrino Physics and Astrophysics*, (eds. K. Huitu, K. Enqvist, and J. Maalampi). pp. 14–24. World Scientific.

HAMPEL, W., ET AL. (GALLEX COLLABORATION) 1996 *Phys. Lett. B* **388**, 384.

SUZUKI, Y. (SUPERKAMIOKANDE COLLABORATION) 1998, In *Neutrino 98, Proceedings of the XVIII International Conference on Neutrino Physics and Astrophysics.* (eds. Y. Suzuki and Y. Totsuka.) To be published in *Nucl. Phys. B (Proc. Suppl.).*

Asteroseismology

By TIMOTHY M. BROWN

High Altitude Observatory/National Center for Atmospheric Research†

1. Introduction

This review is an attempt to explain why asteroseismology is a subject worth pursuing, and to give a brief description of some of the places that this pursuit has led in the recent past. The reader should be aware that the following account is less even-handed than is commonly the case in review papers. In particular, the observations described in the last two sections are largely work with which I have been directly associated, and the background explication in sections 1–3, which deals with subject matter that has not changed very much in the last few years, is taken with little modification from the review of asteroseismology that Ron Gilliland and I wrote for Reviews of Astronomy and Astrophysics in 1994 (Brown & Gilliland 1994). That older review is more complete than this one; anyone desiring greater detail is encouraged to consult it.

Asteroseismology is commonly understood to mean the study of normal-mode pulsations in stars that, like the Sun, display a large number of simultaneously excited modes. The idea of learning about a physical system by examining its oscillation modes is of course an old one in physics, but it is only fairly recently that data of sufficient quality have become available to apply this technique to stars.

The Sun is (and will likely remain) the outstanding example of the progress that can be made using seismological methods. Seismic studies of the Sun have succeeded in mapping the variation of sound speed with depth in the Sun, and the variation of angular velocity with both depth and latitude; they have measured the depth of the Sun's convective envelope, and they have begun to be used to estimate the helium abundance in the convection zone and to reveal at least some of the subsurface structure of solar activity.

Regardless of the type of star or the mechanism driving its pulsations, we will not in the foreseeable future have as much pulsation information about other stars as we have about the Sun. A very large majority of the 10^7 modes seen in the Sun have horizontal wavelengths that are a small fraction of a solar radius. When averaged over the solar disk (as a distant observer would do), the perturbations due to these modes average to zero, rendering them undetectable. It is only in special circumstances that modes with angular degree greater than about 3 are observable on distant stars; the number of modes that may be observed is therefore likely to be at most a few tens. Nevertheless, since oscillation mode frequencies are arguably the most precise measurements relating to a star that we can make, a few tens of such frequencies may still be of great importance to our understanding of stellar structure and evolution.

2. Physical background

The stars that display pulsations may be described with reasonable accuracy as spheres. For this reason it is possible and convenient to write the pulsation eigenmodes as the

† The National Center for Atmospheric Research is sponsored by the National Science Foundation

product of a function of radius and a spherical harmonic. The spatial and temporal variation of a perturbation to the star's mean state are then

$$\xi_{nlm}(r,\theta,\phi,t) \;=\; \xi_{nl}(r)\,Y_l^m(\theta,\phi)e^{-i\omega_{nlm}t} \tag{2.1}$$

Here ξ is any scalar perturbation associated with the mode (e.g. the radial displacement); r,θ,ϕ,t are the radial coordinate, the colatitude, the longitude, and time, respectively. The mode's *radial order* n is usually identified with the number of nodes in the eigenfunction that exist between the center of the star and its surface. Since it deals with the depth structure, n is not accessible to direct observation. The *angular degree* l is the product of the stellar radius R_* and the total horizontal wavenumber of the mode; modes with large values of l display many sign changes across a stellar hemisphere, and hence are usually unobservable on distant stars. The *azimuthal order* m is the projection of l onto the star's equator; it is therefore restricted to be less than or equal to l in absolute value. The mode frequency ω_{nlm} generally depends on n and l in complicated ways, depending on the restoring forces responsible for the pulsation and on the structure of the star. In particular, there is generally no simple harmonic relation between the frequencies of modes with (for instance) given l and successive values of n. Mode frequencies resulting from theoretical calculations are often expressed as angular frequencies ω_{nlm}, as in Eq. (2.1). The results of observations are more commonly written in terms of the circular frequency $\nu_{nlm} \equiv \omega_{nlm}/2\pi$. For stars that are truly spherically symmetric, mode frequencies depend only upon n and l, and are independent of m. This occurs because m depends on the choice of position for the pole of the coordinate system, which is arbitrary for a spherical configuration. Any condition that breaks the spherical symmetry (such as rotation about an axis, or the presence of magnetic fields) can lift this frequency degeneracy.

Observations of stellar pulsations usually involve either the photometric intensity or the radial velocity. The perturbations in intensity and in velocity are related, of course. The displacements of the stellar plasma cause Doppler shifts directly; the accompanying compressions or displacements from equilibrium height also cause temperature changes, resulting in perturbations to the observed intensity.

Although mode frequencies depend in complicated ways on the stellar structure, there is a useful limit (that in which $n \gg l$) in which simple asymptotic formulae give useful approximations to the true frequency behavior (Vandakurov 1968, Tassoul 1980, Christensen-Dalsgaard 1988a). For p-modes, one finds

$$\nu_{nl} \;=\; \Delta\nu_0\left(n+\frac{l}{2}+\epsilon\right) \;-\; \frac{A(L^2)}{(n+l/2+\epsilon)} \;, \tag{2.2}$$

where $\Delta\nu_0$, A, and ϵ are parameters that depend on the structure of the star, and $L^2 \equiv l(l+1)$. If the parameter A were zero, one would therefore find p-mode frequencies to fall in a regular picket fence pattern with frequency spacing $\Delta\nu_0/2$: modes with odd l would fall exactly halfway between modes with even l, and modes with different n at a given l would always be separated in frequency by multiples of $\Delta\nu_0$. The parameter $\Delta\nu_0$, termed the *large separation*, is simply related to the sound travel time through the center of the star:

$$\Delta\nu_0 \;=\; \left(2\int_0^{R_*}\frac{dr}{c}\right)^{-1} \;, \tag{2.3}$$

where c is the local sound speed and R_* is the stellar radius. Consideration of the virial theorem (Cox 1980, Gough 1990) shows that this travel time is related to the mean density of the star, so that

$$\Delta\nu_0 \cong 135 \left(\frac{M_*}{R_*^3}\right)^{1/2} \mu\text{Hz} \ , \tag{2.4}$$

where M_* and R_* are the stellar mass and radius in solar units. Eq. (2.4) holds exactly for homologous families of stars, but it is obeyed quite closely even for stars that are not homologous, such as stars of different mass along the main sequence (Ulrich 1986). The large separation is thus easily interpreted in terms of the stellar structure, and moreover it is likely to be straightforward to observe, even in noisy stellar oscillation data.

Parameters A and ϵ in Eq. (2.2) have to do with the structure near the center of the star and near the surface, respectively. Modes with different degree l penetrate to different depths within the star. Modes with $l = 0$ have substantial amplitude even at the center; those with higher values of l avoid a region in the stellar core that grows in radius as l increases. This difference in the region sampled by modes with different l leads to the second term on the right-hand side of Eq. (2.2), removing the frequency degeneracy between modes that differ by (say) -1 in n and $+2$ in l. This effect is often parameterized in terms of the *small separation*, defined as $\delta_{nl} \equiv \nu_{n+1,l} - \nu_{n,l+2}$. The small separation may be written as an integral analogous to that in Eq. (2.3) (Däppen et al. 1988):

$$\delta_{n,l} = \Delta\nu_0 \frac{(l+1)}{2\pi^2 \nu_{nl}} \int_0^{R*} \frac{dc}{dr} \frac{dr}{r} \ . \tag{2.5}$$

The small separation is thus sensitive to sound speed gradients, particularly in the stellar core. Since these gradients change as nuclear burning changes the molecular weight distribution in the star's energy-producing region, the small splitting contains information about the star's evolutionary state.

2.1. *The solar example*

The Sun provides the best-developed example of seismological inference, and observations of the Sun as a star at once illustrate the phenomena that are observable and motivate the search for similar phenomena on other stars. It is important to remember, however, that many of the successes of helioseismology rest on observation of modes in the range $5 \leq l \leq 100$, which will be inaccessible on most distant stars.

Figure 1 shows the power spectrum of solar p-modes as measured with the IPHIR full-disk photometer while *en route* to Mars on the Soviet Phobos spacecraft (Toutain & Frölich 1992). The IPHIR instrument measured the brightness in several colors, integrated over the visible disk of the Sun, using silicon diode photometers. Except for their low noise level, these observations are thus closely analogous to normal photometric observations of stars. Figure 1 illustrates several important aspects of the solar p-modes. First, the mode frequencies are very well defined, with typical quality factors Q of several thousand. Second, the mode amplitudes are large only within a restricted frequency range, between roughly 2500 and 4000 μHz. Within that range, the low-degree modes to which IPHIR is sensitive are indeed almost evenly spaced in frequency, and in spite of the compressed frequency scale of this figure, many close pairs of modes (corresponding to $l = 0, 2$ or to $l = 1, 3$) may be seen. The separation between pairs of modes turns out to be roughly 68 μHz $= \Delta\nu_0/2$; the separation between modes making up a given $l = 0, 2$ pair is about 9 μHz. The amplitudes of the pulsations are quite small: the largest peaks near 3000 μHz have power corresponding to amplitudes $\delta I/I$ of only about 3×10^{-6}. A similar power spectrum obtained by measuring the disk-integrated solar velocity (see, e.g. Claverie et al. 1984) has virtually identical mode structure, somewhat

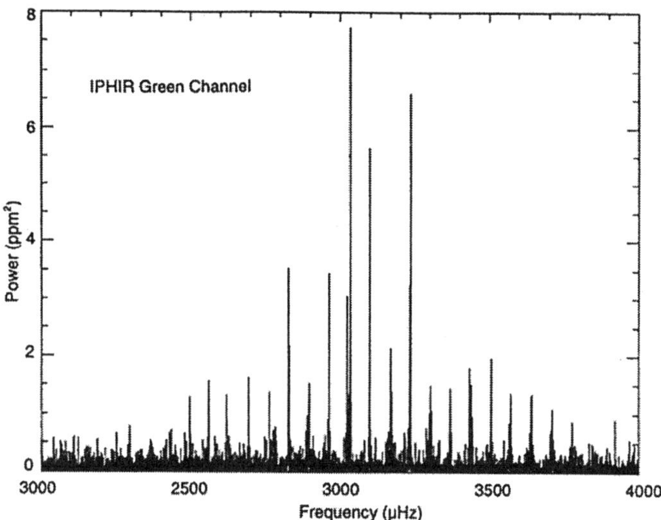

FIGURE 1. Power spectrum of disk-integrated solar intensity measured by the green channel of the IPHIR experiment. From Toutain & Frölich (1992).

smaller background power relative to the mode power, and peak mode amplitudes of roughly 15 cm s^{-1}.

A great deal of information about the Sun's interior has been obtained from the measured pulsation frequencies, including accurate estimates of the variation of sound speed (Christensen-Dalsgaard et al. 1985, Vorontsov 1989) and angular velocity (Duvall et al. 1984, Brown 1985, Brown et al. 1989, Libbrecht 1989) with depth and latitude, and a precise estimate of the depth of the adiabatically stratified region of the solar convection zone (Christensen-Dalsgaard et al. 1991). All of these inferences draw upon observations of modes with a substantial range of l, however; results of useful precision cannot be obtained with disk-integrated observations alone. Several other inferences do not share the requirement for high-l data.

Perhaps most notable among these is the ability to test at least some explanations for the solar neutrino deficit. Since the low-l p-modes penetrate close to the solar core, their frequencies (in particular the small frequency separation) may be used to test for the presence of physical effects that might account for the observed paucity of ^{8}B neutrinos from the Sun. Two astrophysical models to explain the neutrino deficit involve lowering the temperature of the solar core, either by transporting heat by means of Weakly Interacting Massive Particles (WIMPs; see e.g. Steigman et al. 1978, Spergel & Press 1985), or by mixing fresh fuel into the Sun's center (Schatzman et al. 1981). Models of these processes (Gilliland & Däppen 1988, Christensen-Dalsgaard 1991,1992, Lebreton et al. 1988, Cox et al. 1990) yield small frequency separations δ_{02} that are incompatible with the observations.

Another important issue addressable with such data is the frequency splitting of modes with $l > 0$ by the solar rotation (see, e.g. Hansen et al. 1977, Brown et al. 1989, Libbrecht and Morrow 1991). In the simple case in which angular velocity is independent of latitude, the frequencies of modes within a multiplet are given by

$$\nu_{nlm} = \nu_{nl0} + \frac{m\beta_{nl}}{2\pi} \int_0^{R_*} \Omega(r)K_{nl}(r)dr \ , \tag{2.6}$$

where β_{nl} is a correction factor of order unity accounting for Coriolis forces, $\Omega(r)$ is the solar angular velocity, and K_{nl} is a unimodular kernel that is roughly proportional to the local energy density in the mode. Thus, the splitting of low-l multiplets (sets of modes with the same n and l but different m) depends somewhat on the angular velocity near the Sun's center, where only eigenmodes with small l have substantial amplitude.

2.2. *Estimates of information content*

Useful observations of multi-mode pulsations on other stars require heavy commitments of observing time. It is therefore important to ask whether the results one may obtain from seismology are worth the effort required to get them. What exactly can one learn from oscillation data?

The solar case is an exciting and informative example of the power of seismological methods, but it is by no means a perfect predictor of the usefulness of asteroseismology. First, as noted above, our ability to resolve spatial structures on the Sun leads to measurement opportunities that do not exist for other stars. On the other hand, the Sun is only a single star, and its study (even in great detail) is bound to leave many unanswered questions about other stars with different circumstances or histories. It thus seems likely that studying oscillations in a large sample of stars (even restricting the sample to those of roughly solar type) would allow fundamentally new insights. Stars of this sort have been fairly thoroughly studied from a theoretical point of view, since the solar example is so well understood.

The first efforts to estimate the information content of oscillation frequencies for Sun-like stars were made by Ulrich (1986, 1988) and by Christensen-Dalsgaard (1988b). These authors computed the sensitivity of the frequency separations $\Delta\nu_0$ and δ_{nl} to changes in the stellar mass M and age τ; Ulrich (1986) also considered changes in the initial composition parameters Y (the helium abundance), Z (the heavy element abundance), and in the mixing length ratio α. They concluded that if the composition were known, then measurement of the two frequency separations would allow the stellar mass and age to be estimated with useful precision. These results are summarized in the so-called "asteroseismic H-R diagram," shown in Figure 2. If one assumes that individual mode frequencies may be measured with precision comparable to the mode linewidth (unknown for other stars, but typically 1 μHz for the Sun), then Figure 2 suggests that frequency separations could be used to determine stellar masses to within a few percent, and ages to within perhaps 5% of the main-sequence lifetime.

Gough (1987) showed that these estimates are too optimistic, since the frequency separations are also quite sensitive to variations in other parameters, particularly in Z. A realistic uncertainty in Z therefore leads to important uncertainties in mass and age. The generic difficulty Gough illustrated is that most observable properties of stars (both their oscillation frequencies and more standard indices such as luminosity and surface temperature) depend to some extent on all of the parameters of stellar structure. Thus, the two frequency separations can provide two relations among the five structural parameters M, Y, Z, τ, α, but without additional information it generally is not possible to arrive at a definite value for any one of them. Other observational constraints are therefore needed, either in the form of different kinds of oscillation data (e.g. frequencies for individual modes, rather than frequency separations only), or in the form of more traditional astronomical measurements (photometry, astrometry, or spectroscopy). A minor complication is that proper interpretation of some photometric and astrometric data requires that the distance to the star be included as another parameter describing it. Thus, at least six well determined observational properties are required to uniquely define the structural parameters of a field star. More may be necessary if it develops that some

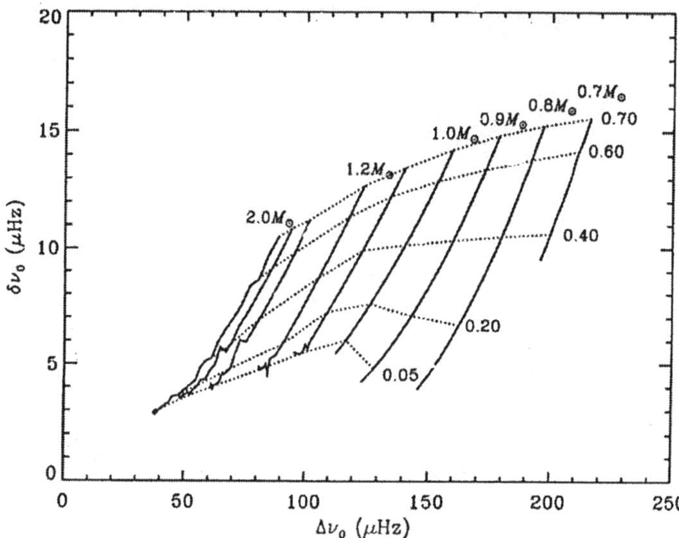

FIGURE 2. The "asteroseismic H-R diagram" of Christensen-Dalsgaard (1988b), showing the variation in large ($\Delta\nu_0$) and small ($\delta\nu_0$) frequency separation with stellar mass and age. Mass is constant along solid lines; age (parameterized by the central hydrogen abundance) is constant along dotted lines.

of the observations are redundant, or if other processes are important in determining the stellar structure, beyond those assumed in this simple 5-parameter description of stars.

Brown et al. (1994) performed a more complete treatment of the problem of estimating model parameters from p-mode frequencies. The approach was to perform a least-squares fit of the model parameters to all of the observations that one might reasonably expect to have, including oscillation frequencies or frequency separations. With reasonable errors ascribed to the various observations, several conclusions emerged. Estimates of the age, mixing length, and mass of field stars can be substantially improved by the addition of oscillation frequencies. Indeed, without the frequency data, parameters such as the age are essentially unconstrained, and must be estimated from more general considerations, such as the age of the galaxy. The relative improvement in errors is greatest for distant stars, for which astrometric data are relatively unreliable. The lowest absolute errors, however, occur for nearby stars with high-quality astrometry. It develops that oscillation frequency data is usually unhelpful in constraining the heavy element abundance Z. In the best field star cases, one should be able to reach errors in mass and mixing length of about 3%, and in helium abundance and age of about 12%. Errors of this size would be interesting from the point of view of galactic evolution if they could be obtained for a good sample of stars near the Sun. They are not, however, small enough to allow tests of the physics of stellar structure theory.

A more interesting situation occurs if mode frequencies can be obtained for both stars in a well-observed visual binary (α Cen, for example). In such a case the two stars may be assumed to have the same age, distance, and initial composition, so that the number of parameters required to describe the system is less than twice that for a single star. Moreover, some new observables (the orbital data) provide fundamentally new sorts of information. One result is that parameter errors become smaller for binaries than for field stars. A more important difference is that, with many more observables than model parameters, one may search for inconsistencies between the observations and the best-fit model. If significant inconsistencies are found, then significant errors must exist in the model of the star system. In this way, it may be possible to detect errors in the physics underlying the calculation of stellar structure. Figure 3 shows an example of the sort of discrepancies that might arise. In this case, observed properties of a binary system (chosen to be similar to the α Cen system) were constructed using LAOL opacity tables (Huebner et al. 1977), but were fit to a model based on OPAL opacities (Rogers & Iglesias 1992). Since the "true" and assumed models of the system employed different physics, no combination of model parameters can match the constructed observations exactly. Figure 3 shows the residuals between the "true" observations and those implied by a best fit to the model using OPAL opacities. These residuals are generally large enough to be detected in spite of observational errors, and the pattern of discrepancies provides clues to the nature of the error in the assumed model. Not all modifications of the input physics result in changes that are as large as those in this example, and the degree to which different physical effects may produce similar sets of residuals is not yet known. Nonetheless, it seems reasonable that oscillation frequencies, if available, would not only allow measurement of the structural parameters of stars, but would also place constraints on at least some aspects of stellar evolution theory.

Star clusters are the natural extension of the progression from single stars to visual binaries; Gough and Novotny (1993a,b) have begun to investigate the utility of cluster observations by considering a group of very similar Sun-like stars, all with the same age, composition, and distance. They found that assuming the stars to be coeval with identical initial compositions causes a partial cancellation in the errors associated with some observables. In effect, measurements that apply to global properties of the cluster (the age and the composition) have their errors reduced by averaging over all of the observed stars. The errors associated with model properties that are peculiar to each star (mass and mixing length), however, take on values that are essentially independent of the number of stars observed. Further work in this area is needed, especially to extend the treatment so that stars throughout the cluster H-R diagram may be included in the solution.

3. Pulsations in Sun-like stars

There is no cause to believe that the Sun is an extraordinary star of its type, so it is reasonable to expect that other stars like the Sun pulsate in much the same way the Sun does. So far, however, observational efforts have yielded results that are equivocal. In this section we shall discuss Sun-like stars mostly from an observational point of view, putting most emphasis on the practical difficulties encountered, ways of dealing with them, and prospects for success. In spite of the observational limitations, considerable progress has been made in predicting the oscillation properties of likely target stars for asteroseismic observations (see e.g.] Demarque & Guenther 1990, Edmonds et al. 1992, Guenther & Demarque 1993).

3.1. *Expected characteristics of the pulsations*

The principal impediment to the observation of pulsations in stars like the Sun is their small amplitudes. As mentioned earlier, the strongest modes in the solar p-mode spectrum show amplitudes of only about 3×10^{-6} in relative intensity or 15 cm s^{-1} in velocity. Unlike the classical stellar pulsations, all of the oscillation modes in the Sun are believed to be intrinsically stable. The reason that pulsations are excited at all (and the reason that the excitation is so indiscriminating, exciting millions of modes to similar amplitudes) is the presence of the solar convection. Motions near the top of the convection

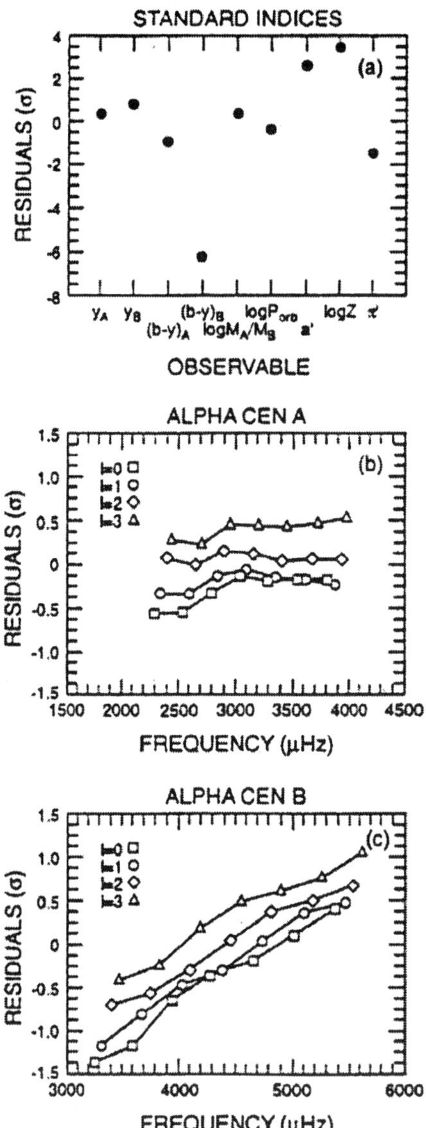

FIGURE 3. Residual errors between the values of observable quantities for the α Cen A/B visual binary system calculated using LAOL opacities and those calculated from a best-fit model using OPAL opacities. The top panel shows errors (in σ) for standard astronomical indices, including Strömgren magnitudes and colors, orbital data, and spectroscopic metal abundance. The bottom two panels show residual mode frequencies for the two stellar components, as functions of l and frequency. From Brown et al. (1994).

zone can reach Mach numbers of order unity (e.g. Nordlund & Stein 1990, Bogdan et al. 1993, Rast & Toomre 1993), providing a source of acoustic noise that is loosely coupled to the part of the Sun in which the p-modes propagate. In the Sun, damping of individual p-modes arises partly from radiative losses, but is most likely dominated by scattering from convectively-generated inhomogeneities in the outer envelope (Goldreich & Kumar 1991). Expected amplitudes therefore depend in a complicated way on the behavior of the stellar convection and on its coupling to the cavity in which the oscillation energy resides. Christensen-Dalsgaard and Frandsen (1983) and Houdek et al. (1995) estimated oscillation amplitudes as a function of T_{eff} and surface gravity. The results suggest a weak dependence of mode surface amplitude on surface gravity. These results are highly uncertain, but remain the best available. The best hope for more accurate theoretical estimates of mode properties in other stars may lie with numerical simulation of the convection-pulsation interaction (Rast & Toomre 1993, Bogdan et al. 1993), but adequate models with the required resolution are probably some years away.

A rough estimate can be made of the frequency range within which oscillations would be seen on distant stars. In the WKB approximation, the reflection of sound waves as they propagate upward through a stellar envelope is governed by the behavior of the *acoustic cutoff frequency*, ω_{ac}. Waves propagate upward until the local value of ω_{ac} becomes greater than the wave frequency, and then they reflect. In the simplest (isothermal layer) approximation, which is adequate for our purposes, one may write

$$\omega_{ac} = \frac{c}{2H} \propto gT^{-1/2} \, , \tag{3.7}$$

where c is the local sound speed, H is the pressure scale height, g is the gravitational acceleration, and T is the temperature. In stellar atmospheres, ω_{ac} reaches a maximum, ω_{ac0}, in the photosphere, where the temperature is minimum. Waves with frequencies above ω_{ac0} never reflect, but rather continue propagating into the tenuous outer parts of the stellar atmosphere. As a result, p-modes with frequencies above ω_{ac0} are not expected to attain significant amplitudes. On the other hand, modes with frequencies much smaller than ω_{ac0} reflect deep in the stellar envelope. This reduces the surface amplitude for a given mode energy, and moreover reduces the coupling between the mode and the near-surface convective driving source. These considerations suggest that maximum p-mode amplitudes should be found at frequencies that are a modest fraction (roughly 0.6, in the Sun) of ω_{ac0}.

From Eq. (3.7), it follows that the expected frequency of maximum p-mode amplitude should scale as $gT_{\text{eff}}^{-1/2}$. If one adopts the scaling appropriate to the Sun, this implies frequencies ranging from about 1 mHz (for F-type subgiants) to about 10 mHz (for M dwarfs). Goldreich and Kumar (1990) have developed a theory explaining the shape of the power envelope of the solar p-modes. This approach offers hope that observations of stellar p-mode amplitudes might provide diagnostics relating to stellar convection and to g.

The foregoing suggests that the most attractive targets for stellar oscillation searches should be stars that have lower surface gravity and are more luminous than the Sun, since such stars should have larger amplitudes and should pulsate with longer periods (simplifying many observational problems). While this conclusion is probably true, the expected frequency separation of p-modes in these stars can work in the other direction.

From Eq. (2.4), $\Delta\nu_0$ for p-modes ranges from roughly 30 μHz (for subgiants) to 400 μHz (for M dwarfs). The frequency spacing one would actually see is $\Delta\nu_0/2$, because of the way in which modes with odd and even values of l interleave. Thus, detecting discrete modes in subgiants requires frequency resolution better than 15 μHz, which is a factor of about 2 better than can be attained from one site in a single night. Simulations show that even if one combines multiple nights of single-site observations, the sidelobes that result from the diurnal cycle may lead to fatal ambiguities when attempting to interpret pulsation time series from such luminous stars (Gilliland & Brown 1992). This is essentially the same problem faced by the solar and white dwarf communities in obtaining observations adequate for their purposes, and the solutions are the same: observe from a network of sites or from space.

4. Observational methods and recent results

4.1. *Photometric and Doppler-shift methods*

The most straightforward approach to observing stellar p-modes is to measure the associated stellar brightness changes, as was done in the IPHIR experiment that produced Figure 1 (Toutain & Fröhlich 1992). The signals to be detected are very small (parts in 10^6, or micro-magnitudes), but even with modest apertures, many stars provide plenty of photons to work with. For instance, with a photon-limited system with 1m aperture, solar-like p-modes should be measurable with 1 month of observing time in stars as faint as about magnitude 10. There is good evidence that, using techniques of ensemble-calibrated CCD photometry (Gilliland & Brown 1992, Gilliland et al. 1993), the necessary photometric precision can be reached. The remaining problem is scintillation originating within the Earth's atmosphere, which causes photometric noise at a level that effectively prohibits detection of stellar pulsations, even using the largest existing telescopes. To bypass this problem, the only solution is to observe from above the atmosphere. Several space missions have therefore been proposed in recent years to do asteroseismology (e.g. Asteroseismology Explorer [Hudson et al. 1986], PRISMA [Appourchaux et al. 1993], and STARS [Badiali et al. 1996]). Other, similar, missions were proposed to search for transits of Earth-sized planets circling distant stars (FRESIP [Borucki et al. 1993], and Kepler [Borucki et al. 1996]); as it turns out, the photometric requirements for these two scientific problems are almost identical. All of these proposed missions involved 1m-class telescopes with large CCD detector arrays, and all of the proposals failed at some point during the selection process. For the present, the seismology community has therefore adopted a strategy involving less ambitious (and much less expensive) missions. This approach is meeting with some success: two small missions have been funded and are expected to fly within the next few years (the CNES mission COROT [Catala et al. 1995] and the Canadian MOST instrument [Matthews 1998, private communication]), and others are under consideration at this writing (The Danish small satellite MONS [Kjeldsen et al. 1998] and an Antarctic balloon experiment BLAST [Buzasi et al. 1998]). Both of the funded experiments will employ apertures of 25 cm or less, and will study only a few bright stars; their choice of targets will be limited, because of the constraints imposed by their low-altitude orbits. In spite of their limitations, these missions will be of singular importance. If successful, they will open the door to the efficient and widespread application of stellar seismic data.

In principle, Doppler measurements of solar-like pulsations have several advantages compared to photometric methods. The Doppler shift measurement process is in essence a differential one, conveying some advantages in the detection of small signals. Also, the contrast between the pulsation signal and the background of stellar convective noise is larger in the velocity than in the intensity signal (Harvey 1988). One pays a price for these advantages, however. Because of the tiny wavelength shifts that are associated with expected pulsation signals (15 cm s^{-1} $\Rightarrow$ $\delta\lambda/\lambda = 5 \times 10^{-10}$), successful measurements require both high spectral resolution and very low noise. If one defines "Sun-like" to mean stars of luminosity class IV or V, with spectral types cooler than F5, then there are only three candidates of roughly first magnitude: α Cen A (G0V) and B (G3V), and Procyon (F5IV). The next brightest star of interest is β Hyi (G2IV), which is 2.3 stellar magnitudes fainter than Procyon. To date, almost all Doppler shift searches for Sun-like pulsations have therefore been aimed at one of these stars.

The most important problem to be solved in pulsation searches to date is that of attaining sufficient Doppler precision to detect the pulsation modes. To see what is involved, let us assume that the largest p-modes may have Doppler amplitudes of 15 cm s^{-1}, and that in one night one can obtain 400 observations with exposure times of roughly 60s each (a rapid cadence is required to sample the pulsation time scale). To obtain a 4σ detection of the largest modes, one therefore requires the noise associated with each observation to be roughly 75 cm s^{-1}. The fundamental limit to the achievable precision results from photon counting statistics (Connes 1985, Brown 1990). To sufficient accuracy, it may be written as

$$\delta v_{rms} = \frac{cw}{\lambda d(N_{pix}N_{lines}I_c)^{1/2}} \ , \tag{4.8}$$

where c is the speed of light, w is the width of the spectrum line (including both instrumental and stellar line broadening processes), λ is the center wavelength of the line, d is the fractional line depth, N_{pix} is the number of wavelength samples obtained across the line width, N_{lines} is the number of spectral lines observed, and I_c is the continuum intensity in the measurement, expressed as the number of detected photons. Putting in plausible values for an echelle spectrograph at a 2m telescope, with an exposure time of 60s, one finds δv_{rms} for a single spectrum line of moderate strength to be roughly 10 m s^{-1}. To obtain observations that are suitable for pulsation studies, this noise level must evidently be improved by more than an order of magnitude. This may be done by (a) improving the instrumental resolution, (b) observing many lines, or (c) getting more light through the system. In practice, it is relatively easy to make the instrumental contribution to the line width smaller than the stellar contribution; beyond this point further improvements are of no use. Increasing N_{line} means using a wider bandwidth. This is a practical strategy; with cross-dispersed echelle spectrographs, hundreds or thousands of lines may be measured simultaneously. Finally, increasing I_c requires larger telescopes or more efficient optical systems, or both.

4.2. *Procyon*

In the search for Sun-like p-mode oscillations in distant stars, the subgiants Procyon (α CMi, F5 IV-V, $m_V = 0.4$) has played a prominent role. Doppler observations of Procyon have been carried out by Gelly et al. (1986), by Brown et al. (1991), and by Innis et al. (1991), and an extensive campaign to measure variations in Balmer line equivalent widths has been undertaken by Bedding, Hill, and coworkers (Hill 1997, private communication). There are several reasons for this star's popularity in this connection. Procyon is the brightest star of approximately solar type visible from the northern hemisphere. Moreover, as explained above, theories of p-mode excitation suggest that, because of their vigorous surface convection zones and low surface gravity, p-modes in subgiants should be excited to amplitudes perhaps as large as 1 m s^{-1}, several times greater than mode amplitudes observed on the Sun (Houdek et al. 1995). The frequencies of p-modes in Procyon are expected to be about 1 mHz, as compared to 3 mHz in the Sun, because lower surface gravity leads to a smaller value for the photospheric acoustic cutoff frequency. This implies that the observational sampling rate for this star can be slower than is practical for the Sun. All of these advantages arising from subgiant status are counterbalanced by a significant observational problem, namely that the characteristic frequency separation between pulsation modes in subgiants should be only about 30 μHz, or about half the separation seen in the solar p-modes (e.g. Guenther & Demarque 1993). Resolving modes with this frequency spacing requires observations spanning at least 9 hours, or preferably twice this long. This means that individual mode frequencies will be difficult or impossible to isolate using data from a single mid-latitude observing site, where the maximum length of a night's observing run is typically about 8 h.

Procyon continues to be of interest, because none of the observations cited above produced unambiguous evidence that it harbors p-modes. The most persuasive observations to date are those by Brown et al. 1991. These included significant coverage on 6 contiguous nights, and achieved a Doppler noise level of about 3.5 m s^{-1} in each 120 s exposure. The average of the 6 nightly power spectra derived from these data is displayed in the top panel of Figure 4. It shows an excess of power in the frequency range between 0.5 and 1.5 mHz, about where p-mode signals should occur in Procyon. The phased spectrum shows in the same frequency range a forest of narrow peaks, each with its attendant sidelobes resulting from the diurnal interruption of the observations. But an extensive analysis of these spectra was not able to prove conclusively that either the broad power excess or the narrow peaks of which it was composed arose from p-modes. Non-oscillatory broad-band processes (such as turbulent convection in the star's photosphere, or instrumental noise) remained possible sources for the observed power. A particular concern was the instrumental drift in Doppler zero point, which sometimes exceeded 200 m s^{-1} during a night. To reduce the effects of this long timescale drift, the time series were high-pass filtered before computing the power spectra shown in Fig. 4. As pointed out by Kjeldsen & Bedding (1994), one might therefore interpret the power excess below 1.5 mHz as simply a high-pass filtered red noise component of either stellar or instrumental origin.

In hopes of resolving these uncertainties, or at least of improving on the earlier results, the Advanced Fiber Optic Echelle (AFOE) spectrograph has been used to make further Doppler measurements of Procyon. The techniques used were refinements of those employed in the earlier works. Better instrumental and analysis methods yielded some improvement in the Doppler noise level above 0.5 mHz, and greatly improved the zero point stability. The resulting power spectra of Procyon (shown in the lower panels of Fig. 4) are of comparable quality to those in Brown et al. (1991) (except at frequencies below 0.3 mHz, where they have much lower noise); they show very similar features, and suffer from the same difficulties of interpretation because of their gapped time coverage. Taken alone, the new data are therefore only slightly more informative than the previous ones. We believe, however, that comparison of the features in the various data sets suggests conclusions that cannot be drawn from any set taken alone.

4.3. *Observations and analysis*

Spectra of Procyon were obtained using the AFOE spectrograph (Brown et al. 1994) at the 1.5 m Tillinghast telescope of the F.L. Whipple Observatory at Mt. Hopkins, AZ. The spectrograph was configured to give resolution $R = \lambda/\delta\lambda = 50000$. Wavelength calibration was provided by a ThAr lamp, whose light was brought into the spectrograph through a second fiber, and was recorded simultaneously with the starlight. Integration times for these spectra were usually 120 s, giving typical signal-to-noise ratio of about 300 in the continuum near 550 nm.

Figure 4 shows the weighted average of the nightly power spectra for each of the 4 Procyon observing runs that have been obtained to date, weighted in proportion to the number of good spectra obtained during each night. Comparing these spectra to one another, one can see similarities in both the shapes and typical power levels. Moreover, runs with more nights of data tend to be more similar to one another than do the shorter runs; this behavior is characteristic of any stochastic process, including randomly-excited p-modes. The frequency resolution of the spectra in Fig. 4 is only about 30 μHz, however, which is too poor to resolve even the large frequency splitting in a subgiant such as Procyon. In hopes of identifying individual modes, we have also computed the phased multi-night power spectra for each of the data sets. Many groupings of peaks may be

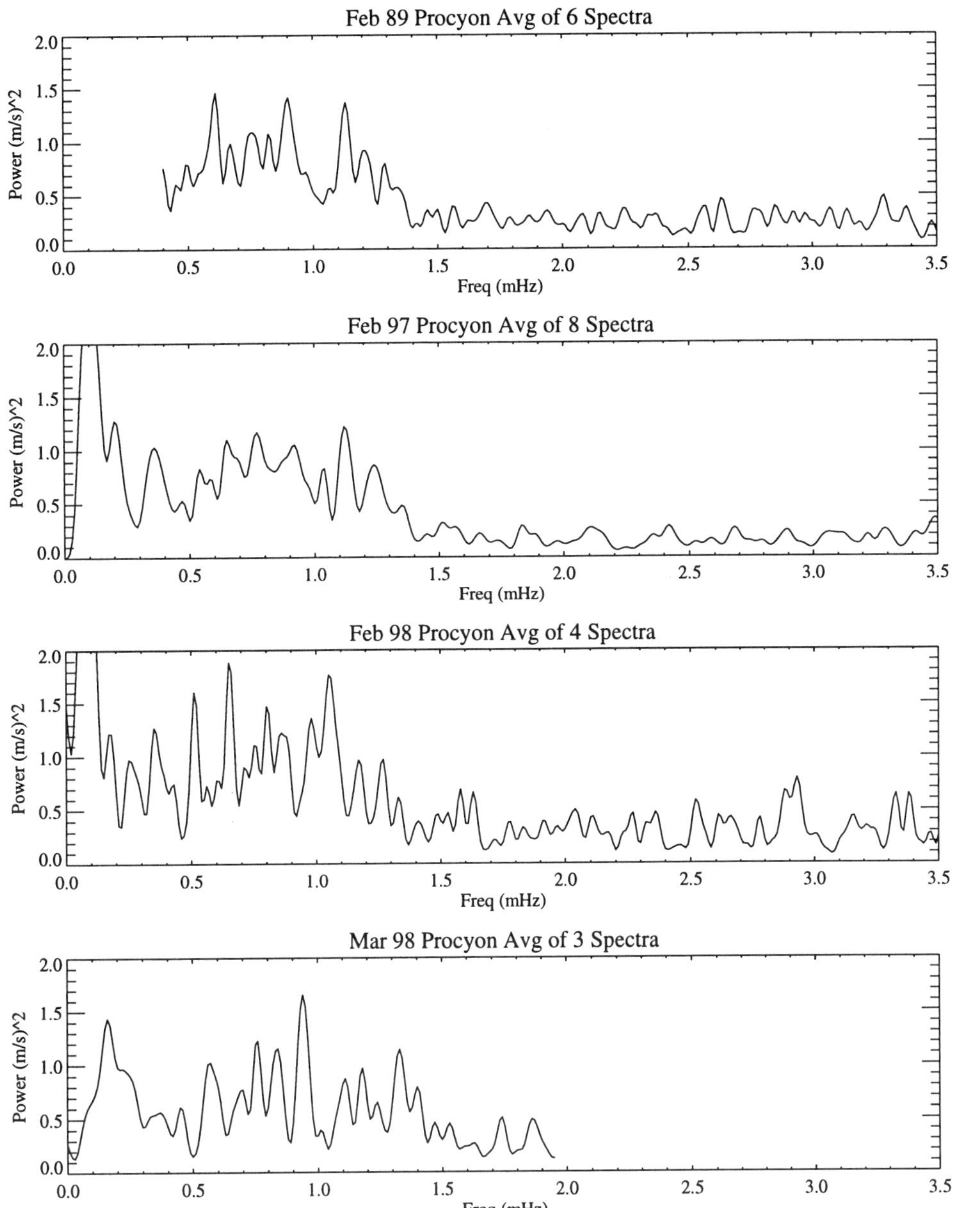

FIGURE 4. Averages of single-night spectra for the 4 Procyon observing runs. The Feb. 1989 spectrum is truncated at $\nu \leq 400$ μHz because low frequencies have been filtered to suppress drifts in the data. The Mar. 1998 spectrum terminates at $\nu = 1.9$ mHz because longer-than-usual exposures decreased the Nyquist frequency to 2 mHz.

seen in these spectra, reproducing (more or less) the pattern of peaks in the spectra of the window functions corresponding to each spectrum. Some features repeat among the 4 phased spectra and many do not; determining whether the repeating features are a signature of p-modes, or whether they may result from random noise, will require further analysis.

The similar power spectrum shapes (and especially the common tendency to converge to a white noise background at about 1.5 mHz) are evidence that the same process was being observed on all four occasions. But what process? Three possibilities need to be considered: non-oscillatory stellar processes, artifacts of the instrument or of the Earth's atmosphere, and p-modes. Stellar noise processes (most plausibly stellar convection—the stellar analog to solar granulation) require serious consideration. Both theory and line bisector observations suggest that Procyon has a vigorous (albeit thin and inefficient) surface convection zone. Perhaps the low-frequency excess power is merely the Doppler manifestation of this convection. A counter-argument is that the Procyon spectrum shows a fairly well-defined maximum, located near 0.8 mHz. Such a maximum agrees well with expectations for p-modes, but is contrary to the behavior of the solar granulation—the granulation power decreases monotonically with frequency, even for ν much smaller than ν_{ac0}. But lacking both a reliable theory of convection and true mode identifications, a convective process cannot be wholly dismissed. Instrumental noise sources seem less likely. The instrumental setup for the 1989 observations was similar in concept to that for the later observations, but it was much different in detail. It seems implausible that noise spectra that are so similar in shape and amplitude would result from two systems that had not a single piece of hardware or analysis software in common. Moreover, observations of other bright stars (Arcturus and η Boo) do not show similar noise signatures. This leaves p-modes as the most likely explanation for the power excess; happily, the predicted mode amplitudes and approximate frequency range are in good agreement with the observed ones. This is reassuring, but more work is needed to reveal whether this likely explanation is in fact the correct one.

If one does interpret p-modes as the cause of Procyon's power excess, then the shape of the power envelope raises a physically interesting question. This is illustrated in the bottom panel of Figure 5, which shows the grand average nightly Procyon spectrum overlaid with the envelope of solar low-l p-mode power, scaled to have the same amplitude and peak frequency. Clearly the solar power envelope is much narrower relative to its central frequency than is that of Procyon. The shape of the Sun's p-mode power envelope is determined on the high frequency side by the interaction of driving and damping processes, and on the low frequency side by the increasing evanescence of modes whose upper turning points lie increasingly far below the photosphere. Both of these tendencies are related to the thermal and velocity structure in the stellar convection zone. An interesting speculation is that the large width of Procyon's power envelope may be related to the shallowness of its convection zone.

5. Conclusions

Asteroseismology promises an entirely new way to probe the interiors of stars, allowing one to determine stellar physical parameters that are different from (and often complimentary to) those that can be observed by more traditional means. In many respects the theory of stellar pulsations is very well developed—in particular, oscillation frequencies may be computed with considerable precision, given a stellar configuration, and there are reasonably reliable methods for inferring many stellar properties, given a list of mode identifications and frequencies. These computations and methods should be particularly satisfactory when applied to stars that are somewhat like the Sun, because the solar case has already been studied in great detail. Other parts of the theory are not so successful, especially those dealing with the nonlinear processes involved in mode excitation and damping. Thus, mode amplitudes and lifetimes on other stars are subjects about which

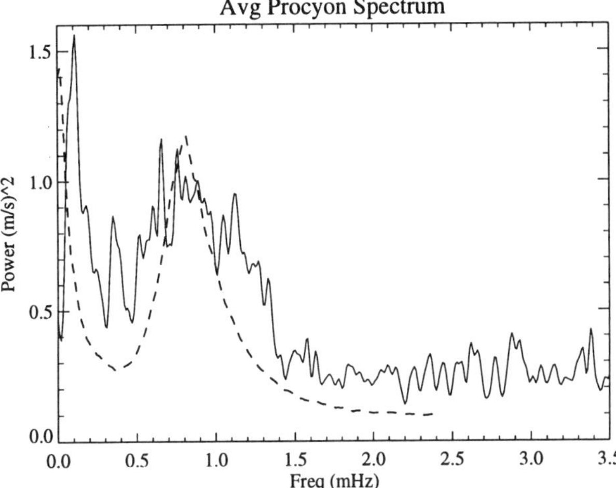

FIGURE 5. Weighted average of the 4 run-averaged spectra shown in Fig. 4. The top panel shows the averaged spectrum alone; the bottom panel shows superposed on this spectrum a version of the solar p-mode spectrum, scaled to have the same peak ν and amplitude. Note the difference in relative widths in frequency.

we know very little; significant improvement in this situation is perhaps unlikely until unambiguous observations become available to constrain the theories.

The bottleneck in the seismic study of Sun-like stars remains the observations. Because of the very small pulsation amplitudes involved, convincing observations of p-modes on other such stars have not yet been obtained. In the short term, the best prospects for success seem to lie with Doppler shift observations, conducted with high-resolution cross-dispersed echelle spectrographs. Recent observations of Procyon, described above, make a fairly good circumstantial case that p-modes have probably been detected in data take with the AFOE spectrograph. If the interpretation in terms of p-modes holds up, then the existing data already allow tests of theories of mode excitation, as well as new estimates of parameters such as the acoustic cutoff frequency in Procyon's photosphere. On the other hand, these data have not yet provided believable mode frequencies, which are necessary for most kinds of seismic analysis, and which must be present to support a definite claim of p-mode detection. Further analysis (and likely more observations) will be needed to be sure that p-modes are indeed responsible for the features in Procyon's power spectrum.

In the longer run, ground-based spectroscopic observations will probably yield to photometric observations made from space. The oscillation process, by its nature, requires long periods of observing time on each star. If each usable stellar power spectrum requires weeks of time on a large telescope, then asteroseismic information will never be a generally useful tool for understanding stellar structure. Photometric methods, with their ability to multiplex the signals from many stars onto a single detector, are capable of extending seismic studies to a much larger sample of stars. But the constraints imposed by Earth's atmosphere demand that such photometry be performed from space. As a result, recent years have seen a number of proposed seismology satellite missions; future years will doubtless see many more. With the benefit of improving technology and experience, we can look forward (soon, I hope) to a generation of survey missions that will, for the first time, take the pulse of all our neighboring stars.

REFERENCES

APPOURCHAUX, T., ET AL. 1993 *PRISMA: Report on the Phase-A Study*. ESA SCI(93)3.

BADIALI, M., ET AL. 1996 *STARS: Seismic Telescope for Astrophysical Research from Space. Report on the Phase-A Study*. ESA SCI(96)4.

BOGDAN, T. J., CATTANEO, F., MALAGOLI, A. 1993 *ApJ* **407**, 316.

BORUCKI, W. J., COCHRAN, W. D., DUNHAM, E. W., KOCH, D. G., & REITSEMA, H. 1993 *Frequency of Earth-Sized Inner Planets: FRESIP*. NASA Study Document.

BORUCKI, W. J., ET AL. 1996 *Kepler: A Search for Habitable Planets*, Proposal to NASA.

BROWN, T. M. 1985 *Nature* **317**, 591.

BROWN, T. M. 1990 In *CCDs in Astronomy* (ed. G. H. Jacoby). ASP Conference Series 8, p. 335. Astronomical Society of the Pacific.

BROWN, T. M., CHRISTENSEN-DALSGAARD, J., WEIBEL-MIHALAS, B. & GILLILAND, R. L. 1994 *ApJ* **427**, 1013.

BROWN, T. M., CHRISTENSEN-DALSGAARD, J., DZIEMBOWSKI, W., GOODE, P., GOUGH, D.O. & MORROW, C. A. 1989 *ApJ* **343**, 526.

BROWN, T. M., GILLILAND, R. L. NOYES, R. W. & RAMSEY, L. W. 1991 *ApJ* **368** 599.

BROWN, T. M. & GILLILAND, R. L. 1994 *Annu. Rev. Astron. Astrophys.* **32**, 37.

BUZASI, D., BROWN, T. M., CHEN, P. & MURPHY, G. 1998 *BLAST: Balloon-Launched Asteroseismic Telescope*. Proposal to NASA.

CATALA, C., ET AL. 1995. In *Fourth SOHO Workshop: Helioseismology* (eds. J. T. Hoeksema, V. Domingo, B. Fleck & B. Battrick), v. 2, p. 549. ESA SP-376.

CHRISTENSEN-DALSGAARD, J. 1988a. In *Advances in Helio- and Asteroseismology*, (ed. J. Christensen-Dalsgaard and S. Frandsen), IAU Symposium No. 123, p. 3. Reidel.

CHRISTENSEN-DALSGAARD, J. 1988b. In *Advances in Helio- and Asteroseismology*, (ed. J. Christensen-Dalsgaard and S. Frandsen), IAU Symposium No. 123, p. 295. Reidel.

CHRISTENSEN-DALSGAARD, J. 1991. In *Challenges to Theories of the Structure of Moderate-Mass Stars*, (eds. D. Gough and J. Toomre), p. 11. Springer-Verlag.

CHRISTENSEN-DALSGAARD, J. 1992 *ApJ* **385**, 354.

CHRISTENSEN-DALSGAARD, J., DUVALL, T. L. JR., GOUGH, D. O., HARVEY, J. W. & RHODES, E. J. 1985 *Nature* **315**, 378.

CHRISTENSEN-DALSGAARD, J. & FRANDSEN, S. 1983 *Solar Phys.* **82**, 469.

CHRISTENSEN-DALSGAARD, J., GOUGH, D. O. & THOMPSON, M. J. 1991 *ApJ* **378**, 413.

CLAVERIE, ET AL. 1984 *Mem. Soc. Astron. Ital.* **55** 63.

CONNES, P. 1985 *Astrophys. Sp. Sci.* **110**, 211.

COX, A. N., GUZIK, J. A. & RABY, S. 1990 *ApJ* **353** 698.

COX, J. P. 1980. In *Theory of Stellar Pulsation*, Princeton University Press.

DÄPPEN, W., DZIEMBOWSKI, W. A. & SIENKIEWICZ, R. 1988. In *Advances in Helio- and Asteroseismology*, (ed. J. Christensen-Dalsgaard and S. Frandsen), IAU Symposium No. 123, p. 233. Reidel.

DEMARQUE, P. & GUENTHER, D. B. 1990. In *Progress of Seismology of the Sun and Stars* (ed. Y. Osaki and H. Shibahashi, p. 405. Springer-Verlag.

DUVALL, T. L. JR., DZIEMBOWSKI, W., GOODE, P. R., GOUGH, D. O., HARVEY, J. W. & LEIBACHER, J. W. 1984 *Nature* **310** 22.

EDMONDS, P., CRAM, L., DEMARQUE, P., GUENTHER, D. B. & PINSONNEAULT, M. H. 1992 *ApJ* **394**, 313.

GELLY, B., GREC, G. & FOSSAT, E. 1986 *Astron. Astrophys.* **164**, 383.

GILLILAND, R. L. & BROWN, T. M. 1992 *Publ. Astron. Soc. Pac.* **104**, 582.

GILLILAND, R. L. & DÄPPEN, W. 1988 *ApJ* **324**, 1153.

GILLILAND, R. L., ET AL. 1993 *Astron. J.* **106**, 2441.

Goldreich, P. & Kumar, P. 1990 *ApJ* **363**, 694.

GOLDREICH, P. & KUMAR, P. 1991 *ApJ* **363**, 694.

GOUGH, D. O. 1987 *Nature* **326**, 257.

GOUGH, D. O. 1990. In *Astrophysics—Recent Progress and Future Possibilities*, (eds. B. Gustaffson and P. E. Nissen), **42**, p. 13. The Royal Danish Academy of Sciences and Letters, *Mat. Fys. Medd.*

GOUGH, D. O. & NOVOTNY, E. 1993a. In *Inside the Stars*, (eds. W. Weiss & A. Baglin), ASP Conference Series 40, p. 550. Astronomical Society of the Pacific.

GOUGH, D O. & NOVOTNY, E. 1993b. In *GONG 1992: Seismic Investigation of the Sun and Stars*, (ed. T. M. Brown), ASP Conference Series 42, p. 355. Astronomical Society of the Pacific.

GUENTHER, D. B. & DEMARQUE, P. 1993. *ApJ* **405**, 298.

HANSEN, C. J., COX, J. P. & VAN HORN, H. M. 1977 *ApJ* **217** 151.

HARVEY, J. W. 1988. In *Advances in Helio- and Asteroseismology*, (eds. J. Christensen-Dalsgaard and S. Frandsen, IAU Symposium No. 123, p. 497. Reidel.

HOUDEK, G., ROGL, J., BALMFORTH, N. & CHRISTENSEN-DALSGAARD, J. 1995. In *GONG '94: Helio- and Asteroseismology*, (eds. R. K. Ulrich, E. J. Rhodes & W. Däppen), p. 641.

HUEBNER, W. F., MERTS, A. L., MAGEE, N. H. & ARGO, M. F. 1977 *Astrophysical Opacity Library*, Los Alamos Scientific Library Report LA-6760-M.

HUDSON, H., BROWN, T. M., CHRISTENSEN-DALSGAARD, J., COX, A. N., DEMARQUE, P., ET AL. 1986 *A Concept Study for an Asteroseismology Explorer*, Proposal submitted to NASA.

INNIS, J. L., ISAAK, G. R., SPEAKE, C. C., BRAZIER, R. I. & WILLIAMS, H. K. 1991 *MNRAS* **249**, 643.

KJELDSEN, H. & BEDDING, T. 1994 *Astron. Astrophys.* **293**, 87.

KJELDSEN, H., ET AL. 1998 *MONS: Measuring Oscillations in Nearby Stars* A Proposal to the Danish Research Council.

LEBRETON, Y., BERTHOMIEU, G. & PROVOST, J. 1988. In *Advances in Helio- and Asteroseismology*, (eds. J. Christensen-Dalsgaard and S. Frandsen), IAU Symposium No. 123, p. 95. Reidel.

LIBBRECHT, K. G. 1989 *ApJ* **336**, 1092.

LIBBRECHT, K. G. & MORROW, C. A. 1991. In *Solar Interior and Atmosphere*, (eds. A. N. Cox, W. C. Livingston, and M. S. Matthews), p. 479. University of Arizona Press.

NORDLUND, A. & STEIN, R. F. 1990 *Comp. Phy. Com.* **59**, 119.

RAST, M. P. & TOOMRE, J. 1993. In *GONG 1992: Seismic Investigation of the Sun and Stars*, (ed. T. M. Brown), ASP Conference Series 42, p. 41. Astronomical Society of the Pacific.

ROGERS, F. J. & IGLESIAS, C. A. 1992 *ApJS* **79**, 507.

SCHATZMAN, E., MAEDER A., ANGRAND, F. & GLOWINSKI, R. 1981 *Astron. Astrophys.* **96**, 1.

SPERGEL, D. N. & PRESS, W. H. 1985 *Astrophys. J.* **294**, 663.

STEIGMAN, G., SARAZIN, C. L., QUINTANA, H. & FAULKNER, J. 1978 *Astron. J.* **83**, 1050.

TASSOUL, M. 1980 *ApJS* **43**, 469.

TOUTAIN, T. & FRÖLICH, C. 1992 *Astron. Astrophys.* **257** 287.

ULRICH, R. K. 1986 *ApJL* **306**, 37.

ULRICH, R. K. 1988. In *Advances in Helio- and Asteroseismology*, (eds. J. Christensen-Dalsgaard and S. Frandsen), IAU Symposium No. 123, p. 299. Reidel.

VANDAKUROV, Y. V. 1968 *Soviet Astr.* **11**, 630.

VORONTSOV, S. V. 1989 *Sov. Astron. Lett.* **15**, 21.

Input physics and parameters: Accretion and losses of mass and angular momentum

By ANDRÉ MAEDER

Geneva Observatory

1. Introduction

Among the unsolved problems in stellar evolution, the changes $\frac{dM}{dt}$ and $\frac{d\mathcal{L}}{dt}$ of stellar mass and angular momentum during evolution are among the most uncertain quantities, although they have a very high impact on all outputs of stellar models: tracks, lifetimes, masses of the remnants, yields, etc. I shall therefore examine some of these problems, both for pre-MS evolution and post-MS evolution.

In spite of some 40 years of calculations in the field of stellar evolution a number of significant discrepancies between models and observations are still present, which shows the need for substantial improvements of the input physics. For instance, for star formation many important questions are unanswered concerning the pre-MS evolution of massive stars. What are the accretion rates? Are massive stars formed by the accretion scenario or by coalescence of intermediate masses? Where is the turning point from a positive to a negative value of the global mass change $\frac{dM}{dt}$ during massive star evolution? What are the main effects which shape the IMF?

Among the problems and discrepancies in post-MS evolution we can mention: 1. The extended cluster main-sequences (Meynet et al. 1993); 2. The high N/C and ^{13}C/^{12}C in red giants of all masses (cf. Charbonnel et al. 1998); 3. The origin of ON-stars (cf. Walborn 1988); 4. The evidences of the He- and N-excesses in fast rotating O-stars (cf. Herrero et al. 1998); 5. The He- and N-excesses in B, A, F supergiants (cf. Gies and Lambert 1992; Fitzpatrick and Bohannan 1993; Lennon 1994; Venn 1995, 1998); 6. The excesses of He and N in the envelope of SN 1987A (cf. Fransson et al. 1998); 7. The evidence for Boron depletion in rotating B-stars (cf. Venn et al. 1996; Fliegner et al. 1996); 8. The existence of Wolf-Rayet stars with a transition spectrum WN/WC (cf. Langer 1991); 9. The difficulties in explaining the B/R ratio of blue-to-red supergiants at various metallicities Z in the Galaxy, the LMC and SMC (cf. Langer and Maeder 1995); 10. The enormous relative N-excess (factor of 10) in A-type supergiants of the SMC, while in the Galaxy the relative excess is quite moderate (Venn et al. 1998); 11. Finally, we may wonder why the SMC and LMC contain much larger fractions of Be stars among B-type stars than the Galaxy (cf. Grebel et al. 1996). A recent analysis of clusters in the Milky Way, the LMC and SMC is showing us that the fraction of Be stars is clearly increasing with the local metallicity Z, (Maeder and Mermilliod, in prep.) Points 10 and 11 may lead us to wonder whether the average stellar rotation is higher at low metallicities. This list, which is of course not exhaustive, shows that in the field of stellar evolution there are many further progresses awaiting in front of us.

Below, several physical effects are examined which will hopefully bring some clarifications to the questions raised above. Section 2 deals with pre-MS evolution of massive stars, Section 3 with mass loss in post-MS evolution, Sections 4 with rotation in the interior and at the stellar surfaces.

2. Pre-MS evolution of massive stars with mass accretion

Several works have shown the interest to study accretion processes during pre-MS evolution (cf. Palla and Stahler 1993; Beech and Mitalas 1994; Bernasconi and Maeder 1996). In the recent studies the accretion rates, rather than assumed to be constant, are obtained from the model of a spherically collapsing cloud sustained by thermal and turbulent pressure. The ^{2}H and CN-burning are treated in detail with a time-dependent convection. The main results are:

(1) An accreting star follows a track called "birthline" (cf. Fig. 1), which in the HR diagram moves upwards; it originates from a faint red object and is reaching the usual zero-age sequence near 5 to 8 $M_\odot$ and then is moving upwards along the zero-age sequence. Let us consider the case of an evolution leading to a 5 $M_\odot$ star. As long as the star is accreting some mass at a sufficient rate, it will follow the upwards track on the birthline. If accretion stops, the star will then essentially follow an horizontal track close to that of constant mass evolution leading to the zero-age main sequence. This shows that for low and moderate stellar masses the birthline should be an upper envelope of the observed stellar distribution in the HR diagram.

(2) For stars with M $\leq 8\,M_\odot$ (on the MS) the accretion timescale t_{accr} is shorter than the Kelvin-Helmholtz timescale t_{KH}, therefore accretion will be completed before central contraction might have progressed to the point of nuclear ignition. For such masses evolution is not drastically different from objects with constant mass.

(3) For M $\geq 8\,M_\odot$, one has $t_{\mathrm{accr}} > t_{\mathrm{KH}}$ and central nuclear ignition may occur before accretion is finished and before the star might have reached its final mass. Thus massive stars are likely to stay hidden in molecular clouds during their pre-MS evolution and maybe even during the early stages of their MS evolution.

(4) The lifetimes $t_{\mathrm{pre-MS}}$ of pre-MS evolution in the accretion scenario are much longer for massive stars than the pre-MS lifetimes obtained with constant mass. As an example, for a 60 $M_\odot$ star $t_{\mathrm{pre-MS}}$ would be 2.8×10^4 yr, which would amount to several 10^5 yr in the accretion scenario.

Figure 1 shows a family of birthlines with rates equal to (from bottom to top) 0.1, 0.15, 0.2, 0.3, 0.5, 0.75, 1.0, 1.25, 1.75, 2.5, 3.5, 5.0 times the values obtained by Bernasconi and Maeder (1996). For a given value of T_{eff}, the luminosity of the birthlines is larger for larger accretion rates. Constant mass pre-MS evolutionary tracks for 2, 3, 5 and 9 $M_\odot$ are shown (dashed lines) as well as post-MS tracks for 15 to 85 $M_\odot$ stars (Schaller et al. 1992). Observations of pre-MS stars are shown as well, together with indication of their accretion rates (cf. Bernasconi and Maeder 1996). This figure reveals the importance of accretion in determining the location of the birthline, which forms the upper envelope of the stellar distribution of low and intermediate mass stars in the HR diagram. On the other hand, it shows that an estimate of the accretion rate is possible by comparing the upper envelope of pre-MS stars with the birthline in the HR diagram. We choose to write the accretion rate as follows:

$$\dot{M}_{\mathrm{accr}}(M) = \dot{M}(0.7 M_\odot) \left(\frac{M}{M_\odot}\right)^\alpha \tag{2.1}$$

Norberg et al. (1999) show that the best fit is obtained for a value of $\dot{M}\,(0.7\,M_\odot)$ of $10^{-5}\,M_\odot\,\mathrm{yr}^{-1}$ and an exponent α between 1.0 and 1.5. This "best" fit is illustrated in Fig. 2. These comparisons show the very important result that the accretion rate during pre-MS evolution is growing quickly with the already accreted stellar mass. Churchwell (this meeting) shows that the mass outflows are also growing almost linearly with the stellar mass, and that a fraction of about one third of the infalling matter is accreted.

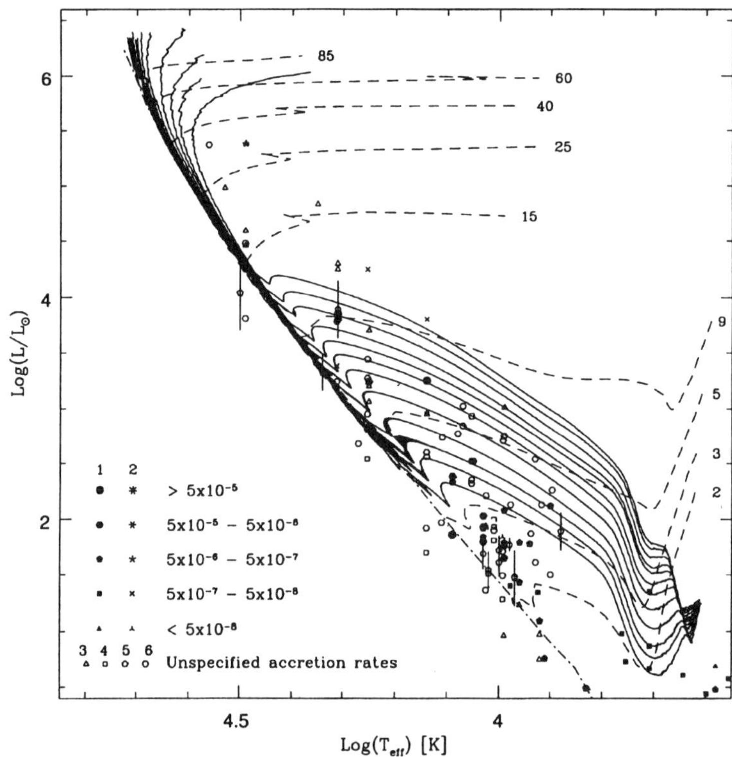

FIGURE 1. Family of birthlines in the HR diagram for different values of the accretion rates from Norberg et al. 1999 (continuous lines). Constant mass pre-MS evolutionary tracks are shown for stars in the range 2–9 $M_\odot$ (dashed lines), and some post-MS evolutionary tracks for massive stars (dashed lines).

Originally the accretion scenario had been applied to low and intermediate mass stars. For massive stars, it is usually considered (cf. Bonnell et al. 1998; Stahler, this meeting) that the accretion onto massive stars is inhibited by their high luminosity. Wolfire and Cassinelli (1987) have studied the conditions for the formation of massive stars. In particular, if the accretion rates are high enough, typically of the order of 10^{-3}, the formation by accretion is possible with account of both the Eddington luminosity of the star and of the Eddington luminosity of the shock produced by the accreted matter. Very interestingly, accretion rates growing as suggested by expression (1) well correspond to the permitted domain for massive stars. Thus, we think that the accretion scenario can also be applied to massive star formation as suggested by Bernasconi and Maeder (1996), if the accretion rates are growing fast enough with the already accreted stellar mass.

The origin of the upper stellar mass M_{up} finds a new interpretation in the context of the accretion scenario for massive stars. In the accretion scenario M_{up} is closely related to the value of the accretion rates and to its dependence on the stellar mass. As an example, a constant $\dot{M}_{\mathrm{accr}}$ of say 10^{-5} $M_\odot$ yr^{-1} would take 10^7 yr to form a 100 $M_\odot$ star, i.e. much longer than the MS lifetime. Such a rate is therefore not possible. A massive star is likely not to have burnt too much of its hydrogen when becoming visible, say for example less than 10% (of course, a better condition for fixing the end of the accretion process should still be found). However, even with this simple condition we need a sufficiently fast growing accretion rate in order to be able to form massive stars

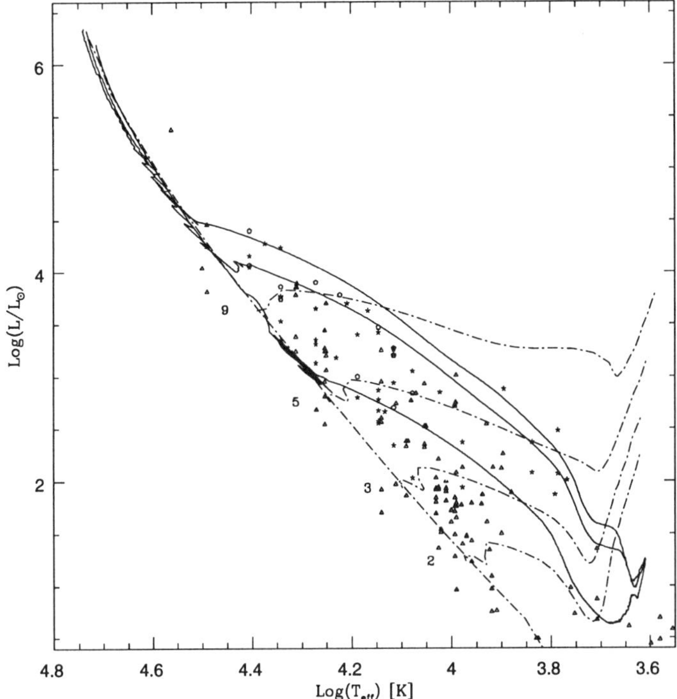

FIGURE 2. The fit of birthlines (continuous lines) and observations with accretion rates given by expression (1) and a slope of 1.5. The top line corresponds to $\dot{M}(0.7M_\odot) = 10^{-5}M_\odot$ yr^{-1}, the second to $5 \times 10^{-6}M_\odot$ yr^{-1} and the third to $1 \times 10^{-6}M_\odot$ yr^{-1}. Dashed-dotted lines show some constant mass evolutionary tracks. The observations also include results by de Winter et al. (1997) and The et al. (1990).

within a limited fraction of their MS lifetimes. Numerical tests show that expression (1) with the numerical parameters given above would allow massive star formation of about 100 $M_\odot$ within about 5% of their MS lifetimes. Lower $\dot{M}_{\mathrm{accr}}$ is leading to stars centrally too much evolved at the time of their completion, while higher values would lead to higher values of the maximum stellar mass, which are not observed.

This approach gives new insight into massive star formation and shows the critical role of $\dot{M}_{\mathrm{accr}}$ not only for pre-MS evolution, but also for the determination of the value of the maximum stellar mass and for its contribution to shaping the IMF.

3. Mass loss in post MS evolution

The general effects of mass loss rates $\dot{M}$ on the evolution of massive stars have been reviewed by Chiosi and Maeder (1986) and by Maeder and Conti (1994). Here we shall address only a few recent problems regarding the $\dot{M}$-rates and their role. Stellar wind theories lead to $\dot{M}$-rates (cf. Pauldrach et al. 1986; Kudritzki et al. 1989; Puls et al. 1996), which agree with observations within a factor of two. Observations however are still uncertain. For many years stellar evolutionary models were based on the compilations by de Jager et al. (1988). New expressions were obtained by Lamers and Cassinelli (1996) for O-type stars, they give rates corresponding roughly to twice the $\dot{M}$-rates (de Jager et al. 1988). This is rather interesting because it was found that our Geneva models

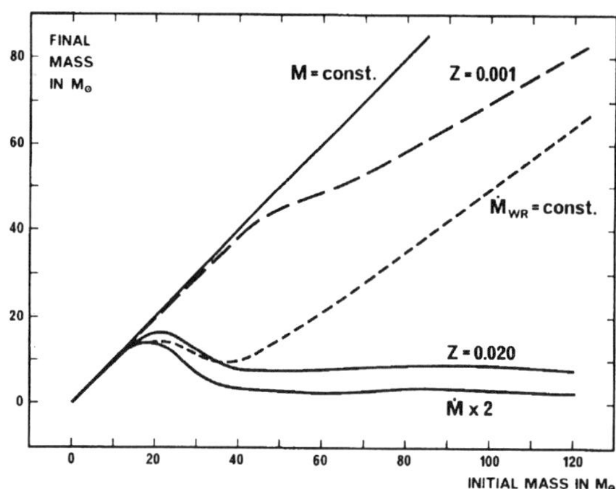

FIGURE 3. Relation between the final and initial masses for different metallicities Z (cf. Maeder 1992). We notice the mass convergence for the high initial stellar masses at solar metallicity.

(cf. Meynet et al. 1994), based on twice the $\dot{M}$-rates by de Jager et al., were fitting observations best.

So far, so good and the matter could be considered as solved. However, it has been recently emphasized that the real rates should be about a factor of two smaller than usually determined, because clumpiness in the wind is generally not accounted for (cf. Hamann 1998). This argument seems convincing, but if so there may be some difficulties in forming a number of Wolf-Rayet (WR) stars large enough to meet observation requirements. In this context we think that rotation, by producing internal mixing and also enhancement of the mass loss, may favour the formation of WR stars, thus compensating the lower values of the $\dot{M}$-rates when taking the clumpiness effect into account.

The mass loss rates of WR stars are also an essential aspect of massive star evolution. Hamann (1994) found that WR stars without hydrogen (generally WNE stars, i.e. early WN stars) follow a relation of the form $\dot{M} \sim L^{1.5}$. If we take the mass-luminosity relation of such WR stars into account (cf. Maeder 1983), which is of the form $L \sim M^{1.73}$, we obtain a mass dependence $\dot{M} \sim M^{2.6}$. This is very much in agreement with the result by Langer (1989)

$$\dot{M}_{WR} = (0.6 - 1.0) \times 10^{-7} \left(\frac{M}{\mathrm{M}_\odot} \right)^{2.5} \tag{3.2}$$

The low numerical coefficient concerns WNE stars and the high one applies to WC stars. This relation leads to very high mass loss when the star enters the WN stage without hydrogen. Numerical models show that these rates can be as high as $10^{-2} \, \mathrm{M}_\odot \, yr^{-1}$ at the beginning of the WNE stage and could thus lead to the formation of a visible shell. Expression (2) produces a remarkable mass convergence (cf. Langer 1989), i.e. massive stars with an initial mass above 30 $\mathrm{M}_\odot$ (at least at solar compositions) will finish their life with very low final masses (cf. Fig. 3).

These low final masses play a very important role in nucleosynthesis. In fact, most of the new Helium and Carbon is ejected during the WR stages before further nuclear processing, and this considerably influences the stellar yields (cf. Maeder 1992). Thus, explosive nucleosynthesis occurs in smaller cores of different supernova types (possibly SN Ib). These low final masses are also critical for the mass limit between neutron star and black hole formation.

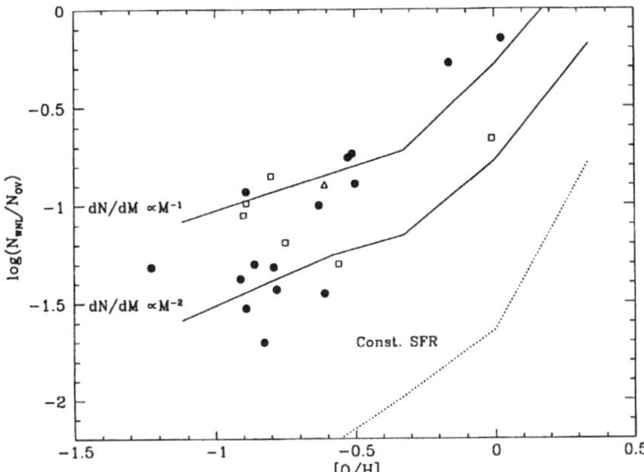

FIGURE 4. Relation between the WNL/O number ratio and metallicities as given by O/H. The continuous lines give the relation for instantaneous starburst and for two different slopes of the IMF. The dotted line corresponds to a constant star formation rate. The observations of starburst regions are from Vacca and Conti (1992) and the number counts were derived from WR lines observed in the integrated spectrum of these galaxies. As shown by Schaerer (1996), the observed number ratios were shifted downwards by a factor of two in order to account for the aging of the nebulae, which lowers the degree of ionisation of the nebulae and thus decreases the intensity of the nebular $H\beta$ lines, which are used to estimate the number of O-type stars.

The way in which mass loss rates depend on stellar metallicity Z has important consequences for stellar evolution. Indeed, for massive stars, the Z-effects essentially stem from the effects of photospheric and wind opacities, which strongly influence the $\dot{M}$-rates. For O-stars, according to Kudritzki et al. (1989) this dependence is expressed as follows:

$$\frac{\dot{M}(Z)}{\dot{M}(Z_\odot)} = \left(\frac{Z}{Z_\odot}\right)^\alpha \tag{3.3}$$

where $\alpha \simeq 0.8$. This means that at low Z mass loss is of little influence on stellar evolution. However, the exponent α is probably not constant with luminosity, the dependence on Z being weaker at high luminosities. At the present time the Z-dependence of the $\dot{M}$-rates is unknown for blue and red supergiants. (Current models are using the same dependence as that for O-stars, which is not very satisfactory). The Z-dependence of the $\dot{M}$-rates leads to an increase of WN star formation with higher ratios of WR/O and WC/WN stars at higher metallicities. Detailed comparisons were made by Maeder and Meynet (1994) and Maeder and Conti (1994), showing that this effect is responsible for the big difference in the WR populations of nearby galaxies (cf. dotted line in Fig. 4).

Massive stars are the tracers of star formation in galaxies and starburst regions. As a matter of fact, number ratios such as those for WR/O stars, not only depend on the local metallicity, but also on the age of the starburst, the duration of the burst and the IMF. Figure 4 shows the relation between the number ratio of WNL (late WN) to O-stars as a function of metallicity, as well as for different IMF. Also, observations of these numbers of massive stars give us access to the important parameters of starburst regions.

For further analysis of star formation in highly redshifted galaxies it is necessary to have models of massive stars which correctly predict all kinds of star number ratios, with the appropriate nucleosynthetic yields. In this respect it would be useful to find an explanation for the particularly high N content in the supergiants of the SMC as well as in starbursts and QSOs.

4. Rotation and evolution

Most of the unsolved problems listed in the introduction suggest the presence of additional mixing processes in massive star evolution. Among the potential sources of additional mixing in massive stars, rotation is of course one of the possible factors. Together with G. Meynet we are investigating this problem in more detail. Here are some of the physical problems encountered.

4.1. *Basic set of equations for shellular rotation*

The study of the various instabilities in differentially rotating stars (Zahn 1992) revealed that vertically the main sources causing instability are shears, while horizontally it is a strong turbulence, analogous to geostrophic circulation. This horizontal turbulence essentially has two main effects: a) it limits the size of the horizontal shears $d\Omega/d\vartheta$, where Ω is the angular velocity, so that rotating stars globally follow rotation laws of the form $\Omega = \Omega(r)$, called "shellular rotation" (Zahn 1992). b) The horizontal turbulence with a diffusion coefficient D_h limits the vertical mixing by meridional circulation, so that the vertical diffusion of chemicals is $D_{\mathrm{eff}} = (rU)^2/(30D_h)$, where $U(r)$ is the vertical velocity of meridional circulation.

Kippenhahn and Thomas (1970) devised a very useful method to include the hydrostatic effects of rotation into 1-D stellar evolutionary codes. This method was subsequently used by all authors, even though it is only valid for solid body or cylindrical rotation. Meynet and Maeder (1997) develop an extension of this method to the case of shellular rotation $\Omega(r)$. The resulting equations were written for the interior, the envelope and the atmosphere, they allowed us to deal with non-uniformly rotating stars, which are the general case.

Evolutionary tracks were then calculated for 9, 20, 40 and 60 $M_\odot$ by Meynet and Maeder (1997) with the following results: 1) The hydrostatic effects are quite modest; for example, they increase the lifetimes by a few percent only. 2) If the usual Richardson criterion is applied to shear instabilities, we observe that the μ-gradients are always strong enough to prevent mixing, (cf. also Chaboyer et al. 1995). Thus we can conclude that either rotation produces no mixing or that the Richardson criterion should not be applied in its usual form.

4.2. *The concept of semiconvective shear zones*

The classical Richardson criterion is written (e.g. Zahn 1992; Maeder and Meynet 1996)

$$\frac{1}{4}\left(\frac{dV}{dz}\right)^2 \geq \frac{g\delta}{H_p}\left(\nabla_{ad} - \nabla + \frac{\varphi}{\delta}\nabla_\mu\right)\frac{vl}{K} \qquad (4.4)$$

where $\frac{dV}{dz}$ is the vertical gradient of the horizontal velocity, v the vertical velocity of the turbulent element with mixing length l, K the thermal diffusivity, g the local gravity, H_p the pressure scale height, δ and φ the usual thermodynamic coefficients. Equation (4) says that shear mixing could occur if the excess energy in the shear is larger than the work necessary to overcome the T- and μ-gradients (with account of the heat losses by the term $\frac{vl}{K}$).

Numerical models by Meynet and Maeder (1997) show a situation often met in rotating massive stars. The following inequalities were both satisfied simultaneously:

$$\frac{1}{4}\left(\frac{dV}{dz}\right)^2 > \frac{g\delta}{H_p}\left(\nabla_{ad} - \nabla\right) \qquad (4.5)$$

$$\frac{1}{4}\left(\frac{dV}{dz}\right)^2 < \frac{g\delta}{H_p}\left(\nabla_{ad} - \nabla + \frac{\varphi}{\delta}\nabla_\mu\right) \tag{4.6}$$

This means that the energy excess in the shear is large enough to overcome the stable thermal gradient, however it is unable to overcome the stable T- and μ-gradients. We call a zone satisfying both (5) and (6) a "semiconvective shear zone." Indeed, in absence of a shear, i.e. $\frac{dV}{dz} = 0$, (5) will be the Schwarzschild criterion and (6) the Ledoux criterion. In this last case, the zone considered is merely a semiconvective zone subject to semiconvective diffusive mixing (cf. Langer et al. 1983). This means that in a semiconvective shear zone, even if the shear does not contain enough energy to produce the complete overturn of the medium, there may be some degree of partial mixing. Indeed, criterion (4) ignores the other sources of turbulence such as semiconvection and horizontal turbulence, which might enable part of the energy present in the shear to be injected in the spectrum of turbulence and thus contribute to mixing. Therefore, we introduced (Maeder 1997) the working hypothesis that some fraction of the local energy excess in the shear is degraded by turbulence, thus changing the local entropy gradient and subsequently the T- and μ-gradients. We assumed that the typical velocity required for mixing is the velocity of thermal adjustments in an inhomogeneous medium (cf. also Kippenhahn and Weigert 1990) and obtained the following diffusion coefficient for shear mixing:

$$D_{\text{shear}} = \frac{4K}{(\frac{\varphi}{\delta}\nabla_\mu + \nabla_{ad} - \nabla)}\left[\frac{H_p}{4g\delta}\left(\frac{dV}{dz}\right)^2 - (\nabla_{ad} - \nabla)\right] \tag{4.7}$$

Interestingly, at high rotation this coefficient is the same as that obtained by Zahn (1992), and for zero rotation it is equivalent to the diffusion coefficient of semiconvective mixing (Langer et al. 1983). The main consequence is a slight progressive erosion of the μ-barriers in rotating stars, where criterion (4) would not predict any mixing at all. Furthermore, this coefficient allows us to understand why evidences of mild mixing are more frequent in the most massive stars. The characteristic time for mixing is of the order of

$$\tau_{\text{mix}} \simeq \frac{R^2}{D_{\text{shear}}} \simeq \frac{R^2}{K\left(\frac{dV}{dz}\right)^2} \tag{4.8}$$

and for a given degree of shear, it scales like

$$\tau_{\text{mix}} \simeq \frac{1}{M^{1.8}} \tag{4.9}$$

while the MS lifetime would be

$$\tau_{\text{MS}} = 10^{7.9}\left(\frac{M}{M_\odot}\right)^{-0.72} \text{yr} \tag{4.10}$$

We see that τ_{mix} decreases much faster than τ_{MS} and reaches the same value for the most massive stars.

4.3. *Meridional circulation revisited*

The problem of meridional circulation has been studied for 3/4 of a century, with major contributions by Eddington, Öpik, Mestel, Kippenhahn and Zahn among others. In particular, Zahn (1992) has reconsidered the problem taking into account the shellular rotation laws. It might seem surprising that progress is still needed, in spite of the numerous extensive studies already done in the past. However, the following four points lead us to follow up again this old problem (cf. Maeder and Zahn 1998).

(1) This first point is the easiest one. Since usually only the case of perfect gas is studied, a more general equation of state is needed, in particular for massive stars.

(2) Generally, the approximation of a stationary circulation is made which certainly is valid for MS stars. However, this approximation cannot be applied to the shell H-burning phase, the He-burning and subsequent phases.

(3) Horizontal turbulence (cf. Zahn 1992) also contributes to the horizontal thermal flux, a contribution which appears to be quite significant.

(4) The effects of the vertical μ-gradient must be introduced consistently. Mestel (1953) took into account the horizontal difference in μ created between the upwards polar motion and the downwards equatorial motion, but omitted the fact that the μ-gradient also enters the entropy gradient. Maeder and Zahn (1998) showed that the μ-gradient, when taken into account properly, may reduce the vertical component of the meridional circulation $U(r)$ by one or two orders of magnitude compared to current expressions. The final expression is then

$$U(r) = \frac{P}{\rho g C_p T (\nabla_{ad} - \nabla + \frac{\varphi}{\delta}\nabla_\mu)} \left[\frac{L}{M_\star}(E_\Omega + E_\mu) + \frac{C_P}{\delta}\frac{\partial\Theta}{\partial t} \right] \qquad (4.11)$$

The term in brackets is explicated by Maeder and Zahn (1998), however the most important change is the presence of the term ∇_μ in the denominator of expression (11). While the usual expressions for the meridional circulation predict an infinite velocity at the edge of a radiative and a semiconvective zone and an inverted circulation in a semiconvective zone, this expression offers a continuity of the solutions for the circulation.

4.4. Transport of chemicals and angular momentum

Both the turbulent diffusion and the meridional circulation contribute to the transport of chemicals and angular momentum. The equation expressing the radial conservation of the angular momentum in Lagrangian coordinates is (cf. Zahn 1992; Maeder and Zahn 1998)

$$\rho\frac{d}{dt}(r^2\Omega)_{M_r} = \frac{1}{5r^2}\frac{\partial}{\partial r}(\rho r^4\Omega U(r)) + \frac{1}{r^2}\frac{\partial}{\partial r}\left(\rho D_{shear} r^4\frac{\partial\Omega}{\partial r}\right) \qquad (4.12)$$

It shows that the change of angular momentum of a mass element results from the (downwards) advection by meridional circulation and from the (upwards) diffusion by shears. It also shows that meridional circulation cannot be treated as a diffusion process. The effects of contraction or expansion are automatically contained in this Lagrangian expression.

The transport of chemical elements also obeys an advection-diffusion equation. However, based on Chaboyer and Zahn (1992) we could express the time variation of the horizontal fluctuation of μ in terms of meridional circulation and of horizontal diffusion. The advection-diffusion equation could then be transformed into a diffusion equation only, and that for any species of mass fraction X_i.

$$\rho\left(\frac{dX_i}{dt}\right)_{M_r} = \frac{1}{r^2}\frac{\partial}{\partial r}\left[\rho\,(D_{shear} + D_{eff})\,r^2\frac{\partial X_i}{\partial r}\right] \qquad (4.13)$$

where D_{eff} has been defined in Sect. 4.1 above.

4.5. *The von Zeipel theorem, the Eddington factor and the $\Omega\Gamma$ limit*

The von Zeipel theorem, in its usual form, says that the radiative flux on the surface of a rotating star is locally proportional to the effective gravity (von Zeipel 1924). This is an essential ingredient for models, ingredient which allows us to know the distribution of temperatures, i.e. $T_{\text{eff}} \sim g_{\text{eff}}^{1/4}$ on the surface of a rotating star. It determines the spectrum, the mass loss and all the surface properties of a rotating star. We extended the von Zeipel theorem by including a shellular rotation law $\Omega(r)$ and found (cf. Maeder 1998) a small additional term $\zeta(\vartheta)$, which in fact contributes to a further increase of the radiative flux at the pole and its decrease at the equator.

The Eddington factor Γ, which in its usual form is

$$\Gamma = \frac{L\kappa}{4\pi cGM} \tag{4.14}$$

needs to be redefined more precisely for the case of a rotating star, since both the flux and the gravity are varying over the stellar surface. Considering a local $\Gamma(\vartheta)$, where ϑ is the colatitude, taken as the ratio of the local stellar flux to the local limiting flux, we find (Maeder 1998)

$$\Gamma(\vartheta) = \frac{\kappa(\vartheta)L}{4\pi cGM\left(1 - \frac{\Omega^2}{2\pi G\rho_m}\right)}(1 + \zeta(\vartheta)) \tag{4.15}$$

where ρ_m is the mean density inside the shell mass considered. We note three differences between expressions (14) and (15). 1) The ζ-term favours a high Γ at the pole. 2) The opacity $\kappa(\vartheta)$ applies only locally, which means that if κ grows for lower T, Γ is larger at the equator. 3) Γ depends also on rotation. These results are very useful in relation with the concept of an "Ω-limit" recently introduced by Langer (1997) and Langer and Heger (1998). The central idea is that if one has, for the critical break-up velocity

$$v_{\text{crit}}^2 = \frac{GM}{R}(1 - \Gamma) \tag{4.16}$$

the value of v_{crit}^2 would then tend towards zero for a star approaching the Eddington limit. Thus, Langer's conclusion was that for any non-zero rotation a star would reach its critical rotation velocity (the "Ω-limit") before the formal Eddington limit is reached.

Indeed, one has, for the total gravity taking into account von Zeipel's theorem,

$$\vec{g_{tot}} = \vec{g_{\text{grav}}} + \vec{g_{\text{rot}}} + \vec{g_{\text{rad}}} = \vec{g_{\text{eff}}}\left[1 - \frac{\kappa(\vartheta)L(1 + \zeta(\vartheta))}{4\pi cGM\left(1 - \frac{\Omega^2}{2\pi G\rho_m}\right)}\right] \tag{4.17}$$

The break-up velocity occurs when rotation is such that $\vec{g_{tot}} = 0$ at the equator. We see that expression (17) has formally two roots, one given by $\vec{g_{\text{eff}}} = 0$ and the other obtained by nulling the bracket term in (17). The condition $\vec{g_{\text{eff}}} = 0$ at the equator gives

$$v_{\text{crit}}^2 = \Omega^2 r_{eb}^2 = \frac{GM}{r_{eb}} \tag{4.18}$$

where r_{eb} is the equatorial radius at break-up velocity. Thus we see that expression (16) is not correct since the current expression v_{crit}^2 does not contain any Γ. The reason for this is simply that close to break-up the effective gravity goes down to zero, and so does the radiative flux according to von Zeipel's theorem. A closer look at the second root of expr. (17) reveals that most stars near the Eddington limit reach the Ω- and Γ-limits simultaneously. For instance, at 95% of the Eddington limit, v_{crit}^2 is still equal to 41% of its usual value (18). Thus, as long as a detailed stability analysis of the situation, which may be called the $\Omega\Gamma$-limit, is not made we cannot say which effects dominate.

4.6. *Mass loss and rotation*

According to the radiative wind theory (cf. Pauldrach et al. 1986; Kudritzki et al. 1989; Puls et al. 1996) the mass loss rates are essentially scaling like

$$\dot{M} \sim (k\alpha)^{1/\alpha} \left(\frac{1-\alpha}{\alpha}\right)^{\frac{1-\alpha}{\alpha}} F(\vartheta)^{1/\alpha} g_{tot}^{1-\frac{1}{\alpha}}(\vartheta) \tag{4.19}$$

where k and α are the force multiplier parameters, k representing the number of lines with strengths above a critical value and α the slope of the line strength distribution. $F(\vartheta)$ is the local flux as given by the revised von Zeipel theorem. The values of k and α are changing with T_{eff} (cf. Pauldrach et al. 1986); for $T_{\mathrm{eff}} = 50'000, 40'000, 30'000$ and $20'000$ K one has respectively $k = 0.124, 0.124, 0.17, 0.32$, and $\alpha = 0.64, 0.64, 0.59, 0.565$. Then, taking into account the expression for the flux with the revised von Zeipel theorem, as well as g_{tot} we obtain, for the mass flux,

$$\dot{M}(\vartheta) \sim (k\alpha)^{\frac{1}{\alpha}} \left(\frac{1-\alpha}{\alpha}\right)^{\frac{1-\alpha}{\alpha}} \left[\frac{L(P)}{4\pi G M_\star(P)}\right]^{\frac{1}{\alpha}} \frac{g_{\mathrm{eff}} (1 + \zeta(\vartheta))^{\frac{1}{\alpha}}}{(1 - \Gamma(\vartheta))^{\frac{1}{\alpha}-1}} \tag{4.20}$$

$M_\star$ is the effective mass as explicated at the denominator of expression (15). This relation expresses the dependence of the mass loss rates on colatitude. If α and k are constant in latitude (as normally expected in O-stars), we see that $\dot{M}(\vartheta)$ mainly depends on g_{eff} (cf. also Owocki et al. 1996, 1998). This means that in a rotating hot star the mass loss rates by unit surface are much larger over the polar caps than at the equator. The terms $\zeta(\vartheta)$ and $\Gamma(\vartheta)$ in (20) slightly reinforce the polar mass loss.

As mentioned above k and α vary with T_{eff}, these variations being larger for lower T_{eff}. This means that over the surface of a rotating star k and α also vary. The term $(k\alpha)^{\frac{1}{2}} \left(\frac{1-\alpha}{\alpha}\right)^{\frac{1-\alpha}{\alpha}}$ increases by a factor of three from $T_{\mathrm{eff}} = 50'000$ K to $20'000$ K. The term in brackets in (20) is also larger for lower α-values. The term containing Γ increases with the value of Γ, this growth being much faster in case of lower α-values. On the whole the following picture emerges:

(1) In hot, rotating stars the mass flux is higher at the poles and lower at the equator (the respective surface areas must of course be accounted for in numerical models). Let us call this the "g_{eff}-effect" in rotating stars.

(2) This behaviour is also present near the $\Omega\Gamma$-limit where the mass flux is strongly increased as shown by (20).

(3) In B and later type stars the enhanced polar ejection is progressively compensated by effects of larger line opacities (higher k and lower α), which favour progressively larger mass flux in the cooler equatorial regions. We shall call this the "κ-effect" in rotating stars.

(4) For B and later type stars near the $\Omega\Gamma$-limit the mass flux is strongly enhanced, particularly in the equatorial regions.

A so-called bistability of stellar winds has been found by Lamers (1997) in non-rotating stars: near a T_{eff} of $20'000$ K and also close to $10'000$ K, large and rather abrupt changes of the force multiplier parameters k and α modify the relations between v_∞ and v_{esc} and the mass loss rates. These important changes of k and α should also occur in B- and A-type rotating stars, as a result of the decrease of T_{eff} between the pole and the equator; furthermore they should produce the corresponding strong changes in the mass loss regime.

The HST pictures of η Carinae show two broad polar ejections and an equatorial skirt (Davidson 1997). η Carinae is clearly a hot star close to the $\Omega\Gamma$-limit. Among the various explanations possible for the observed geometry of the ejections from η Carinae, we might point out the possibility that polar ejections result from the "g_{eff}-effect" in (20) while the equatorial skirt is more likely to stem from the " κ-effect."

The complex structure around SN 1987 A consists of a bright elliptical inner ring and of two outer rings moved away from the central ring. Currently the two outer rings are interpreted as real rings and not as rings due to the limb brightening of an hour glass shell (cf. Burrows et al. 1995). Their location and CNO composition suggest (cf. Panagia et al. 1996) that they were ejected at an earlier stage of their evolution than the inner ring, perhaps when the SN progenitor was a blue supergiant. The bright inner ring is generally associated to the red supergiant stage in view of its composition, location and timescales involved (cf. Burrows et al. 1995 and references therein). In absence of numerical models we think that an equatorial ring could be consistent with the "κ-effect" in cool stars while symmetrical and possibly also polar ejections better correspond with polar ejection resulting from the "g_{eff}-effect" in hot stars. These comparisons are of course still speculative and numerical 2D models already in progress will hopefully allow quantitative comparisons.

Anisotropic stellar winds remove selectively the angular momentum. For example, winds passing through polar caps in O-stars remove very little angular momentum. Thus an excess of angular momentum is retained by these stars, which is rapidly redistributed by horizontal turbulence. These excesses might lead some Wolf-Rayet stars to be fast spinning objects. We show how anisotropic ejection can be treated in numerical models when the outer boundary conditions for the transport of angular momentum are properly modified.

On the whole, among the unsolved problems in the field of stellar evolution are the role and relative importance of the effects discussed above, such as shear mixing, meridional circulation, mass loss and loss of angular momentum. Numerical models in progress will show the interplay of these various physical effects.

REFERENCES

BEECH, M. & MITALAS, R. 1994 *ApJS* **95**, 517.

Bernasconi, P. A. & Maeder, A. 1996 *A&A* **307**, 829.

BONNELL, I. A., BATE, M. R., & ZINNECKER, H. 1998 *MNRAS* **298**, 93.

BURROWS, C. J., KRIST, J., HESTER, J. J. ET AL. 1995 *ApJ* **452**, 680.

CHABOYER, B., DEMARQUE, P., & PINSONNEAULT, M. H. 1995 *ApJ* **441**, 865.

CHABOYER, B. & ZAHN J.-P. 1992 *A&A* **253**, 173.

CHARBONNEL, C., BROWN, J. A., & WALLERSTEIN, G. 1998 *A&A* **332**, 204.

CHIOSI, C. & MAEDER, A. 1986 *Ann. Rev. Astron. Astrophys.* **24**, 329.

DAVIDSON, K. 1997 *Bull. American Astron. Soc.* **188**, 50.04.

DE JAGER, C., NIEUWENHUIJZEN, H., & VAN DER HUCHT, K. A. 1988 *A&AS* **72**, 259.

DE WINTER, D., KOULIS, C., THE, P.S., VAN DEN ANCKER, M. E., PEREZ, M. R., & BIBO, E. A. 1997 *A&AS* **121**, 223.

FITZPATRICK, E. L. & BOHANNAN, B. 1993 *ApJ* **404**, 734.

FLIEGNER, J., LANGER, N., & VENN, K. A. 1996 *A&A* **308**, L13.

Fransson, C., Cassatella, A., Gilmozzi, R., Kirshner, R.P., Panagia, N. et al. 1989 *ApJ* **336**, 429.

Gies, D. R. & Lambert, D. L. 1992 *ApJ* **387**, 673.

Grebel, E. K., Roberts, W. J., & Brandner, W. 1996 *A&A* **311**, 470.

Hamann, W. R. 1994 *Space Sci. Rev.* **66**, 237.

Hamann, W. R. 1998. In *Variable and Non-spherical Stellar Winds in Luminous Hot Stars* (Eds. B. Wolf et al.) IAU Coll. 169. Springer-Verlag, in press.

Herrero, A., Villamariz, M. R., & Martin, E. L. 1998. In *Boulder-Munich II: Properties of Hot, Luminous Stars*, ASP Conf. Ser. 131. p. 159. ASP.

Kippenhahn, R. & Thomas, H.-C. 1970. In *Stellar Rotation* (Ed. A. Slettebak), IAU Coll. 4, p. 20.

Kudritzki, R.P., Pauldrach, A., Puls, J., & Abbott, D. C. 1989 *A&A* **219**, 205.

Lamers, H. 1997. In *ASP Conf. Ser. 120*. p. 76. ASP.

Lamers, H. & Cassinelli, J. P. 1996. *ASP Conf. Ser. 98*. p. 162. ASP.

Langer, N. 1989 *A&A* **220**, 135.

Langer, N. 1991 *A&A* **248**, 531.

Langer, N. 1997. *ASP Conf. Ser. 120*. p. 83. ASP.

Langer, N. 1998. *ASP Conf. Ser. 131*. p. 76. ASP.

Langer, N. & Maeder, A. 1995 *A&A* **295**, 685.

Langer, N., Sugimoto, D., Fricke, K. J. 1983 *A&A* **126**, 207.

Lennon, D. 1994 *Space Sci. Rev.* **66**, 127.

Maeder, A. 1983 *A&A* **120**, 113.

Maeder, A. 1992 *A&A* **264**, 105.

Maeder, A. 1997 *A&A* **321**, 134 (paper II).

Maeder, A. 1998 *A&A*, in press.

Maeder, A. & Conti, P. S. 1994 *Ann. Rev. Astron. Astrophys.* **32**, 227.

Maeder, A. & Meynet, G. 1994 *A&A* **287**, 803.

Maeder, A. & Meynet, G. 1996 *A&A* **313**, 140.

Maeder, A. & Zahn, J. P. 1998 *A&A* **334**, 1000 (paper III).

Mestel, L. 1953 *MNRAS* **113**, 716.

Meynet, G., Maeder, A., Schaller, G., Schaerer, D., Charbonnel, C. 1994 *A&AS* **103**, 97.

Meynet, G. & Maeder, A. 1997 *A&A* **321**, 465 (paper I).

Meynet, G., Mermilliod, J.-C., & Maeder, A. 1993 *A&AS* **98**, 477.

Norberg, P., Maeder, A., & Meynet, G. 1999 *A&A*, in prep.

Owocki, S. P., Cranmer, S. R., & Gayley, K. G. 1996 *ApJ* **472**, L115.

Owocki, S. P., Gayley, K. G., & Cranmer, S. R. 1998. In *Boulder-Munich II: Properties of Hot, Luminous Stars*. ASP Conf. Ser. 131. p. 237. ASP.

Palla, F. & Stahler, S. W. 1993 *ApJ* **418**, 414.

Panagia, N., Scuderi, S., Gilmozzi, R., Challis, P. M., Garnavich, P. M., & Kirshner, R. P. 1996 *ApJ* **459**, L17.

Pauldrach, A., Puls, J., & Kudritzki, R. P. 1986 *A&A* **164**, 86.

Puls, J., Kudritzki, R. P., Herrero, A. et al. 1996 *A&A* **305**, 171.

Schaerer, D. 1996 *ApJ* **467**, L17.

Schaller, G., Schaerer, D., Meynet, G., & Maeder, A. 1992 *A&AS* **96**, 269.

The, P. S., de Winter, D., Feinstein, A., & Westerlund, B. E. 1990 *A&AS* **82**, 319.

Vacca, W. & Conti, P. S. 1992 *ApJ* **401**, 543.

VENN, K. A. 1995 *ApJ* **449**, 839.

VENN, K. A., LAMBERT, D. L., & LEMKE, M. 1996 *A&A* **307**, 849.

VENN, K. A., McCARTHY, J. K., LENNON, D. J., & KUDRITZKI, R. P. 1998. In *ASP Conf. Ser. 131*. p. 177. ASP.

WALBORN, N. 1988. In *Atmospheric Diagnostics of Stellar Evolution* (Ed. K. Nomoto). IAU Coll. 108. p. 70. Springer-Verlag.

WOLFIRE, M. G. & CASSINELLI, J. P. 1987 *ApJ* **319**, 850.

VON ZEIPEL, H. 1924 *MNRAS* **84**, 665.

ZAHN, J. P. 1992 *A&A* **265**, 115.

Why stars evolve to giant radii

By **PETER P. EGGLETON**

Institute of Astronomy, Madingley Rd, Cambridge CB3 0HA, UK

I would certainly not have put forward this topic in a conference on 'unsolved problems in astrophysics,' were it not that our host very kindly insisted on giving it to me. For me, the unsolved problem in this area is why people still publish papers on it. And even there I can give a partial answer: *Astrophysical Journal* sends the papers to me to referee, and then ignores my advice. But in fairness to *Astrophysical Journal* I should say that both *Astronomy & Astrophysics* and *Monthly Notices* have also ignored my advice.

As a student I was lectured to by Prof. Tayler and by Prof. Hoyle, who both worked independently on stellar modelling in the days when 'computer' meant a person who computes, using tables or a machine whose handle had to be turned. Both told their classes that evolution towards giants is due to

(a) the molecular-weight gradient

(b) the concentration of energy production into a shell away from the centre.

There have been several occasions subsequently when I have questioned this, at least to myself. However in 1991 Robert Cannon, then a Ph.D. student, and I came up with a theorem which actually proves it (we were thinking about something else at the time). The theorem does not say everything that one would hope to know: it is a necessary condition for a giant, but not a sufficient condition. In other words, if a computed model is giant-like (in the sense of being highly centrally condensed), the theorem allows one to identify the region of the star that is responsible; but it does not, unfortunately, say that a star with such-and-such physics in it must, or must not, become a red giant.

Although it has been known in general terms for about 50 years why stars change their structure from dwarf-like to giant-like as they evolve, it is still common today—in fact, more common than it used to be—for incorrect answers to be given. This is not to say that there exists a totally correct answer, but even in the absence of a totally correct answer it is quite possible to say that another answer is wrong. Usually, this is because one can find a counter-example.

The following potential explanations have all been offered at least in informal discussion:

(c) the gravitational energy released by core contraction causes the envelope to expand

(d) it is because the radiative part of the envelope is polytropic with $n \sim 3$

(e) it is because the envelope has to lie on the Hayashi track

(f) it comes from the virial theorem

(g) it is due to a thermal instability in the envelope.

None of these is correct. Even the correct (a) and (b) are not *sufficient* conditions, since there has to be 'enough' in some sense of one or the other, or both. I will not deal at any length with (e) to (g), although I have heard them being bandied around in discussion, and even seen some publications claiming to support them. Answers (c) and (d) need more substantial refutation.

I believe that (c), which is based on a misreading of Sandage & Schwarzschild (1952), is quite the most prevalent mistake. Indeed, it seems to be dogma. Sandage & Schwarzschild (1952) certainly showed that as the core contracted, the envelope expanded, but they did *not* claim cause and effect. Their calculation did not in fact include the absorption of thermal/gravitational energy in the envelope, only the release of it in the core. They argued entirely reasonably that as it was only a small fraction ($\lesssim 4\%$) of the nuclear luminosity from the shell, it would make little difference to the envelope. For low-mass stars like the Sun, the transition from MS to RG is not even on a thermal timescale. The models evolve virtually in thermal equilibrium throughout. But even if the contraction of the core contributed substantially to the star's luminosity, it is not possible for the envelope to 'know' which bit of energy comes from gravitational contraction and which from the much more copious source of nuclear burning. Thus although energy does indeed flow out of the core, and energy is indeed absorbed in the envelope—but not in the computations of Sandage & Schwarzschild (1952), and not in any event necessarily in equal amounts—this is a *consequence* of giantward evolution, and not at all the *cause* of it.

Answer (d) has been raised by two or three people or groups in the last 15 years. A reasonable counter-example to this is that one can construct a very simple *analytic* giant-like model (Eggleton, Faulkner & Cannon 1998) that has no region anywhere near $n \sim 3$. It consists of an $n = 5$ core, an $n = 1$ envelope, an interface between them with a molecular-weight jump by a factor of 3, and a core mass/total mass ratio of $2/\pi - \delta$. As $\delta \to 0$ the ratio envelope radius/core radius $\to \infty$. A more detailed refutation is given towards the end of this article; the essence is that two nearby singularities in the U, V plane (e.g. Schwarzschild 1958) are being confused.

An interesting point to bear in mind is the evolution of *helium* stars (Paczyński 1971). For some masses they evolve into red giants, but for either low masses or high masses they do not. At all masses evolved He stars show all of (a), (b) and (d); how do they *avoid* becoming giants? Helium stars also show half of (c)—the contracting core—but only some of them show the other half—the expanding envelope.

We need, of course, to be reasonably clear about what we mean by a red giant (RG). We cannot just define an RG as a star which is cool but luminous, because there exists a class of cool luminous stars, pre-main-sequence Hayashi-track stars, which are homogeneous in composition. Their largeness is interesting, but it is not a consequence of their having evolved by nuclear burning. The chief characteristic, which I shall take as the defining characteristic, of an *evolved* RG is that the central condensation C is large, where

$$C \equiv \text{central density / mean density}. \tag{1}$$

As Sandage & Schwarzschild (1952) pointed out, evolved RGs must be centrally condensed, because their cores must be at a temperature to support hydrogen burning, and yet the virial theorem applied to a reasonably uniform large star would give a low temperature (such as one finds in pre-MS red giants). On the MS (and also in pre-MS stars, including contracting homogeneous red giants) $C \sim 10\text{--}10^4$, whereas in evolved RGs $C \sim 10^6\text{--}10^{18}$. Why does the burning of hydrogen to helium have such a drastic effect on C?

A rather definite answer emerges from the following very general theorem (Eggleton & Cannon 1991):

Let $n(r)$, the local polytropic index, and also a quantity $s(r)$, a 'softness index' directly related to $n(r)$, be defined by

$$\frac{n}{n+1} \equiv s \equiv \frac{d\log\rho/dr}{d\log p/dr} \quad . \tag{2}$$

Further, let s_{max} be the maximum value of $s(r)$ throughout the star.

Then the theorem states that if $s_{\mathrm{max}} < 5/6$ ($n_{\mathrm{max}} < 5$), the star *cannot* be more centrally condensed than a polytrope of index n_{max}. Using the radii of numerical models of polytropes, which are easily constructed, or which can be obtained from Chandrasekhar (1939), we obtain the result

$$C \leq C_{\mathrm{poly}}(n_{\max}) \approx \frac{5.4(n_{\max}+1)^3}{(5-n_{\max})^3} \quad , \quad 3 \lesssim n_{\max} < 5 \quad . \tag{3}$$

This result has a very strong consequence. If we have a red giant with $C = 10^9$, say, then $n_{\max}$ *cannot* be less than 4.99: there must be some point in the star where $n(r)$ takes at least this value. Although the theorem does not say this, it seems very likely that since n is normally substantially less than 4.99 throughout the bulk of an RG (e.g. $n = 1.5$ in the degenerate part of the core, and in the convective part of the envelope), n should probably exceed 4.99 by quite a wide margin in some region. Hence if we can explain this relatively large value in terms of the basic physics, we have in effect explained why the star is a giant.

Dr. Iben, in his otherwise excellent summing-up (this conference), dismissed this discussion briefly by saying 'giants have $n = 1.5$ cores and $n = 1.5$ envelopes,' and then shrugging, as if to imply that they had no other region of interest that could influence their structure. But the point is that they *must* have a region with $n \gtrsim 5$, and indeed they do, as I am sure he would find if he plotted n (or better s). This can be seen in Fig. 1, which I come to shortly. The region need not be large, and is indeed usually rather small, but it is critical. With no such region, a star cannot be an evolved red giant, i.e. a star with a high degree of central condensation. But I emphasise that this is *necessary*, not *sufficient*, and many models with an $n \sim 5$ region are not red giants.

There are people who feel that polytropes are not adequate approximations to real stars, and that therefore any reference to polytropes can only at best be an approximation. It should not be necessary to point out, but I suspect that it is, that *every* stellar model that has ever been computed is locally polytropic, that is to say a value for $n(r)$ can be calculated from Equn (2) at each value of r—whether by differencing in tabular output or analytically is immaterial. In our theorem, actual polytropes are only used as a comparison: an actual star has a value of C that is compared with C for a polytrope.

I shall use the word 'soft' to describe a situation where the density changes rapidly with pressure, and 'hard' to mean the opposite. An incompressible fluid is extremely hard, an isothermal gas is moderately soft (but it is possible to have situations which are even softer than isothermal gas: see below). Note, however that softness or hardness are not necessarily statements about the intrinsic nature of the material. In a star n is usually determined by the energy transport process, being ~ 3 or 3.25 in regions that are stable to convection, and ~ 1.5 in unstable regions.

In hydrostatic equilibrium, a circumstance that is assumed in virtually all stellar modelling, $p(r)$ must decrease monotonically outwards and so the denominator of the right-hand member of Equn (2) does not go through zero, or even reach zero. This does not mean that n or s are always finite, because the *numerator* can be infinite if there is a jump discontinuity in the molecular weight. The bilinear nature of the relation between n and s means that n can behave in an apparently rather perverse way, passing from positive to negative values through the value $\pm\infty$ which corresponds to $s = 1$. On the other hand s behaves in a relatively simple way, perhaps passing through zero and perhaps also attaining $\pm\infty$ at jump discontinuities ($-\infty$ if μ jumps downwards going inwards). But only in very contrived situations does s, as distinct from n, pass 'through' $\pm\infty$ from positive to negative values or vice versa. A consequence of the bilinearity of the $s(n)$ relation is that $s > 5/6$ is not precisely the same as $n > 5$, unless one excludes the possibility of $s > 1$. To be almost pedantically careful, one should make statements about s,

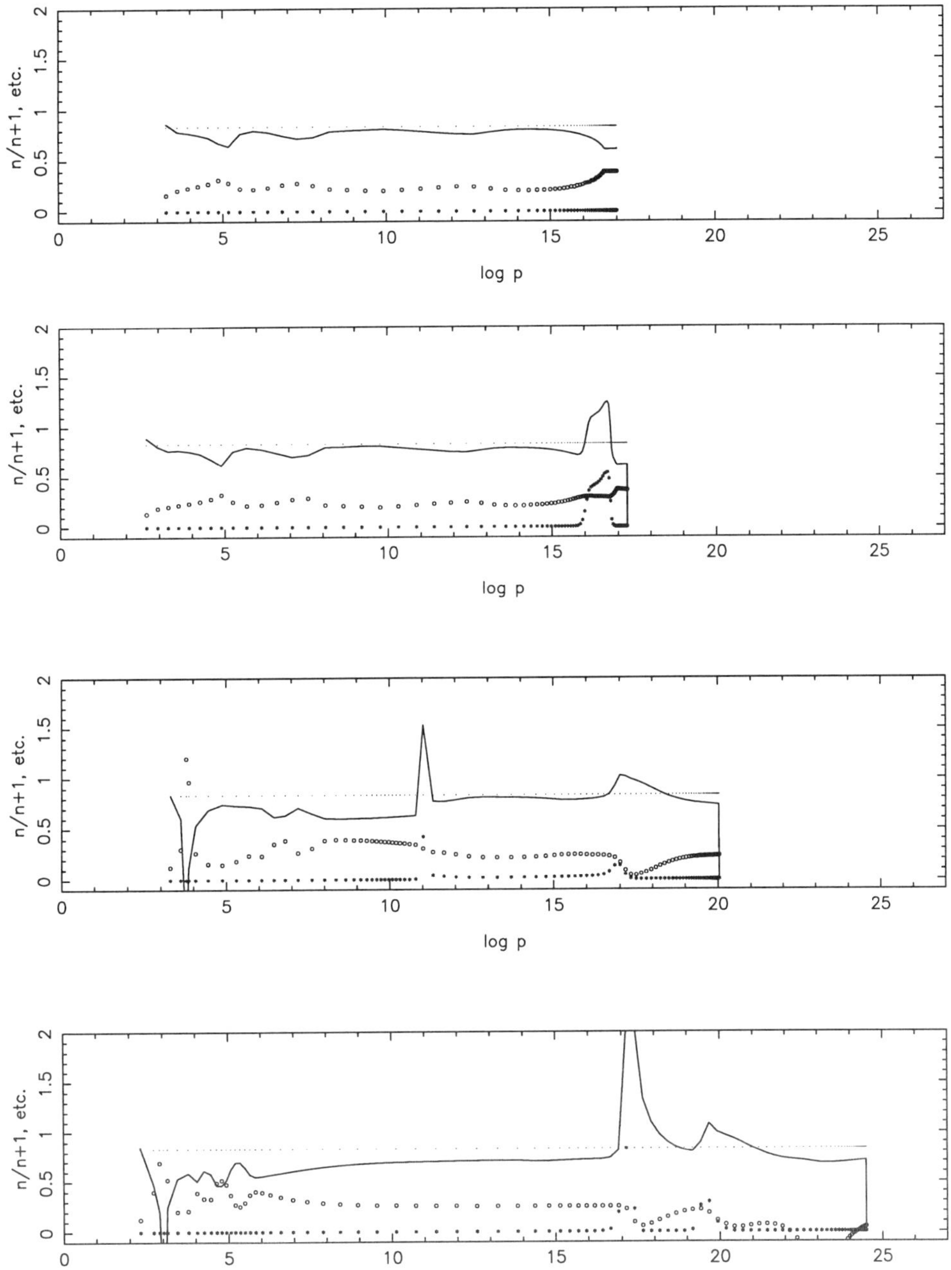

FIGURE 1. The softness parameter $s \equiv n/(n+1)$ (continuous line) through a $4\,M_\odot$ star as a function of $\log_{10}(\text{pressure})$, for four different evolutionary stages: (a) ZAMS ($C = 57$), (b) near TMS ($C = 800$), (c) just before He ignition ($C = 2 \times 10^6$), (d) AGB, core masses (He, and C/O) $\sim 0.94\,M_\odot$ ($C = 10^{15}$). Also shown are (a) the critical softness (5/6: dots), (b) the molecular weight gradient $n_{\max}$ (asterisks), and (c) the temperature gradient ∇_T (circles). Since the mesh is rather coarse (100 meshpoints), and since some gradients were computed directly, and others by differencing on this rather coarse mesh, the gradients do not necessarily satisfy Equn (4) *exactly*. In particular, some spikes in s, and particularly in $n_{\max}$, must be much narrower, and perhaps somewhat higher, than shown (not necessarily by the same amount in s and $n_{\max}$). Apparently erratic fluctuations near the left are due to ionisation zones and the dissociation of molecular hydrogen.

but I shall follow the more usually route of making statements about n. Equalities are preserved, but inequalities have to be looked at cautiously.

By any reasonable interpretation of the word 'explain', it must be the soft region or regions required by the theorem, and no other region, which explains why the star has such a large value of C. The theorem *demands* such a soft region in any star which is highly centrally condensed (but if there are two or three such regions it does not of course identify which is most important). The theorem proves necessity, but not sufficiency. A star with a soft region need not be a giant, but a star without a soft region *cannot* be a giant. Thus if an evolved star is a giant, we can explain this by first identifying the soft region, and then explaining the softness.

In all models I compute which are giants (say, $C \gtrsim 10^6$), I find a soft region—see Fig. 1—as the theorem requires, and the softness is very clearly due primarily to (a) and (b), as I show shortly. In a few models, however, one can see an additional region with n quite close to 5, and even slightly greater, in the upper layers of the radiative envelope close to the base of the surface convection zone. This is due to the nature of the opacity law: Kramers' law ($\kappa \propto \rho/T^{3.5}$) predicts $n \sim 3.25$, but a law like $\kappa \propto \rho^{0.5}/T^{4.5}$ would predict $n \sim 5$. This may be roughly how the opacity behaves at some intermediate temperatures and densities. One might certainly suppose that it contributes to giantness, but since it is absent in some giants whereas the major softness due to (a) and (b) is always present, I do not think it can be seen as the dominant contribution. And of course with the usual Milne-Eddington approximation to the boundary condition at optical depth $2/3$ we get $n = 7$ at the photosphere, but dropping very rapidly inwards. I am sure that there is sufficiently little mass and radius in the photosphere that its softness makes little difference.

Fig. 1 shows how $n(r)$, or rather the 'softness index' $s(r) \equiv n/(n+1)$, varies through a $4\,M_\odot$ star at four stages in its evolution. The stages are (a) the ZAMS (b) near the TMS (c) near the tip of the FGB (i.e., near He ignition), and (d) near the tip of the AGB (where the core mass approaches $\sim M_\odot$: no mass loss was built into these models). In the two stages where C is large, we see a marked local maximum of s with s substantially in excess of unity, i.e., a soft region. On the ZAMS where C is not so large, s is well below the critical value $5/6$ except for a tiny region at the photosphere. Fig 1b illustrates the fact that the condition, though necessary, is not sufficient. There is a substantial soft region, and yet the star is not really giant-like. I suspect that to be a giant a star not only has to have a substantial soft region, but also that this region must be not too near the centre, and not too near the surface, in some sense.

Fig. 1 shows two other homology variables, $\nabla_T \equiv d\log T/d\log p$ and $\nabla_\mu \equiv d\log \mu/d\log p$. Note that for a simple perfect gas with $\rho \propto p\mu/T$ we have

$$s \equiv \frac{n}{n+1} = 1 + \nabla_\mu - \nabla_T \qquad \text{(perfect gas } \textit{only}\text{)}. \tag{4}$$

It is easy to see that spikes of ∇_μ, and/or troughs of ∇_T, make for an increase of softness (although when degeneracy sets in the core, ∇_T ceases to influence s, and s becomes $3/5$). Obviously spikes of ∇_μ are caused by molecular weight gradients; slightly less obviously, troughs of ∇_T are caused by 'strong' burning shells. By 'strong' I mean that L is much lower just beneath the shell than it is just above. Normally ∇_T above a shell is dictated by the opacity law there, and has a value $1/4.25$ for Kramers' opacity, say. If the luminosity drops instantaneously by 90%, going inwards through the shell, then obviously ∇_T would drop by the same amount; if the luminosity dropped by only a few per cent (a 'weak' shell) then so would ∇_T. By Equn (4) a drop in ∇_T causes a rise in s. In both cases, ∇_T picks up and returns towards either its opacity-driven value approaching the next shell, or else towards the convective value ($\nabla_T = 2/5, n = 3/2$) approaching a central burning core. Thus a strong burning shell causes a soft region *below* the burning shell, while a burning core causes no soft region. In Figs. 1c and 1d one can distinguish easily between the contributions of μ-gradient and shell-burning, i.e., of ∇_μ and ∇_T, to the soft region or regions. One cannot say, at least on the basis of the theorem, that one is more important than the other.

Figs. 1c and 1d show a moderately complicated distribution of softness s over the star. One can wonder whether all of this complexity is necessary. But we can show that it is not. The very simple model (Eggleton et al. 1998) referred to earlier consists of an $n = 5$ ($s = 5/6$) core, a molecular weight jump by a factor of 3 (a δ-function in s), and an $n = 1(s = 0.5)$ envelope. The fact that this model shows an infinite degree of central condensation at a specific core mass-fraction that is well away from either zero or unity shows that it is not necessary for 'giantness' that the envelope have a complicated n-structure. A soft region somewhere is necessary, as the theorem says, and in the analytic example it is the molecular weight jump that provides this (by way of a δ-function spike), since the $n = 5$ core is exactly marginal. But the outer envelope has constant softness, and at $n = 1$ is a great deal harder than $n \sim 3$.

In Fig. 1 some minor apparent discrepancies can be noted, which are due to the following circumstances:

(a) the evolution code used to generate (for other purposes) the models from which those displayed were selected produces $s, X, Y, p, T, \ldots$ at 200 meshpoints, every 300 timesteps, and prints the results only for every *second* meshpoint, and only to either 3 or 4 significant figures

(b) the quantities ∇_μ, ∇_T were computed within a plotting routine which used the previous output, taking ratios of differences of the appropriate (logged) variables. I approximated μ by $1/\mu = 1.5X + 0.25Y + 0.5$, corresponding to full ionisation with all metals having two baryons per electron.

Thus the figure does not show the actual rise in μ in the surface layers, due to recombination of ions (and of molecular hydrogen in some models). Also Equn (4) is not satisfied exactly, even in regions that are close to perfect gas, because ∇_T and ∇_μ are differenced over two meshpoints whereas s was not differenced. Because ∇_μ in particular is often quite sharply peaked, mild discrepancies show up at these peaks.

What about the helium stars of Paczyński (1971)? The reason that the *low*-mass models do not become giants is that the density-temperature conditions for helium burning are normally a lot nearer to degeneracy than those of hydrogen. Thus the isothermal region below the burning shell is almost completely degenerate, and therefore $n \sim 3/2$ right away. In hydrogen-shell burning stars the shell is further from degeneracy, and so the isothermal region below the shell has $n \sim \infty$ for a few pressure scale heights. For the *high*-mass helium stars that do not become giants, the reason is partly that the molecular weight jump is not as large as in hydrogen stars, and partly that the core luminosity is larger relative to the shell luminosity than in intermediate mass helium stars. Thus the shell, though present, is not strong enough to force the transition to giant-like character.

Fig. 2 shows the behaviour of a metal-poor ($Z = 0.001$) star of $10\,M_\odot$. Dr. Renzini (after this talk) alluded to the fact that such stars tend to spend most or all of their helium-burning lives on the blue side of the Hertzsprung gap, whereas solar-metallicity stars move to the yellow side at least, if not entirely to the red edge. One would hope to see a difference particularly in Figs. 1c and 2c (shortly before, and, as it happened, shortly after helium ignition, respectively). The $4\,M_\odot$ Pop I star was a red giant ($T \sim 4300\,\mathrm{K}$) at this stage and the $10\,M_\odot$ Pop II star was quite blue ($T \sim 16000\,\mathrm{K}$). Although the spike in ∇_μ in the former, at the base of the convection zone, might be supposed to make a difference, the spike would not have developed until the star had already almost reached the giant branch. The hump further in, at and just below the burning shell in each case, though different in detail is about as prominent in each.

This is consistent with the fact that the theorem alluded to above is a *necessary* condition, and not a *sufficient* condition. A star which is highly centrally condensed must have a soft region, but a star with a soft region does not have to be highly centrally condensed (although the model in Fig. 2c is in fact quite centrally condensed, at $C \sim 1.4 \times 10^6$). It would be nice to have a sufficient condition, but I suspect it would be quite complicated. One might hope that it would depend only on, for example, the size of the hump in s, but I suspect that it depends on the shape, and perhaps even more importantly on the location. The humps in Figs. 1b and 2b, both near the terminal MS, are about as substantial as those in Figs. 1c and 2c, but are a little 'nearer' the centre (in terms of $\log p$). The hump in Fig. 2c appears to me to be also fractionally nearer the centre, in terms of $\log p$, than in Fig. 2b, but one would certainly wish for a clearer distinction.

I would not recommend anyone, particularly if they are in a temporary research position, to waste any time looking for a sufficient condition. I suspect that even if an iron-clad one were found, it could only be shown to be satisfied by computing the model, in which case one can easily see whether it is centrally condensed or not without any analysis. But a sufficient condition already tells us a great deal: when a star *is* centrally condensed, one can hope to identify the cause rather clearly, and to rule out other hypothetical causes as clearly. For realistic stellar models it points unambiguously to the μ-gradient and to the nearly-isothermal non-degenerate zone below the burning shell. However, if one enlarges the investigation to include 'quasi-stars', stars in which the physics is artificially altered (for example, a different opacity law), then other processes can easily come in to play and cause giantness even without (a) or (b).

The reason, I think, why several people have come mistakenly to the conclusion that the $n \sim 3$ radiative part of the envelope is necessary is that they are confusing two different singularities that occur in the classic U, V plane (Chandrasekhar 1939, Schwarzschild 1958). In the U, V plane there is always one singularity on the V-axis, and if $n > 3$ there is another in the positive quadrant (there is a third, on the U axis, and a fourth at the origin, which are not relevant to this discussion). The two relevant ones coincide at exactly $n = 3$, and the nature of the solution becomes rather complicated in that case. For the off-axis singular point there is a singular solution

$$\rho^{\frac{1}{n}} = \theta = \left[\frac{2(n-3)}{(n-1)^2 r^2} \right]^{\frac{1}{n-1}} \quad , \quad n > 3 \quad . \tag{5}$$

For the on-axis singular point there is no singular solution, but solution curves emerging from there have

$$\theta \sim \frac{\text{const.}}{r} \quad , \tag{6}$$

which corresponds to the case that there is a finite point mass at the origin. Either singular point therefore gives $\theta \sim 1/r, \rho \sim 1/r^3$ for $n \sim 3$ (I use a twiddles sign here to mean 'is approximately proportional to,' which includes the possibility 'is exactly proportional to'). It is helpful, but not necessary, to have $\rho \sim 1/r^3$ in order to be able to construct a large radially-extended envelope containing rather little mass. However, *any* surface solution which starts (at its base) with $U \ll 1$ ($U \equiv d\log m / d\log r$) and a modest V ($V \lesssim 1$, where $V \equiv -\{n+1\}d\log\theta / d\log r$) will obviously have the giant-like property that r increases by a large amount while m increases by a small amount, without θ decreasing by a large amount. Many solutions which emerge from the on-axis singularity (those which head downwards initially) have this property. This is why for

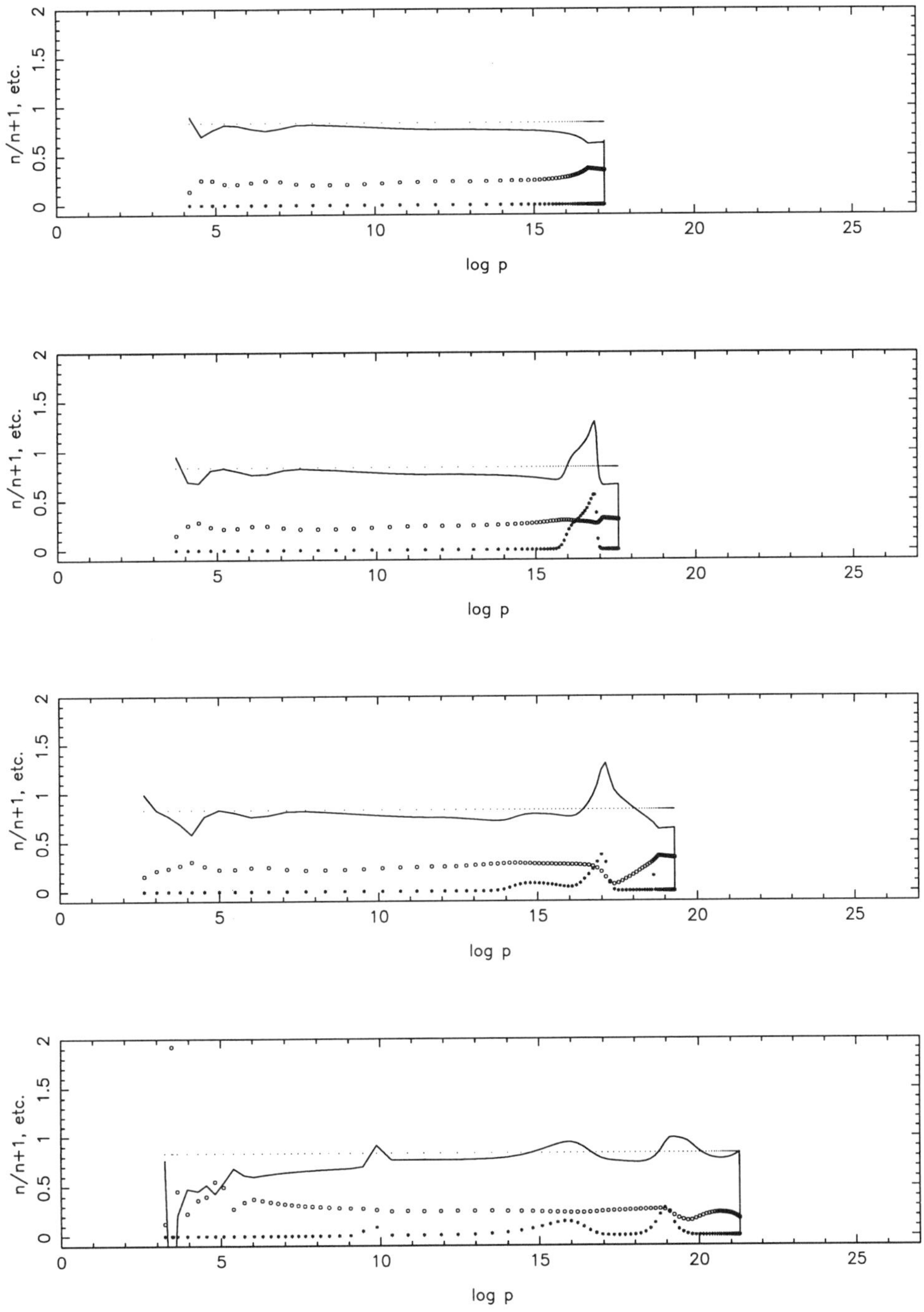

FIGURE 2. As for Fig. 1, but for a $10\,M_\odot$ star of low metallicity ($Z = 0.001$). Such a star ignites He, and burns it, while still relatively blue. The four panels correspond to (a) ZAMS ($C = 25$), (b) near TMS ($C = 10^3$), (c) just after He ignition ($C = 1.4 \times 10^6$), (d) AGB, core (He) $3.1\,M_\odot$ ($C = 3.2 \times 10^{11}$).

example the analytic model cited above with the $n = 1$ envelope is giant-like. Such behaviour is not confined to solutions which pass near the off-axis singularity, although many will pass near anyway since it is nearby. It is clear that the fact that the off-axis singularity is close to the on-axis singularity when $n \sim 3$ is a source of confusion here.

REFERENCES

CHANDRASEKHAR, S. 1939 *An Introduction to the Study of Stellar Structure*. Univ. Chicago Press.

EGGLETON, P. P. & CANNON, R. C. 1991 *ApJ*, **383**, 757.

EGGLETON, P. P., FAULKNER, J. & CANNON, R. C. 1998 *MNRAS*, in press.

SANDAGE, A. R. & SCHWARZSCHILD, M. 1952 *ApJ*, **116**, 463.

SCHWARZSCHILD, M. 1958 *The Structure and Evolution of the Stars*. Princeton Univ. Press.

PACZYŃSKI, B. 1971 *Acta Astr.*, **21**, 1.

Abundance anomalies in stars:
A 30-minute tour of the HR Diagram

By C. A. PILACHOWSKI

National Optical Astronomy Observatories,† PO Box 26732, Tucson, AZ 85726-6732;
email: catyp@noao.edu

1. Introduction

The HR diagram is home to an astounding diversity of odd stars. To cover all these
anomalous stars in a modest review paper is, of course, impossible. And if I tried to do
justice to all the anomalous stars in the heavens, I'd probably leave you in confusion about
the differences among the beta Cephei stars, the BY Draconis stars, the SX Phoenecis
stars, the RV Tauri stars, and the RS CVn (or AM CVn!) stars. To begin, we need a
definition of what an abundance anomaly is. I suggest the following: a stellar abundance
anomaly is a difference in the stellar surface composition from the general galactic composition at the time the star formed. Given the relative simplicity of the general picture
of stellar evolution, the extraordinary variety of real stars is truly amazing.

I will approach the subject of abundance anomalies in stars not as a visit to the stellar
zoo, but rather by reviewing how the physical processes described in many of the other
lectures of this Symposium are manifest in the surface abundances of stars. Mass loss,
element segregation, mixing, mass transfer, and other processes can all have subtle to profound effects on stellar surface abundances. Understanding the patterns these processes
impose on stellar surface abundances can help us identify the cause of peculiar abundances, and the observed abundances, in turn, constrain the physical processes. Studying
abundance anomalies in stars provides a laboratory for modeling physical processes, and
provides clues for us to follow the step-by-step evolution of stars from one stage to the
next. Abundance anomalies also illuminate the complexities of stellar populations and
galactic chemical enrichment. I suspect, however, that the underlying motivation for the
study of abundance anomalies by stellar astronomers is primarily both the challenge of a
puzzle to be solved and the confrontation with standard theory—the unsolved problems
of stellar evolution.

2. Mass loss

Nature provides many ways for a star to lose its envelope to expose material processed
by nuclear reactions to our view. Radiatively driven winds from massive stars, "su-
perwinds" in stars on the AGB, and envelope ejection through mass transfer, common
envelope evolution, or stellar collisions can all strip off the outer layers of a star to show
us what's inside. What we find there is often more complex than simple mass loss would
suggest.

2.1. *Evolved massive stars*

Radiatively driven massive winds, discussed by Maeder and by Kudritzki in this volume,
remove the surface layers of massive stars so quickly that stellar evolution codes must

† Operated by the Association of Universities for Research in Astronomy, Inc. (AURA) under
cooperative agreement with the National Science Foundation

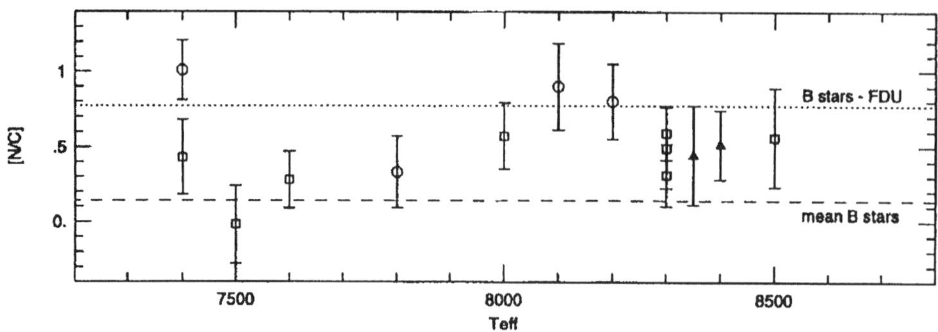

FIGURE 1. [C/N] ratios in A supergiants as a function of effective temperature. The dashed line is the mean [N/C] ratio for B stars and the first dredge-up prediction for a 10 $M_\odot$ star is shown by the dotted line (Venn 1995, Figure 11).

struggle to keep up. As massive stars evolve, deeper and deeper layers of the interior are exposed, producing a bounty of abundance anomalies to confound the classification spectroscopist. WC and WN stars, WO, OBN, OBC, luminous blue variables, and so forth all show anomalous abundances of carbon, nitrogen, and oxygen, indicative of the peeling off of the outer layers of the star to show us successively the products of CN-cycle hydrogen burning and other cycles of proton-capture nucleosynthesis, helium burning, and other nuclear processes. As Kudritzki describes in this volume, breathtaking progress has been made in the interpretation of the spectra of early type stars to produce accurate physical models and abundances.

Mixing, a process to which I will return in the section on red giants, also seems to be implicated in the production of abundance anomalies in massive stars. The rapid rotation seen in many massive stars suggests that meridional circulation currents in conjuction with mass loss must be considered in modeling the evolution and explaining the abundances found in massive stars (Langer and Heger 1998). Fitzpatrick & Bohannan (1993) found nitrogen enhancements and oxygen depletions in B supergiants in the Large Magellanic Cloud, suggesting that CNO-processed material has contaminated the stars' surfaces. Venn (1995) investigated the carbon, nitrogen, and nitrogen abundances in A-type supergiants, demonstrating the partially mixing of CN cycle gas (see Figure 1). Takeda & Takada-Hidai (1994) found modest sodium excesses in similar stars, and Venn et al. (1996) found severe depletions of boron. These observations are all consistent with a mixing explanation.

2.2. *Post-AGB stars*

Mass loss on the AGB, driven by thermally pulsing helium and hydrogen burning shells surrounding the stellar core, terminates the red giant evolution of low and intermediate mass stars, and strips off nearly all of the outer envelope of the star, leaving only the carbon-oxygen core, the helium shell, and (in most cases) a thin ($\sim 10^{-2}$ $M_\odot$) hydrogen layer (cf. Iben & Renzini 1983). The process of envelope ejection itself is discussed by Willson and by Bond elsewhere in this volume. For purposes here, we consider only the abundance anomalies observed in the AGB remnants.

The remnants of low and intermediate mass stars crossing the top of the HR diagram include by several classes of supergiants with abundance anomalies, including the post-AGB supergiants, the RV Tauri stars, the R CrB stars, the hydrogen deficient carbon stars, and unusual objects like FG Sagittae and Sakuri's Object, discussed by Gonzalez in this volume. Those with hydrogen atmospheres show clear signatures of CNO processing, and those which are hydrogen deficient show helium atmospheres and products of triple-α burning. Many show enhancements of s-process elements. Their anomalous compositions are primarily attributed to mass loss and nucleosynthesis during thermal pulses (and possible grain formation—see section 3.3).

The post-AGB supergiants also provide an interesting case study of abundances affected by mass loss. Described by Parthasarathy (1994), these stars are IRAS sources with spectral types ranging from K1 to B1, and they are often found quite far from the galactic plane, some with high radial velocities. Some are seen with significant apparent deficiencies of iron. The stars have recently undergone an episode of strong mass loss and may exhibit dust, H_α emission, or CO millimeter-wave emission. The abundance ratios of carbon, nitrogen, and oxygen suggest these supergiants are evolved, low-mass stars (Luck et al. 1990), but the pattern of abundances of the alpha-process and iron group metals are anomalous. Iron and other refractory metals (Ca, Mg, Si, etc.) are depleted relative to the volatile metals (e.g. C, N, O, S, Zn). Bond (1991, 1992) suggested that the surface compositions of these post-AGB supergiants have been modified by grain formation. Van Winckel et al. (1992) hypothesize that dust must have formed from the refractory elements and been lost during a mass loss episode, and the remaining gas, including the volatile elements, must have fallen back onto the star.

A similar mechanism is invoked by Venn & Lambert (1990) to explain the λ Boo stars. The λ Boo stars are rapidly-rotating, A-type main sequence stars which also show deficiencies of refractory metals. Grain formation and mass loss during a pre-main sequence phase may be responsible for the observed abundances.

The [WC] type central stars of planetary nebulae (CSPN) are probably also examples of stars with anomalous abundances due to mass loss. These central stars, comprising less than 10% (Tylenda 1996) of the CSPN population, are recognized by strong, broad emission lines indicating strong winds. These CSPN are hydrogen deficient, with helium and carbon as their primary constituents; oxygen has an abundance of 5–10% by mass. Leuenhagen and Hamann (1998) also invoke diffusion to explain the compositions of the [WC] CSPH. They note that evolution models from Herwig et al. (1997) incorporate diffuse mixing into the overshoot layer to modify the composition of the intershell region on the AGB. With these models Leuenhagen and Hamann are able to approximate the composition which later appears at the surface of the [WC] central stars.

3. Element segregation

Microscopic diffusion was first suggested by Michaud (1970) to explain the chemical anomalies observed in stars on the upper main sequence. With improved stellar models and atomic data, it is now clear that diffusion processes are active in many, if not all, stars, including the Sun (see Bahcall's paper in these proceedings for a discussion of diffusion in the solar model). Diffusion processes have been seen to produce abundance anomalies not only in stars of the upper main sequence, but in white dwarfs and horizontal branch stars as well, in fact in any region of the HR diagram where stellar surfaces are stable against mixing or mass loss. Gravitational acceleration and radiation pressure can both cause significant segregation of elements in a stable stellar atmosphere. The combination of these effects, particularly in the presence of a magnetic field, can cause highly unusual stellar surface abundances.

3.1. *Chemically peculiar stars of the Upper Main Sequence*

Slowly rotating main sequence stars in the spectral range from the early B stars through the mid-F stars are particularly susceptible to diffusion. Their upper layers are affected neither by significant convection, which sets in at mid-F spectral type, nor by massive stellar winds, which become an important factor in massive O and early B stars. Their main sequence lifetimes are long, providing ample time for diffusion to operate. Meridional circulation in rapidly rotating B, A, and F stars causes mixing which impairs the effectiveness of diffusion.

This region of the main sequence is home to several classes of peculiar stars, including the Am-Fm stars (metallic lined, non-magnetic), the magnetic peculiar A (Ap SrCrEu) and B (Bp Si) stars, and the HgMn stars, as well as the λ Boo stars (mentioned above). The chemical anomalies in these groups of stars have been summarized by Smith (1996) and are reproduced here in Figure 2.

The Am-Fm stars show the least dispersion in their elemental abundances from star to star and also show only modest anomalies relative to their peculiar cousins. Calcium and scandium are characteristically deficient, while the iron group and heavy metals are overabundant by factors up to 100. The abundance patterns of the Am-Fm stars are well described by diffusion models, particularly when coupled with modest mass loss rates (Michaud et al. 1983a, Michaud & Charland 1986, Alecian 1993).

The addition of magnetic fields dramatically complicates the element transport mechanisms in stars, introducing diffusion along field lines as well as in the radial direction. The SrCrEu Ap stars show the most extreme abundance anomalies seen on the main sequence, with enhancement factors for some metals up to 10^6. Some elements (e.g. Ti and Cr) may exhibit striking surface inhomogeneities.

The HgMn stars also show extreme enhancements of some metals, including Be, Mn, Ga, Xe, Pt, and Hg, with striking deficiencies of N, Al, and Zn. The abundances (and patterns) vary wildly from star to star, and even the relative abundances of the isotopes of Hg may be affected (cf. Wahlgren et al. 1995).

Roby and Lambert (1990) determined C, N, and O abundances in a wide sample of normal and peculiar stars in the temperature range 7000–15,000 K. Gonzalez et al. (1995) modeled the diffusion of C, N, and O within this temperature range to produce estimates of the equilibrium abundances as a function of temperature. Agreement is satisfying for carbon, while the models predict deficiencies of nitrogen and oxygen which are orders of magnitude greater than observed. Still, the models do not include hydrodynamical mixing process which compete with diffusion, and which may play a critical role.

3.2. *The lithium dip*

Observations of lithium in F dwarfs in the Hyades Cluster by Boesgaard & Tripicco (1986) revealed a surprising new phenomenon: a narrow dip in the abundance of lithium in stars near 6600 K in temperature (see Figure 3). Stars hotter than 7000 K have lithium abundances which are consistent with the interstellar medium abundance of lithium, and which presumably reflect the initial lithium abundance in Hyades Cluster stars. Stars cooler than 6000 K show a decline of the lithium abundance with temperature typical of G and K dwarfs.

Subsequent work established that the dip is not present in very young clusters, but grows increasingly deep as stars age (Balachandran 1995). Michaud (1986) proposed that the "Boesgaard Dip" could be explained by diffusion. Lithium is supported by radiative acceleration at the bottom of the convective zone in hotter stars, but not in stars cooler than 7000 K. Hence, lithium sinks below the convection zone. In stars cooler than 6400 K, Michaud argued that the convection zone is too deep for diffusion to modify the lithium abundance since the Hyades stars formed.

Talon & Charbonnel (1998) suggest that this picture is too simple for three reasons: a) the observed dip in the Hyades is wider than predicted by pure microscopic diffusion;

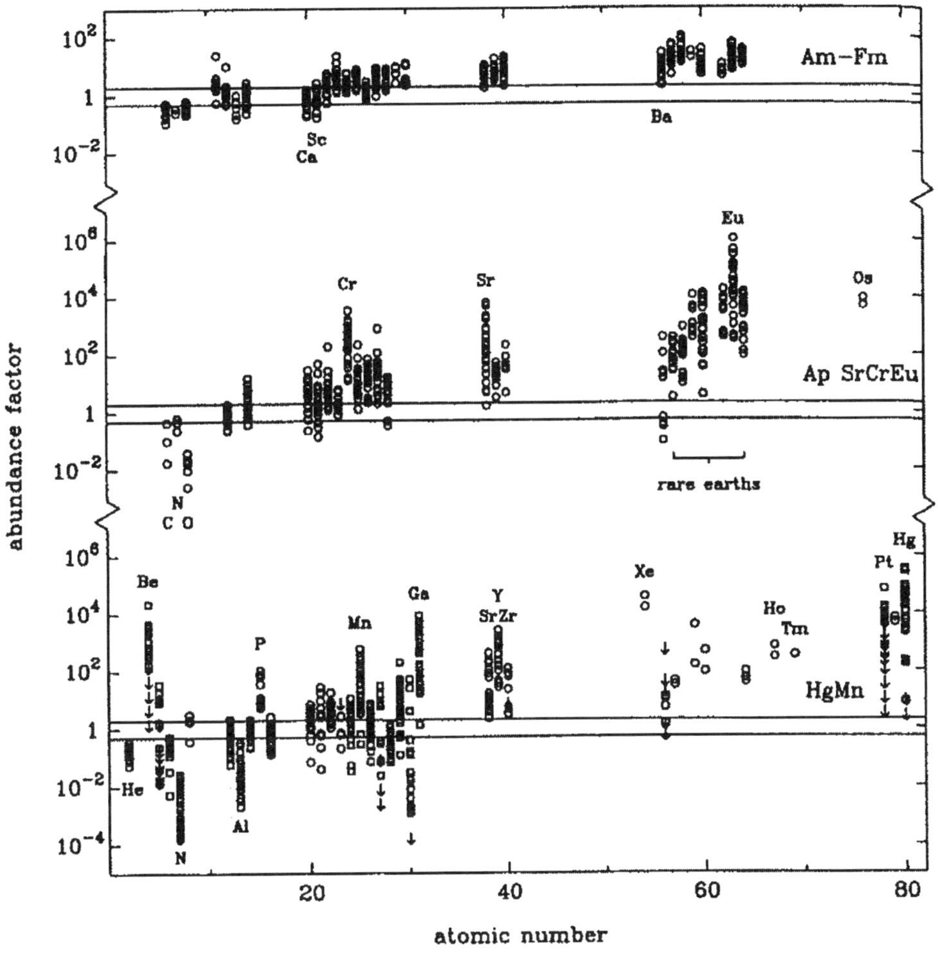

FIGURE 2. Abundances of the elements in three classes of chemically peculiar stars of the upper main sequence (Smith, 1996; Figure 5).

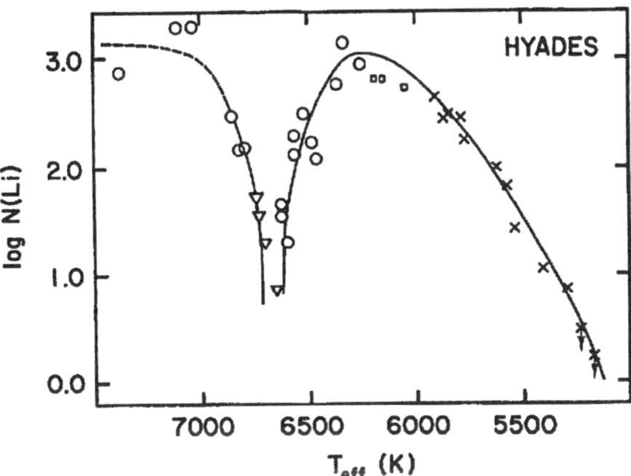

FIGURE 3. Lithium abundances in F dwarfs in the Hyades Cluster (Boesgaard & Tripicco 1986; Figure 2).

b) the diffusion hypothesis would also predict deficiencies of carbon, oxygen, and boron, which are not observed (cf. Michaud 1986, Boesgaard et al. 1998); and c) diffusion would deposit lithium in a buffer zone below the convective zone, from which it could be remixed to the surface when the star becomes a subgiant. Observations show that the lithium does not reappear in the subgiants in M67 (Balachandran 1995), a cluster whose subgiants have evolved from the region of the lithium dip. Talon & Charbonnel apply models including microscopic diffusion, meridional circulation, and shear turbulence to reproduce the hot side of the lithium dip. Lithium is destroyed rather than stored below the convective zone, and their models reproduced the observed beryllium abundance in the Hyades (Boesgaard & Budge 1989). They do not reproduce the red side of the dip, and suggest that some other mechanism of angular momentum transport becomes efficient in this regime, perhaps linked to the growth of the convective zone.

3.3. *Evolved stars of low and intermediate mass*

While diffusion processes produce intriguing abundance anomalies in main sequence stars, they also appear in stars in the late stages of stellar evolution, including blue horizontal branch stars, subdwarf O and B stars, and white dwarfs.

The sdB and extended blue horizontal branch (EHB) stars are well known to exhibit helium deficiencies (Greenstein & Sargent 1974, Heber 1987) attributed to diffusion (Greenstein et al. 1967), but a helium deficiency should not be the only symptom of diffusion at work. Glaspey et al. (1989) examined spectra of two hot horizontal branch stars in the globular cluster NGC 6752, and found an enrichment of a factor of 50 in the abundance of iron, consistent with the predictions of Michaud et al. (1983b).

Diffusion may also play a role in post-AGB stars at the low mass end of the remnant mass distribution. Models suggest that the crossing times of remnants with masses less than $0.55\ M_\odot$ are in excess of 10^5 years (Schönberner 1997). The fossil nebulae ejected at the end of the AGB phase will have long since dissipated before these stellar cores reach high enough temperatures to light up their planetary nebulae. These time scales may be sufficiently long for diffusion to modify the remnant's very thin ($\sim 10^{-2}\ M_\odot$) surface hydrogen layer, although hydrodynamic processes may act to inhibit diffusion.

Diffusion almost certainly plays a role in the surface compositions of white dwarfs as well. Dehner and Kawaler (1995) have investigated quantitatively whether diffusion can provide a credible evolutionary path from the hot PG 1159 stars (with He+C+O surface layers) to the cooler DB white dwarfs. They find that within the time that a PG 1159 star cools from 140,000 K to the DB regime starting at 30,000 K, that diffusion can produce a thin surface helium zone ($\sim 10^{-6}\ M_\odot$) consistent with pulsational models for DBs. The time scales involved are of order 10^7 years. Radiative levitation is invoked to provide opacity in the form of metal ions to explain the short-wavelength flux distributions of white dwarfs hotter thank 50,000 K (cf. Wolff et al. 1998).

Diffusion may also play a role in explaining the "DB gap" in white dwarfs (Wesemael et al. 1985). Among the white dwarfs, some 20% of stars are hydrogen deficient, containing little or no hydrogen in their surface layers. Between the DO white dwarfs, with temperatures down to 45,000 K, and the DB white dwarfs, the hottest of which has temperature of 30,000 K, is a gap. What happens to the cooling DO stars when they reach a temperature of $\sim 45,000$ K? Liebert et al. (1987) proposed that a hydrogen layer is built up by upward diffusion of trace amounts of hydrogen as DO white dwarfs cool below 45,000 K. The stars remain DA white dwarfs until, reaching a temperature of 30,000 K, the thin hydrogen layer is remixed into the massive helium convection zone and the stars become DB white dwarfs.

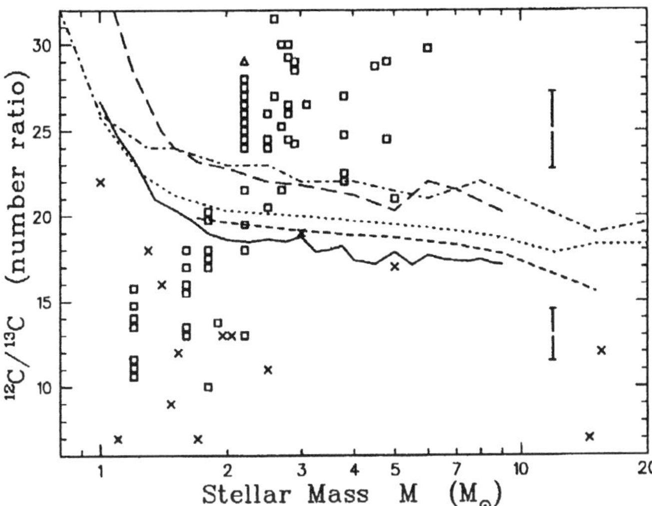

FIGURE 4. The carbon isotope ratio in giant stars as a function of stellar mass (Wasserburg et al. 1995, Figure 1).

4. Nucleosynthesis and mixing

4.1. *First ascent giants of low mass*

The evolution of low mass red giants is discussed in excellent review papers by Iben (1977 and in Wallerstein et al. 1997). As a star leaves the main sequence, a hydrogen shell begins to burn around the inert helium core. The surface convective layer deepens to become a true convective envelope, encompassing nearly half the mass of the star. The envelope reaches deepest into the star near the base of the giant branch, dredging up material processed by the CNO-cycle during early main sequence evolution. (This event is known as the "first dredge-up.") The convective envelope then retreats slightly, leaving behind a discontinuity in the mean molecular weight which is thought to inhibit any mixing between the convective envelope and layers deep enough for nuclear processing. The temperature at the base of the convective envelope itself is too cool for proton-capture nucleosynthesis.

As outlined by Iben (1964), changes in the surface abundances of the elements expected to occur at the first dredge-up include a modest drop in the carbon abundance, an increase in the nitrogen abundance, and a drop in the carbon isotope ratio from an initial value of $^{12}C/^{13}C \sim 90$ to a value between 20 and 30. By the end of the first dredge-up, the dilution of lithium is fully complete, dropping the abundance from its final main sequence value by a factor of 25–50 (Iben 1965, 1967). The resulting lithium abundance in a giant is complicated by main sequence depletion, however. In stars of spectral type later than mid-F, lithium is lost during the main sequence phase due to any of several factors, including diffusion, turbulent mixing. and possibly mass loss.

The observations of lithium and ^{13}C in low mass red giants, however, tell us that the classic first dredge-up is not the whole story. Observations of these species in low mass red giants over a range of metallicities have established that nuclear processing of their convective envelopes continues beyond the first dredge-up throughout the first ascent of the giant branch.

Carbon isotope ratios have been investigated systematically in giants over a wide range of mass in several galactic and open clusters by Gilroy (1989). She found that for masses greater than about 2 $M_\odot$, theoretical models differed only slightly from the observations.

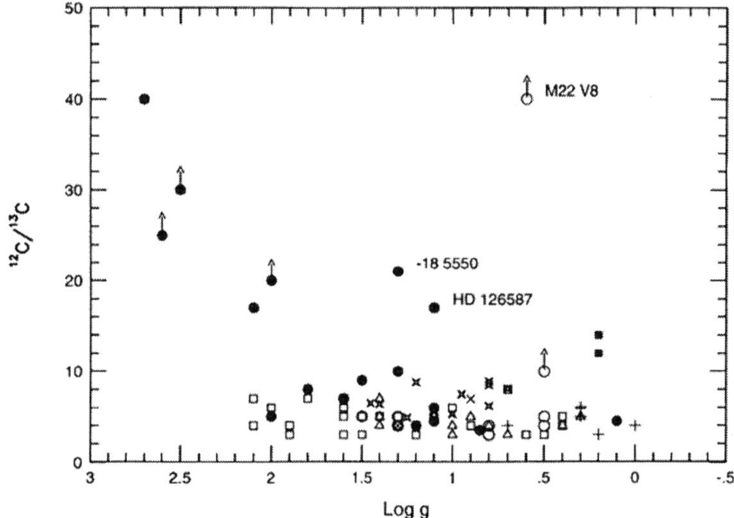

FIGURE 5. Isotope ratios in field halo and globular cluster giants (Pilachowski et al. 1997, Figure 3).

For less massive stars, the giants' carbon isotope ratios drop to values around four, near the CNO equilibrium ratio, while models predicted the stars should have much higher isotope ratios. Her data are included in Figure 4, taken from Wasserburg et al. (1995). Several theoretical models are shown and none correctly follow the decline in the isotope ratio in low mass giants.

Carbon isotope ratios have also been explored in halo field giants by Sneden et al. (1986) and by Pilachowski et al. (1997), and in old disk giants by Shetrone et al. (1993) and Charbonnel et al. (1998). All of these authors note that giants below a luminosity of $M_{Bol} \sim +0.5$ (corresponding to log $g \sim 2$) have isotope ratios above ~ 20, but that a sharp drop in the ratio occurs as giants brighten past that limit, as shown in Figure 5.

Lithium abundances in halo giants have been investigated by Pilachowski et al. (1993), who surveyed the subgiant regime to examine the convective dilution of lithium. They found that the decline in the lithium abundance due to convective dilution is generally consistent with the original predictions of Iben (1965, 1967b) as shown in Figure 6, but that additional depletion of lithium on the giant branch was necessary to account for the continued decline of the lithium abundances as stars progressed to higher luminosity on the giant branch. A similar result was found by Pilachowski (1986) for giants in the old galactic cluster NGC 7789, in which the decline in the lithium abundance can be tracked clearly to the tip of the giant branch.

Charbonnel (1995) and Wasserburg et al. (1995) have both investigated models which account for these carbon isotope and lithium abundance anomalies in low mass giants. Wasserburg et al. invoked deep mixing currents to transport material from the convective envelope to close to the hydrogen burning shell. Charbonnel found that rotation-induced mixing could account consistently for both the lithium and the carbon isotope changes seen in metal-poor giants, as seen in Figure 7.

4.2. *Abundance anomalies in globular cluster giants*

The cause of abundance variations among stars in globular clusters has been a major question in the field for two decades since Cohen (1978) first noted differences in sodium

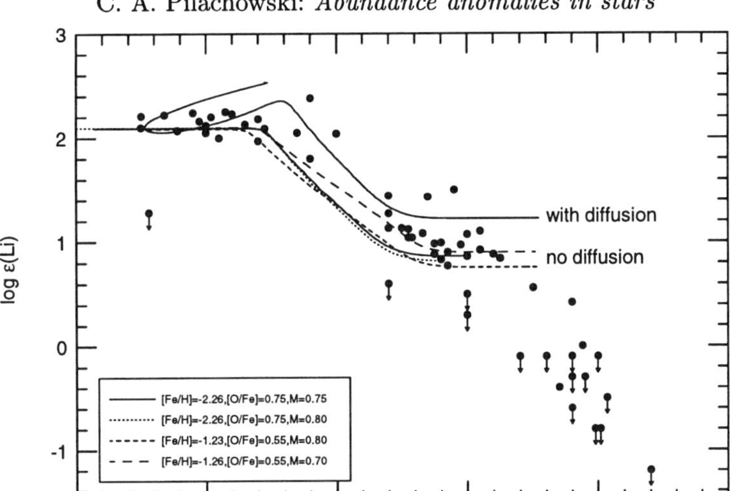

FIGURE 6. Lithium abundances in metal-poor halo subgiants and giant stars (Pilachowski et al. 1993, Figure 8).

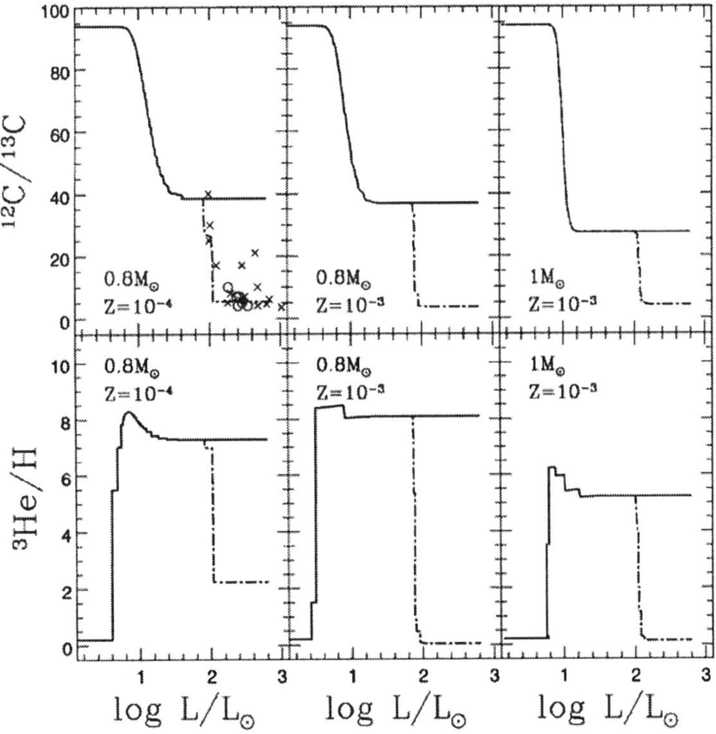

FIGURE 7. Rotation induced mixing can explain the decline in the lithium abundance seen in halo red giants (Charbonnel 1995, Figure 1).

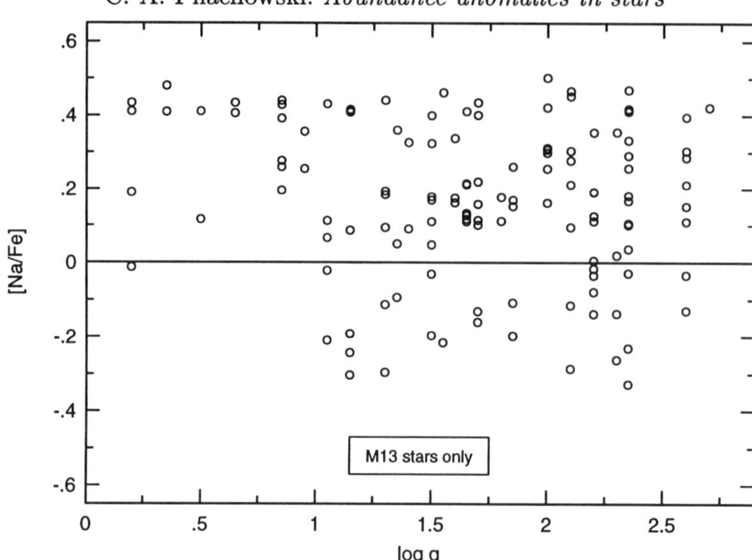

FIGURE 8. The abundance of sodium as a function of temperature in M13 giants (Pilachowski et al. 1996, Figure 7).

abundances in giants in M13. Smith (1987) comprehensively summarized what was known at the time about abundance variations in cluster giants, including the iron-peak and alpha-process metals, sodium, aluminum, the strengths of the CN and CH molecular bands, and correlations among these quantities. Two hypotheses have dominated the discussion. One point of view is that the abundance differences were caused by material from intermediate mass stars in the cluster which evolved and lost mass during the period in which the low mass stars were still forming (Cottrell & Da Costa 1981). The second hypothesis is that the abundance differences are caused by nucleosynthesis and mixing in the stars themselves. Most investigators now agree that a case can be made for invoking both mechanisms to account for all the abundance differences observed in individual clusters.

Analysis of many giants in M13 has shown that proton-capture nucleosynthesis including the hydrogen-burning chains of the CN, ON, NeNa, and MgAl cycles is occurring in stars evolving up the giant branch (Pilachowski et al. 1996, Kraft et al. 1997). These data (see Figure 8) established a clear dependence of the sodium abundance on luminosity: stars near the tip of the giant branch have higher average Na than less luminous giants. The Na abundance also anti-correlates tightly with the O abundance. The Na and Mg abundances demonstrate a partial anti-correlation (i.e. Na-weak stars are always Mg-rich, and some Na-strong stars are Mg-weak). Mg and Al are anti-correlated, and the sum Mg+Al is a constant along the giant branch.

Theoretical calculations by Denisenkov & Denisenkova (1990), Langer & Hoffman (1997), and Cavallo et al. (1996, 1998) provide the mechanisms for proton capture nucleosynthesis on the giant branch. Still, the detailed mechanism of mixing and the phase at which the nucleosynthesis happens remain unclear.

Is M13 unique, or are similar processes at work in other globular clusters? Clearly, M13 shows much more variation from star to star then other clusters of similar metallicity such as M3 (Kraft 1994). The contrasting morphology of the horizontal branches in M3 and M13 is generally attributed to a difference in age. But stars which have undergone extreme deep mixing and nucleosynthesis have higher helium content, driving them to

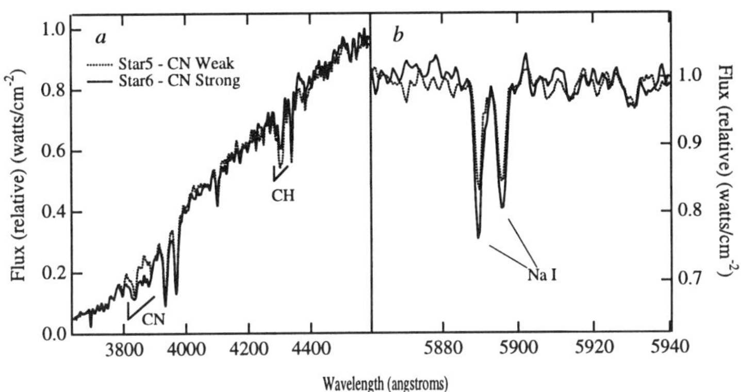

FIGURE 9. Spectra of two stars near the turnoff of the globular cluster 47 Tuc. Note the correlation of the Na D lines with the strength of the CN band at 3880 Å (Briley et al. 1996, Figure 1).

a blue locus on the horizontal branch (Langer & Hoffman 1995). The difference in horizontal branch morphology between M3 and M13 may result, in part, from a difference in the degree of deep mixing. Understanding quantitatively how these two clusters differ in the degree of deep mixing and nucleosynthesis is important for understanding the general question of the ages of the globular clusters systems.

Abundance anomalies in globular clusters aren't restricted only to giant stars, however. In Figure 9, Briley et al. (1996) demonstrate differences in the spectra of two stars near the turnoff of the globular cluster 47 Tucanae. In addition to differences in the strength of the CN and CH molecular bands, the strengths of the Na D lines also very, in correlation with the strength of the CN band. Such differences show that we are still far from understanding abundance anomalies in globular clusters.

4.3. *Peculiar red giants*

In discussions on the origins of abundance anomalies in stars, the combined processes of nucleosynthesis and mixing are what we often think of first. The peculiar red giants provide very lively examples of these phenomena at work. Perhaps the most compelling evidence for nucleosynthesis and mixing in red giants is the discovery of the radioactive element technicium in the atmospheres of some. While the element was discovered in red giant spectra by Merrill (1952, 1955), Peery provided a good illustration in his 1971 article, shown in figure 10. All isotopes of technicium are radioactive, with half-lives short compared to the evolutionary lifetime of a peculiar red giant. Its presence in a star demands its recent creation.

Since Merrill's initial discovery, several investigators have surveyed the various subclasses of peculiar red giants—types M, MS, S, SC, C, to determine in which types of stars it is and isn't found. The presence of technicium is strongly correlated with light variability, with non-variable stars rarely, if at all, showing technicium (Little et al. 1987). Peculiar red giants of spectral type S† without technicium are warmer and intrinsically fainter than those with technicium (Van Eck et al. 1998), based on Hipparcos trigonometric parallaxes. Smith & Lambert (1988) concluded that MS and S stars without technicium are the coolest members of the class of barium stars, to which we will return in a later section.

---

† S stars are luminous red giants whose spectra are dominated by oxides of zirconium and other s-process elements, rather than TiO.

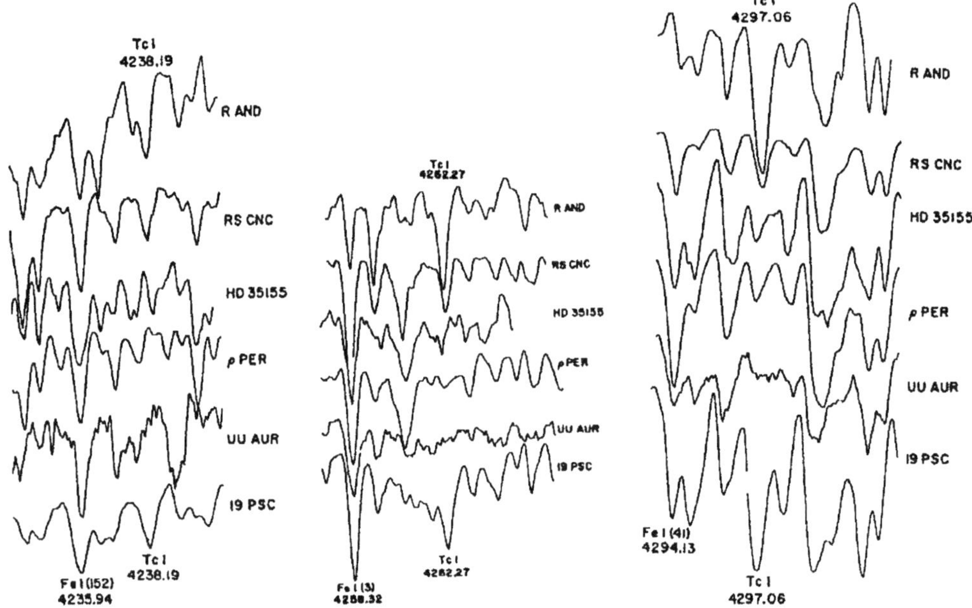

FIGURE 10. Technicium is identified in the spectra of peculiar red giants (Peery 1971).

The technicium stars themselves are, of course, prime examples of stars with abundance anomalies attributed to nucleosynthesis and mixing. In addition to technicium, the peculiar red giants also show enhancements of the s-process elements and carbon. The ratio of carbon to oxygen controls the spectral type of the star: the spectra of stars with more oxygen than carbon are dominated by oxides (TiO, ZrO, YO, etc.) while those with more carbon than oxygen are dominated by carbon molecules (CH, CN, C_2, etc.). The SC stars are an intermediate group of red giants with carbon and oxygen very nearly equal in abundance.

Some of the peculiar red giants in the Milky Way have long been known to show high surface abundances of lithium (Boesgaard, 1970a, b). Pivotal observations by Smith & Lambert (1989) of AGB stars in the Small Magellanic Cloud (SMC) demonstrated that all AGB stars in the SMC above a luminosity of $M_V \sim -6$ show enhanced lithium, shown in Figure 11. (Distances and, hence, absolute magnitudes of galactic S stars were difficult to determine before Hipparcos.) These observations established that lithium is produced, most probably by the Be-transport mechanism (Cameron & Fowler 1971), at the hot base of the convective envelope in the most massive of the AGB stars. Following Scalo (1976), Smith & Lambert noted that AGB stars may be an important source of lithium production in the galaxy.

The abundance anomalies identified in the peculiar red giants (technicium, CNO, lithium) are well understood to result from mixing and nucleosynthesis occurring during thermal pulses at the He- and H-burning shells in AGB stars. Recent papers by Busso et al. (1995) and Gallino et al. (1998) provided detailed stellar models to follow the nucleosynthesis. A small amount of hydrogen from the envelope mixes into the top of the ^{12}C-rich intershell region. When the hydrogen ignites, a ^{13}C-rich zone is formed, providing neutrons via the ^{13}C(alpha, n)^{16}O reaction (the ^{22}Ne(alpha, n)^{25}Mg reaction also playing a minor role). The distribution of s-process elements thus produced cannot be approximated by a simple exponential law of neutron irradiations. A wide range of

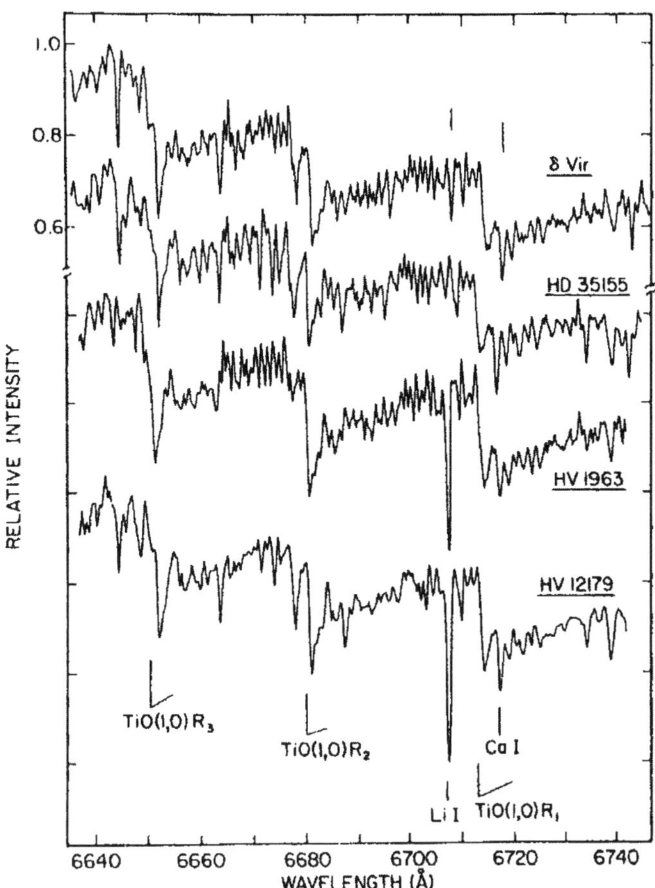

FIGURE 11. Lithium in AGB stars of the SMC. The upper two stars are galactic red giants, and the lower two are SMC giants. (Smith & Lambert 1989, Figure 1).

s-process element distributions can be produced in AGB stars of different metallicities. The C/O ratios are also affected, producing marked differences in the M star/carbon star ratio among stellar populations (cf. Iben and Renzini 1983).

5. Mass transfer

Many of you will remember what a puzzle the barium stars were 20 years ago. Discovered by Bidelman & Keenan (1951), the barium stars are low mass G and K giants rich in barium and other s-process elements, and in carbon. Their origin, as well as the origins of numerous other classes of late type stars with similar anomalies, was hotly debated in the 1970s (cf. Scalo & Miller 1979). While the anomalies in their compositions are reminiscent of the peculiar red giants on the asymptotic giant branch, which also show s-process element enhancements and carbon excesses, the luminosities of the barium stars are too faint for the stars to have experienced the necessary thermal pulses associated with double shell burning. Investigators struggled to find schemes to return AGB stars to lower luminosity by mixing hydrogen into their cores, or to produce extensive mixing and nucleosynthesis at helium core ignition at the tip of the red giant branch.

The problem was neatly solved by McClure et al. (1980), who conducted a survey to monitor the radial velocities of the barium stars. They found that nearly all barium

Stellar Class	Reference
Probable Classes	
Ba II Stars	McClure et al. 1980
Dwarf Ba Stars	Porto de Mello & Da Silva 1997
Pop II CH Stars	McClure 1984, 1997
Dwarf Carbon Stars	Heber et al. 1993
	Green & Margon 1994
Some Tc-Poor MS, S, and C Stars	Jorissen et al. 1998
	Smith and Lambert 1988
Suspected Classes	
N-rich Halo dwarfs	Beveridge & Sneden 1994
Li-depleted Halo dwarfs	Norris et al. 1997

stars were binaries with low mass companions; further surveys revealed that many had degenerate companions. The barium stars' originally more massive companions evolved more quickly, experienced the nucleosynthesis and mixing associated with thermal pulses on the AGB, dumped some of their enriched envelopes onto their less massive siblings, and then faded away.

Today, numerous classes of low mass stars, both giants and dwarfs, with abundance anomalies are known or suspected to result from mass transfer processes. Some have been shown to be binary stars, and for some, white dwarf companions have been identified. In the table above, I summarize the current state of knowledge about anomalous abundances due to mass transfer.

One significant group of stars remains unexplained—the R-type carbon stars. These giants typically show carbon enhancements, but without the s-process enrichments typical of the barium stars (Dominy 1984), and their luminosities are too low to be explained by thermal pulses. McClure (1997) reports 16 years of radial velocity observations of 22 R-type carbon stars, finding no evidence of binary motion in any. The lack of any binaries is particularly surprising since typically 20% of normal late type giants are binaries. McClure suggests that the absence of detected binaries implicates binaries in a formation scenario for R-type carbon stars, but in such a way that the binary is eliminated, possibly through coalescence. The coalescence may cause extra carbon from helium core burning to be mixed outward, where it can be brought to the surface by convection.

6. Coalescence

It is fitting, then, to turn next to abundance anomalies which might arise from stellar coalescence. Two well know classes of stars, the blue stragglers and the anomalous Cepheids might arise from coalescence, either due to binary mergers or collisions. Mateo will discuss the blue stragglers in a later contribution to this volume, and I refer the reader to his paper for a detailed discussion of their origins.

The compositions of field and cluster blue stragglers have been investigated by several authors. Andrievsky et al. (1995) analyzed several field blue stragglers, finding them somewhat metal poor, but with compositions similar to other field dwarfs of similar metallicity. Two M67 blue stragglers analyzed by Mathys (1991) show modest anomalies quite similar to what is seen in Am stars.† Generally the compositions of the two M67

---

† Am stars are metallic-line A stars which may show enhancements of iron group and heavier metals or deficiencies of calcium or scandium. The majority are spectroscopic binaries with relatively short periods (Wolff 1983).

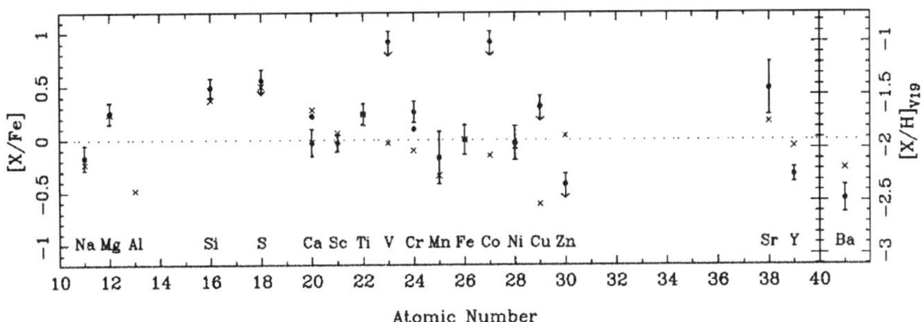

FIGURE 12. Abundances of the elements in the anomalous Cepheid V19 in the globular cluster NGC 5466 (McCarthy and Nemec 1997, Figure 11).

blue stragglers appear normal, without striking CNO or heavy element anomalies. The lithium abundance in blue stragglers is usually deficient compared to normal A stars (Glaspey et al. 1994), possibly because mixing of the outer layers has diluted the surface lithium abundance. One counter example to the low lithium abundance in blue stragglers is the field blue straggler HD 35863 observed by Andrievsky et al.: this star appear to have retained its surface lithium, although its evolutionary history is ambiguous. Models computed by Lombardi et al. (1995) show that the helium cores of colliding stars merge without mixing with the envelopes, but the positions of at least some blue stragglers in the HR diagram suggest they must have mixed. A fuller discussion of this problem can be found by Rasio (1996) and in other papers in that same volume. These arguments suggest that some blue stragglers should have high helium abundances, and, presumably, low carbon isotope ratios.

The abundances of the single anomalous Cepheid† which has been analyzed to date, V19 in the globular cluster NGC 5466, are similarly boring. McCarthy & Nemec (1997) find its composition to be very similar to field giants of comparable metallicity ([Fe/H] $\sim$ -1.9) analyzed by a variety of authors as shown in Figure 12. They do derive a mass for V19 of 1.66 $+0.7/-0.5$ $M_\odot$, consistent with a binary coalescence or significant mass transfer origin.

Neither for the blue stragglers nor for the anomalous Cepheid does coalescence seem to produce significant abundance anomalies beyond the dilution of lithium. Coalescence at later stages of evolution may produce more spectacular results (such as possibly the R-type carbon stars discussed above).

The lithium-rich giants may also be an example of an abundance anomaly caused by coalescence, but with a planet rather than a star. Wallerstein & Sneden (1982) first noted an unusually strong lithium line in the K giant HD 112127; since then, numerous other similar K giants have been identified. Lithium-rich giants comprise about 1% of otherwise normal G and K giants (Brown et al. 1989). These are intermediate to low mass stars which often display IR excesses and circumstellar shells or chromospheric activity (Fekel & Balachandran 1993, de la Reza et al. 1996). Their carbon isotope ratios are normal, demonstrating they have undergone the convective dilution normal stars at this stage of evolution (Brown et al. 1989, da Silva et al. 1995). The presence of significant lithium in low mass giants which have clearly developed normal convective envelopes is a serious inconsistency.

† Anomalous Cepheids are found primarily in dwarf spheroidal galaxies. They have luminosities up to two magnitudes brighter than RR Lyrae stars, and periods < 1.5^d. Their period-luminosity relations differ from the dwarf Cepheids more typically found in globular clusters (McCarthy & Nemec 1997).

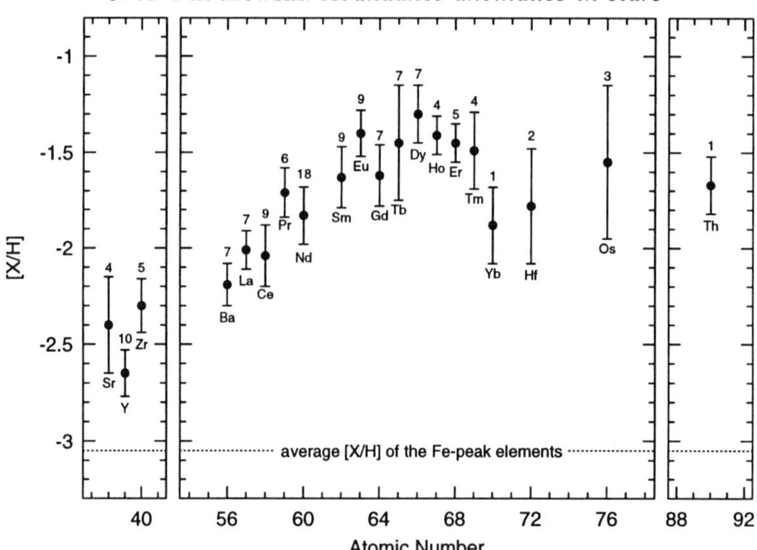

FIGURE 13. Abundances in the ultra-metal-poor, r-process-element-rich star CS 22892–052. The star has an average deficiency of the iron-peak elements of [Fe/H] = −3.1, as indicated by the dashed line. (Sneden et al. 1996, Figure 4)

When Wallerstein and Sneden first discussed the origin of a lithium rich giant they noted a suggestion by Alexander (1967) that stars might engulf planets, thereby raising the abundance of lithium. Since the discovery of "hyper-Jupiter," 51 Peg-type planets, their hypothesis becomes more credible. Siess and Livio (1999) explore this hypothesis in more detail, and find that the merger of a 51 Peg-type planet with its parent star could, with judicious parameter choices, reproduce exactly the observed characteristics of the lithium-rich giants.

7. Star Formation history

The star formation history of a stellar population or an individual star forming event can also produce stellar abundance anomalies. One of the most dramatic examples is the ultra-metal-poor, r-process-element-rich star CS 22892–052 (Sneden et al. 1996). This halo giant has a metallicity of [Fe/H] $\sim$ −3.1, but shows striking enhancements (+0.5 < [X/Fe] < +1.5 dex) of the r-process metals, as shown in Figure 13. The pattern of the abundances of the lanthanide metals matches closely the r-process fractions of the solar system abundances, and is inconsistent with production by the s-process. Generally, halo giants more metal poor than [Fe/H] < −2.0 show a wide dispersion in heavy element abundances (Gilroy et al. 1988, McWilliam et al. 1995) attributed to the influence of small numbers of supernova events.

Another abundance anomaly that can be attributed to star formation history is the recent discovery of alpha-process-element poor stars. Among halo stars, the [Ca/Fe] ratio is typically +0.3 dex with very small scatter (Timmes et al. 1995). Carney et al. (1997) have analyzed the metal-poor subgiant BD +80 245 and find it has a calcium abundance of [Ca/Fe] = −0.3, a factor of four below what is found in "normal" halo stars. King (1997) found that the common proper motion pair HD 134439+HD 134440 also shows modest deficiencies of the alpha-process elements Mg, Si, and Ca, and Nissen & Schuster (1997) identified a handful of additional halo stars with low alpha-process element abundances. Uavane et al. (1996) have explored the compositions to be expected in dwarf spheroidals after a hiatus of star formation. Supernovae of Type I rather than Type II may be more important as sources of chemical enrichment, producing relative deficiencies of alpha-process elements. These interesting objects may then result from the accretion of stars from disrupted dwarf spheroidals by the Milky Way.

A related study by McWilliam & Rich (1994) highlights an even odder abundance pattern in the galactic bulge. They find that the alpha-elements Mg and Ti show a modest enhancement [Mg,Ti/Fe] = +0.3 for stars with near-solar metallicity, while Si and Ca show no enhancement ([Si,Ca/Fe] = 0.0). Clearly, star formation history can produce subtle and very interesting abundance anomalies in stars.

8. Surface activity

Finally, nucleosynthesis occurring in stellar flares has long been suspected to produce some elements in stellar atmospheres. Perhaps the most compelling evidence for nucleosynthesis in flares is the recent discovery by Montes & Ramsey (1998) of the apparent production of ^{6}Li and ^{7}Li by spallation reactions during a long duration flare in the chromospherically active RS CVn† binary 2RE J0743+2 24. The star is a double-lined spectroscopic binary with a K1III primary. During the flare, the Li I photospheric line strength gradually increased by about 40%, and the ^{6}Li/^{7}Li ratio, as measured by the wavelength of the Li I doublet, increased to about 10%.

Nucleosynthesis has also been seen occur in solar flares. Gamma ray emission from excited states of ^{7}Li and ^{7}Be produced by $\alpha\alpha$ reactions in solar flares was found by Murphy et al. (1990) in observations from the Solar Maximum Mission. Kotov et al. (1996) explored the production rate of ^{7}Li from solar flares and concluded that sufficient lithium might be produced to explain the observed surface abundance on lithium on the Sun. Livingston et al. (1997) have searched for enhancement of the Li I λ6706 resonance feature, and particularly the presence of ^{6}Li in sunspots, and find evidence for lithium enhancement (but not ^{6}Li) at one umbral position following a flare. Ritzenhoff et al. (1997) reported similar results from their search. While evidence remains sketchy that stellar abundance anomalies arise from nucleosynthesis in flares, such events may play a role in some stellar environments, such as in dMe stars.

9. Conclusions

For many, if not most, stars with anomalous abundances multiple processes are at work to create unusual abundance patterns we observe—nucleosynthesis and mixing, nucleosynthesis and mass loss, mass loss and diffusion, mass loss and grain formation, etc. Understanding the origin of stellar abundances requires detailed evolution models, especially including the many complexities now incorporated into stellar evolution codes. While we have a general picture of what causes abundances anomalies in stars, we have much to do to match real stars with the detailed predictions of our models. And one of the lessons learned from the study of stellar abundances is that if a problem looks intractable, we should look elsewhere for a solution. So many times in the last two decades astronomers have had to look outside the framework of current understanding to deduce the origin and evolution of particular stars.

This meeting has also highlighted the role of binaries in all phases of stellar evolution, from star formation to star death. Binaries are especially important in understanding

---

† RS CVn stars are chromospherically-active, rapidly rotating late type stars which rotate because of the tidal effect of a relatively nearly binary companion.

stellar abundance anomalies, which can arise from coalescence, mass transfer, collisions, common envelope evolution, and enhanced mass loss. The evolution of binaries is the single most significant "unsolved problem" in stellar astronomy.

10. Apology

Finally, let me close with an apology to all the authors of many interesting papers on stellar abundance anomalies that I have not cited. In some cases I have cited only the first "discovery" paper for particular phenomena, neglecting many subsequent papers which include better data and more complete samples. In other cases, I've selected single papers which highlight particularly well the phenomenon under discussion, and which lead to a clearer understanding of the stellar physics. In no sense is this review comprehensive; it serves only to demonstrate the many complex physical processes at work in stars to produce abundance anomalies, and to show examples of these processes at work. To appreciate fully the wealth of data and models of stellar abundance anomalies, the reader is encouraged to seek out review papers on particular subjects in the proceedings of many other recent topical conferences. Here, I've provided only a taste of the feast to be found in the literature of stellar abundances.

REFERENCES

ALECIAN, G. 1993. In *ASP Conf. Ser. 44*. p. 450. ASP.

ALEXANDER, J. B. 1967 *Observatory* **8**, 238.

ANDRIEVSKY, S. M., CHERNYSHOVE, I, V., & KOVTYUKH, V. V. 1996 *A&A* **310**, 277.

BALACHANDRAN, S. 1995 *ApJ* **446**, 203.

BEVERIDGE, R. C. & SNEDEN, C. 1994 *AJ* **108**, 285.

BIDELMAN, W. P. & KEENAN, P. C. 1951 *ApJ* **114**, 473.

BOESGAARD, A. M. 1970a *ApJ* **161**, 163.

BOESGAARD, A. M. 1970b *ApJ* **161**, 1003.

BOESGAARD, A. M. & BUDGE, K. G. 1989 *ApJ* **338**, 875.

BOESGAARD, A. M. & TRIPICCO, M. J. 1986 *ApJ* **302**, L49.

BOESGAARD, A. M., DELIYANNIS, C. P., STEPHENS, A., & LAMBERT, D. L. 1998 *ApJ* **492**, 727.

BOND, H. E. 1991. In *IAU Symposium 145*. p. 341.

BOND, H. E. 1992 *Nature* **356**, 474.

BRILEY, M. M., SMITH, V. V., SUNTZEFF, N. B., LAMBERT, D. L., BELL, R. A., & HESSER, J. E. 1996 *Nature* **383**, 604.

BROWN, J. A., SNEDEN, C., LAMBERT, D. L., & DUTCHOVER, E. JR. 1989 *ApJS* **344**, 1058.

BUSSO, M., LAMBERT, D. L., BEGLIO, L., GALLINO, R., PAITERI, C. M., & SMITH, V. V. 1995 *ApJ* **446**, 775.

CAMERON, A. G. W. & FOWLER, W. A. 1971 *ApJ* **164**, 111.

CARNEY, B. W., WRIGHT, J. S., SNEDEN, C., LAIRD, J. B., AGUILAR, L. A., & LATHAM, D. W. 1997 *AJ* **114**, 363.

CAVALLO, R. M., SWEIGART, A. V., & BELL, R. A. 1996 *ApJ* **464**, L79.

CAVALLO, R. M., SWEIGART, A. V., & BELL, R. A. 1998 *ApJ* **492**, 575.

CHARBONNEL, C. 1995 *ApJ* **453**, L42.

CHARBONNEL, C., BROWN, J. A., & WALLERSTEIN, G. 1998 *A&A* **332**, 204.

COHEN, J. 1978 *ApJ* **223**, 487.

COTTRELL, P. & DA COSTA, G. 1981 *ApJ* **245**, L79.

DA SILVA, L., DE LA REZA, R., & BARBUY, B. 1995 *ApJ* **448**, L41.

DEHNER, B. T. & KAWALER, S. D. 1995 *ApJ* **445**, L141.

DE LA REZA, R., DRAKE, N. A., & DA SILVA, L. 1996 *ApJ* **456**, L115.

DENISENKOV, P. A. & DENISENKOVA, S. N. 1990 *Sov. Astron. Lett.* **16**, 275.

DOMINY, J. F. 1984 *ApJS* **55**, 27.

FEKEL, F. C. & BALACHANDRAN, S. 1993 *ApJ* **403**, 708.

FITZPATRICK, E. L. & BOHANNAN, B. 1993 *ApJ* **404**, 734.

GALLINO, R., ARLANDINI, C., BUSSO, M., LUGARO, M., TRAVAGLIO, C., STRANIERO, O., CHIEFFI, A., & LIMONGI, M. 1998 *ApJ* **497**, 388.

GILROY, K. K. 1989 *ApJ* **347**, 835.

GILROY, K. K., SNEDEN, C., PILACHOWSKI, C. A., & COWAN, J. J. 1988 *ApJ* **327**, 298.

GLASPEY, J. W., MICHAUD, G., MOFFAT, A. F. J., & DEMERS, S. 1989 *ApJ* **339**, 926.

GLASPEY, J. W., PRITCHET, C. J., & STETSON, P. B. 1994 *AJ* **108**, 271.

GONZALEZ, J.-F., ARTRU, M.-C., & MICHAUD, G. 1995 *A&A* **302**, 788.

GREEN, P. J. & MARGON, B. 1994 *ApJ* **423**, 723.

GREENSTEIN, J. L., TRURAN, J. W., & CAMERON, A. G. W. 1967 *Nature* **213**, 871.

GREENSTEIN, J. L., & SARGENT, A. I. 1974 *ApJS* **28**, 157.

HEBER, U. 1987 *Mitt. Astr. Ges.* **70**, 79.

HEBER, U., BADE, N., JORDAN, S., & VOGES, W. 1993 *A&A* **267**, 31.

HERWIG, F., BLÖCKER, T., SCHÖNBERNER, D., & EL EID, M. 1997 *A&A* **324**, L81.

IBEN, I. JR. 1964 *ApJ* **140**, 1631.

IBEN, I. JR. 1965 *ApJ* **142**, 1447.

IBEN, I. JR. 1967 *ApJ* **147**, 650.

IBEN, I. JR. 1977. In *Advanced Stages of Stellar Evolution* (eds. P. Bouvier & A. Maeder). p. 1. Geneva Observatory.

IBEN, I. JR. & RENZINI, A. 1983 *ARA&A* **21**, 271.

JORISSEN, A., VAN ECK, S., MAYOR, M., & UDRY, S. 1998 *A&A* **332**, 877.

KING, J. R. 1997 *AJ* **113**, 2302.

KOTOV, Y. D., BOGOVALOV, S. V., & ENDALOVA, O. V. 1996 *ApJ* **473**, 514.

KRAFT, R. P. 1994 *PASP* **106**, 553.

KRAFT, R. P., SMITH, G. H., SHETRONE, M. D., SNEDEN, C., LANGER, G. E., & PILACHOWSKI, C. A. 1997 *AJ* **113**, 279.

LANGER, G. E. & HOFFMAN, R. D. 1995 *PASP* **107**, 1177.

LANGER, G. E., & HOFFMAN, R. D. 1997 *PASP* **109**, 244.

LANGER, N. & HEGER, A. 1998. In *ASP Conf. Ser. 131*, p. 76. ASP.

LEUENHAGEN, U., & HAMANN, W.-R. 1998 *A&A* **330**, 265.

LIEBERT, J., FONTAINE, G., & WESEMAEL, F. 1987 *Mem.S.A. It.* **58**, 17.

LITTLE, S. J., LITTLE-MARENIN, I. R., & BAUER, W. H. 1987 *AJ* **94**, 981.

LIVINGSTON, W., POVEDA, A. & WANG, Y. 1997. In *ASP Conf. Ser. 118*. p. 86. ASP.

LOMBARDI, J, C., RASIO, F. A., & SHAPIRO, S. L. 1995 *ApJ* **445**, L117.

LUCK, R. E., BOND, H. E., & LAMBERT, D. L. 1990 *ApJ* **357**, 188.

MATHYS, G. 1991 *A&A* **245**, 467.

MCCARTHY, J. K. & NEMEC, J. M. 1997 *ApJ* **482**, 203.

MCCLURE, R. D. 1984 *ApJ* **280**, L31.

MCCLURE, R. D. 1997 *PASP* **109**, 256.

MCCLURE, R. D. 1997 *PASP* **109**, 536.

McClure, R. D., Fletcher, J. M., & Nemec, J. M. 1980 *ApJ* **238**, L35.

McWilliam, A. & Rich, R. M. 1994 *ApJS* **91**, 749.

McWilliam, A., Preston, G. W., Sneden, C., & Searle, L. 1995 *AJ* **109**, 2757.

Merrill, P. W. 1952 *ApJ* **116**, 21.

Merrill, P. W. 1955 *PASP* **67**, 70.

Michaud, G. 1970 *ApJ* **160**, 641.

Michaud, G. 1986 *ApJ* **302**, 650.

Michaud, G., Tarasick, D., Charland, Y., & Pelletier, C. 1983a *ApJ* **269**, 239.

Michaud, G. & Charland, Y. 1986 *ApJ* **311** 326.

Michaud, G., Vauclair, G., & Vauclair, S. 1983b *ApJS* **45**, 259.

Montes, D. & Ramsey, L. W. 1998 *ApJL*, in press; *BAAS* **192**, #82.03.

Murphy, R. J., Hua, X.-M., Kozlovsky, B., & Ramaty, R. 1990 *ApJ* **351**, 299.

Nissen, P. E. & Schuster, W. J. 1997 *A&A* **326**, 751.

Norris, J. E., Ryan, S. G., Beers, T. C., & Deliyannis, C. P. 1997 *ApJ* **485**, 370.

Parthasarathy, M. 1994. In *ASP Conf. Ser. 60.* p. 261. ASP.

Peery, B. F. Jr. 1971 *ApJ* **163**, L1.

Pilachowski, C. A. 1986 *ApJ* **300**, 289.

Pilachowski, C. A., Sneden, C., & Booth, J. 1993 *ApJ* **407**, 699.

Pilachowski, C. A., Sneden, C., Kraft, R. P., & Langer, G. E. 1996 *AJ* **112**, 545.

Pilachowski, C. A., Sneden, C . Hinkle, K., & Joyce, R. 1997 *AJ* **114**, 545.

Porto de Mello, G. F. & Da Silva, L. 1997 *ApJ* **476**, L89.

Rasio, F. A. 1996. In *ASP Conf. Ser. 90.* p. 368. ASP.

Ritzenhoff, S., Schroter, E. H., & Schmidt, W. 1997 *A&A* **328**, 695.

Roby, S. W. & Lambert, D. L. 1990 *ApJS* **73**, 67.

Scalo, J. M. 1976 *ApJ* **206**, 795.

Scalo, J. M. & Miller, G. E. 1979 *ApJ* **233**, 596.

Schönberner, D. 1997. In *Planetary Nebulae* (eds. H. J. Habing & H. G. L. M. Lamers). p. 379.

Shetrone, M. D., Sneden, C., & Pilachowski, C. A. 1993 *PASP* **105**, 337.

Siess, L. & Livio, M. 1999 *MNRAS*, submitted.

Smith, G. 1987 *PASP* **99**, 67.

Smith, K. C. 1996 *AP&SS* **237**, 77.

Smith, V. V. & Lambert, D. L. 1988 *ApJ* **333**, 219.

Smith, V. V. & Lambert, D. L. 1989 *ApJ* **345**, L75.

Sneden, C., McWilliam, A., Preston, G. W., Cowan, J. J., Burris, D. L., & Armosky, B. J. 1996 *ApJ* **467**, 819.

Sneden, C., Pilachowski, C. A., & VandenBerg, D. A. 1986 *ApJ* **311**, 826.

Takeda, H. & Takada-Hidai, M. 1994 *PASJ* **46**, 395.

Talon, S. & Charbonnel, C. 1998 *A&A*, in press.

Timmes, F. X., Woosley, S. E., & Weaver, T. A. 1995 *ApJS* **98**, 617.

Tylanda, R. 1996. In *ASP Conf. Ser. 96.* p. 101. ASP.

Unavane, M., Wyse, R. F. G., & Gilmore, G. 1996 *MNRAS* **278**, 727.

Van Eck, S., Jorissen, A., Udry, S., Mayor, M., & Pernier, B. 1998 *A&A* **329**, 971.

Van Winckel, H., Mathis, J. S., & Waelkens, C. 1992, *Nature* **356**, 500.

Venn, K. A. 1995 *ApJ* **449**, 839.

Venn, K. A. & Lambert, D. L. 1990 *ApJ* **363**, 234.

VENN, K. A., LAMBERT, D. L., & LEMKE, M. 1996 *A&A* **307**, 849.

WALLERSTEIN ET AL. 1997 *Rev. Mod. Phys.* **69**, 995.

WALLERSTEIN, G. & SNEDEN, C. 1982 *ApJ* **255**, 577.

WASSERBURG, G. J., BOOTHROYD, A. I., & SACKMANN, I.-J. 1995 *ApJ* **447**, L37.

WAHLGREN, G. M., LECKRONE, D. S., JOHANSSON, S. G., ROSBERG, M., & BRAGE, T. 1995 *ApJ* **444**, 438.

WESEMAEL, F., GREEN, R. F., & LIEBERT, J. 1985 *ApJS* **58**, 379.

WOLFF, S. C. 1983 *The A Stars: Problems and Perspectives*, NASA-SP 463.

WOLFF, B., KOESTER, D., DREIZLER, S., & HAAS. S. 1998 *A&A* **329**, 1045.

Winds and mass-loss of hot stars

By ROLF-PETER KUDRITZKI

Institut für Astronomie und Astrophysik der Universität München, Scheinerstr.1, D-81679
München, Germany

Max-Planck-Institut für Astrophysik, D-85740 Garching bei München, Germany

The physical effects of stellar winds on atmospheric structures, energy distributions, ionizing
fluxes, spectra and spectral diagnostics of hot stars are discussed. The *Wind Momentum–
Luminosity Relationship* is introduced as a result of the hydrodynamics of radiation driven
winds and then verified by observations of winds of A-, B- and O-supergiants. The relationship
depends on spectral type. New stellar wind calculations are presented reproducing the observed
spectral type dependence. Simple fit formulae are given for mass-loss rates, wind momenta and
wind power as function of stellar parameters along evolutionary sequences. The observed wind
momentum of the very massive "Pistol Star" in the Galactic Centre is discussed and found to
be in accord with the theory.

Finally, the *Wind Momentum–Luminosity Relationship* is discussed as a tool for the determination of extragalactic distances. Recent results obtained for the Galaxy, LMC, SMC, M33
and M31 are presented. The potential of the method is discussed with the conclusion that it
*may allow independent distance moduli to be obtained with an accuracy of ten percent out to the
Virgo and Fornax clusters of galaxies.*

1. Introduction

Astronomers working on hot stars look at winds from two sides. At the one hand, they
are simply sources of trouble. They corrupt the nice photospheric spectra and introduce
an enormous complication of the model atmosphere theory. On the other hand, stellar
winds are a gift of nature allowing quantitative spectroscopic studies of the most luminous
stellar objects in distant galaxies and, thus, enabling us to obtain important quantitative
information about their host galaxies.

All hot stars have winds. These winds become directly observable in the spectra as
soon as the stars are above certain luminosity borderlines in the HRD. For massive
stars of spectral type O, B, and A the borderline corresponds to 10^4 $L_\odot$. Above this
threshold in luminosity all massive stars show direct spectroscopic evidence of winds
throughout their lifetime (Abbott 1979). Stars of originally intermediate or low masses
($M_{ZAMS} \leq 8$ $M_\odot$) also show signatures of winds, when they evolve through the post-
AGB phases towards the White Dwarf final stage. Here, all objects more massive than
0.6 $M_\odot$ or more luminous than $10^{3.7}$ $L_\odot$ exhibit direct signatures of winds in their spectra
(Pauldrach et al. 1989).

The fact that for stars below these thresholds no direct and clear spectroscopic evidences of winds can be identified (such as broad P-Cygni or emission line profiles, radio-
or IR-excess, thermal X-ray emission) does not mean that they do not have winds. It just
means that because of the lower luminosity the mass loss rates are smaller and, therefore,
the mass and particle density of the wind is too small to produce significant absorption
or emission features in the spectrum. Still, these weak winds might be extremely important. In binary systems they can be partially accreted or can produce shocks of colliding
winds, their momentum and energy might still be sufficient for the dynamics and energetics of the ambient interstellar medium in galaxies or of surrounding gaseous nebulae,
and—most importantly—weak winds are still able to modify the ionizing radiation of
hot stars dramatically (see discussion in section 2.2).

The existence of stellar winds leads to important consequences. They heavily affect

- *the evolution of galaxies* by their input of energy, momentum and nuclear processed material to the ISM,

- *the evolution of stars* by modifying evolutionary timescales, chemical profiles, surface abundances and stellar luminosities,

- *the physics of stellar atmospheres* by dominating the stratification and radiative transfer through the presence of their macroscopic transsonic velocity fields.

In the following section we will discuss how spectral diagnostics and model atmospheres have to be modified to account for the presence of winds. We will give examples demonstrating the present state of the art and will discuss possible future applications of the new generation of model atmospheres on the problem of re-ionization of the universe and on the population synthesis analysis of starbursting galaxies at high redshift.

Then, we will concentrate on the properties of stellar winds as a function of stellar parameters. By a simplified analytical solution of the hydrodynamic equations of radiation driven winds, we will introduce the concept of the *Wind Momentum–Luminosity Relationship (WLR)* (section 3) and then investigate the observed WLR of galactic blue supergiants and Central Stars of PN (section 4). We will present stellar wind models based on a new modified approach for line driven winds and compare with the observed stellar wind properties (section 5). Based on these results, we will provide simple scaling relations for wind momentum, wind power and mass-loss rates as function of stellar parameters in section 6. In section 7 we will present wind models for the "Pistol Star," the extremely massive IR-object in the central region of our Galaxy. Section 8 deals with the WLR as a tool for the determination of extragalactic distances. We discuss wind momenta of A-supergiants in M31, the effects of metallicity based on supergiants in the LMC, SMC and in the outskirts of M33 and the potential of the method in the era of new 8m–class optical telescopes.

2. Effects of winds on atmospheres and spectral diagnostics of hot stars

Winds have drastic effects on the atmospheric structure of hot stars, even when they are weak and not directly detectable in the spectrum. They modify the density stratification, the geometrical extension, the NLTE effects, the emergent energy distributions and the ionizing fluxes. An extensive discussion of all these effects has been given very recently by Kudritzki 1998. Here, we give only a brief overview.

2.1. *Hydrodynamic NLTE model atmospheres with spherical extension*

The spectra of luminous hot stars usually show simultaneously photospheric absorption and emission lines (originating from atmospheric layers, where the outflow velocity is smaller or of the order of the sound speed) and stellar wind lines. For this reason, Gabler et al. 1989 developed their concept of "Unified Model Atmospheres," which has now become the standard treatment for model atmospheres of hot stars with winds. Unified Models are spherically extended and yield the entire sub- and supersonic atmospheric structure taking the mass-loss rate, density and velocity structure from the hydrodynamics of radiation driven winds. They can be used to calculate energy distributions simultaneously with "photospheric" and "wind" lines and, most importantly, they can treat the multitude of "mixed cases," where a photospheric line is contaminated by wind contributions. This concept turned out to be extremely fruitful for the interpretation of hot star spectra and has led to the rapid development of significant refinements and improvements in parallel by several groups. Basic papers describing the status of the

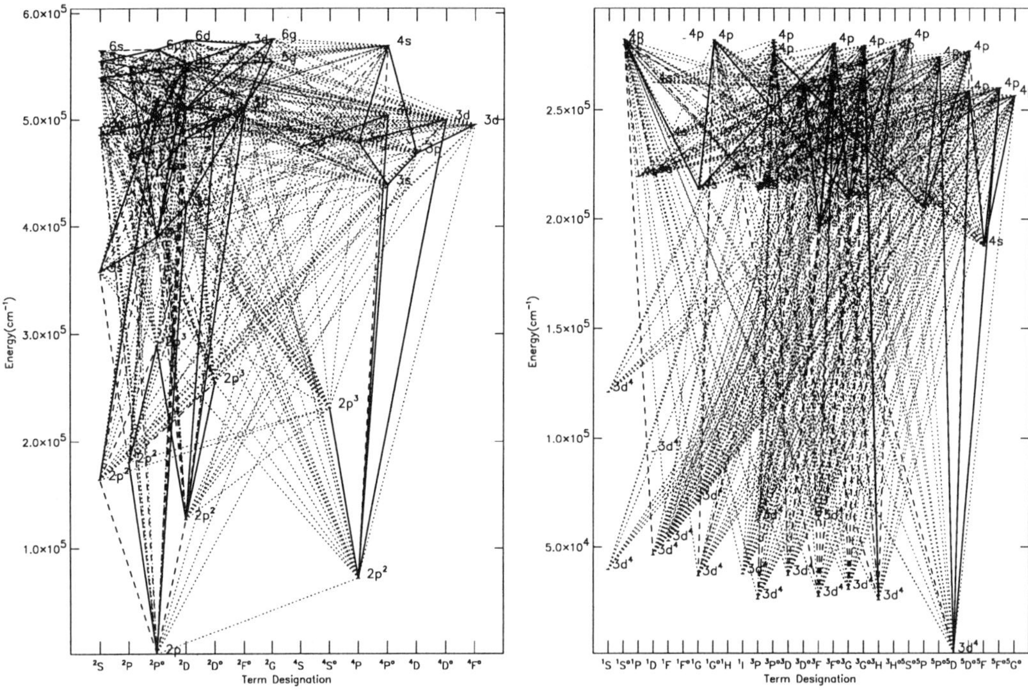

FIGURE 1. Energy level diagram and transition scheme for O IV (left) and Fe V (right) in the Munich model atmosphere code. The transitions are coded by the strengths of their gf-values: solid $gf \geq 1.0$; dashed $1.0 \geq gf \geq 0.1$; dotted $0.1 \geq gf$. In the same way, more than 150 ions are included in the Munich University Observatory stellar wind code developed by A. Pauldrach, J. Puls and collaborators.

work in the different groups are Schaerer & Schmutz 1994, Schaerer & de Koter 1997, Hillier & Miller 1998, Pauldrach et al. 1994, Taresch et al. 1997, Pauldrach et al. 1998, Santolaya-Rey et al. 1997. In the following, we give a few examples of the model atmosphere approach developed at the Munich University Observatory by A. Pauldrach, J. Puls and collaborators. These model atmospheres are fully line-blanketed with millions of spectral lines and solve rate equations for 150 ionic species together with hydrodynamic equations of radiation driven winds to calculate the atmospheric structure and the emergent spectra (Fig. 1 gives examples for level diagrams and transition schemes of two individual ionic models as representatives for the other 150). They allow a detailed spectrum synthesis of UV and optical spectra for the purpose of complete quantitative spectral analyses. Fig. 2 and Fig. 3 give nice examples from the recent papers by Taresch et al. 1997 and Haser et al. 1998.

2.2. *The impact of stellar winds on energy distributions and ionizing fluxes*

Energy distributions are severely affected by the presence of stellar winds, as is demonstrated by Fig. 4. Free-free thermal emission of the spherically extended stellar wind plasma causes a significant IR- and radio-excess at longer wavelengths. But also the ionization edges of bound-free transitions in the UV and EUV are substantially modified. This is caused by two effects: the change in density stratification (see above) and the influence of the velocity induced Doppler-shifts on the optical thickness of resonance lines of ions producing strong absorption edges. A detailed discussion is given by Gabler et al. 1989 and Najarro et al. 1996. Basically, the optical thickness in the

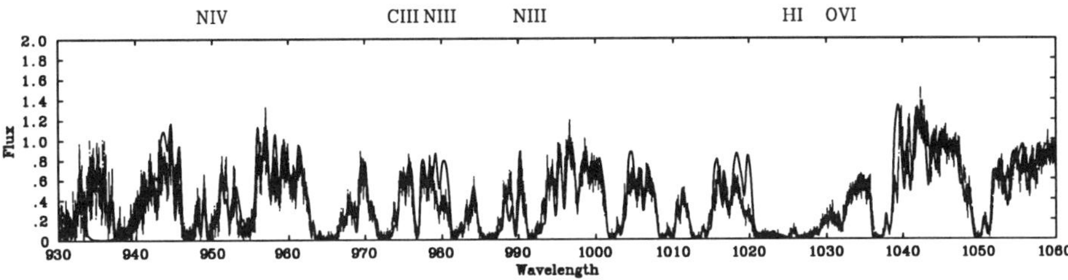

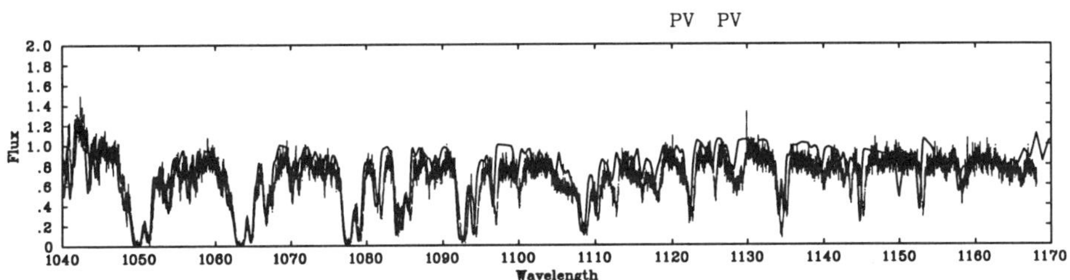

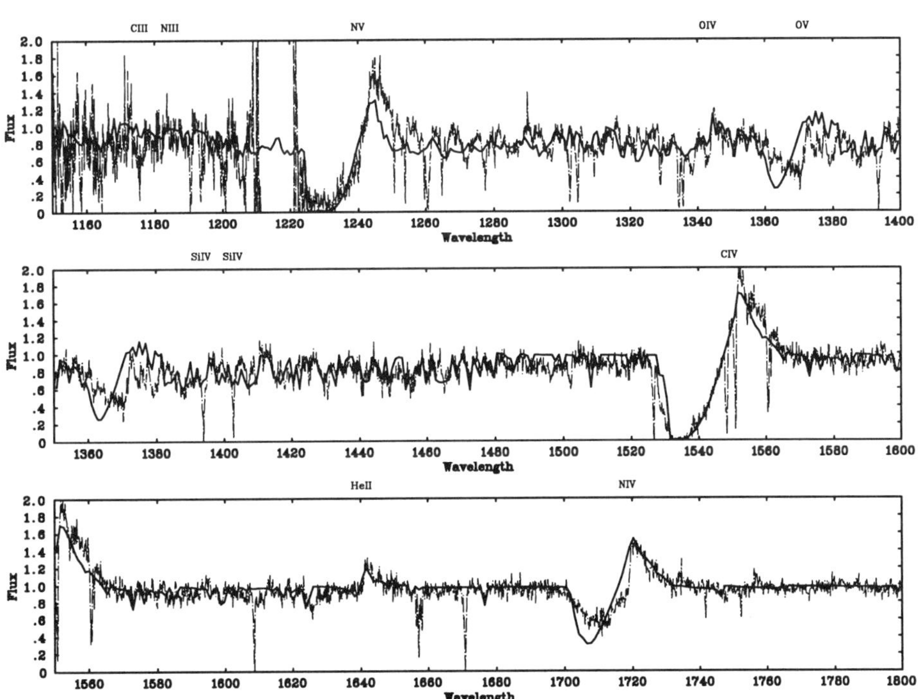

FIGURE 2. Comparison of observed and calculated spectrum of the O3 Iaf+ supergiant HD 93129A. Upper two panels: FUV (ORFEUS), lower three panels: UV (IUE). The OR-FEUS spectrum contains many interstellar molecular lines of H_2 and HD which are included in the spectrum fit. The narrow sharp lines in the IUE spectrum are of atomic or ionic interstellar origin and are not included in the spectral fit. From Taresch et al. 1997.

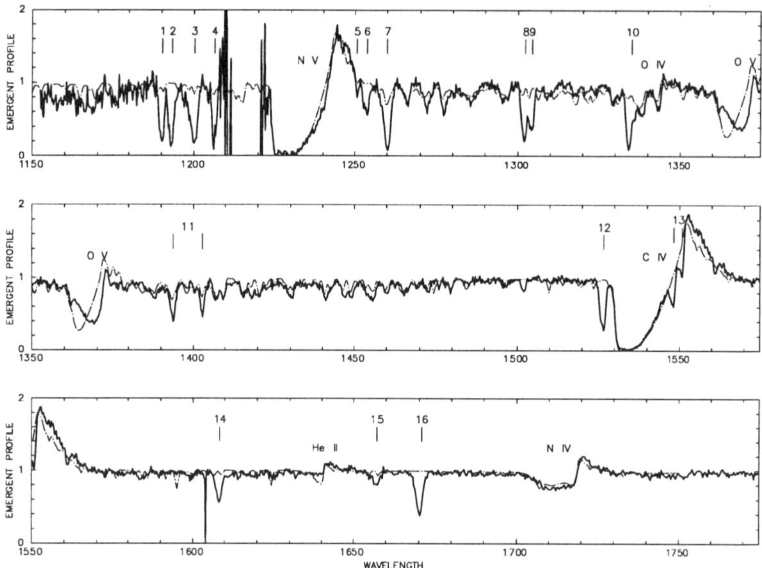

FIGURE 3. Final model atmosphere fit of the HST UV spectrum of the O3-star Sk−68°137 in the LMC adopting a metallicity of $z = 0.4z_\odot$. Interstellar lines are indicated by vertical bars with numbers. From Haser et al. 1998.

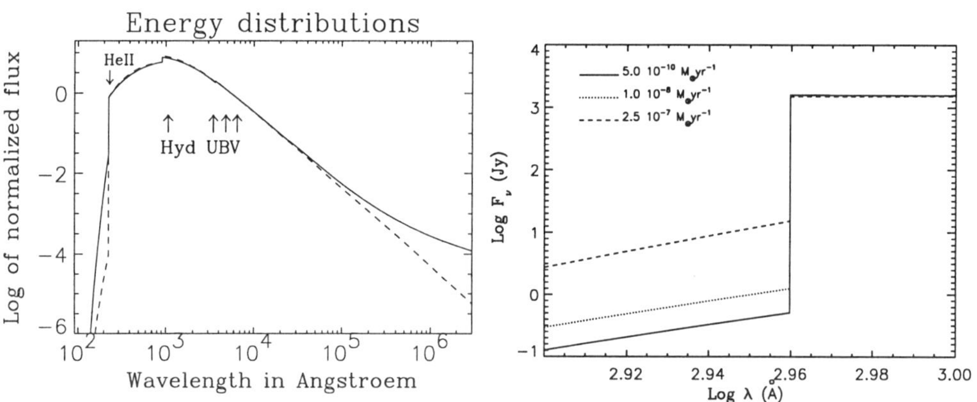

FIGURE 4. Left: Energy distribution of a hot ($T_{\rm eff} = 47500$ K) O-supergiant calculated for a hydrostatic model (dashed) and a hydrodynamic model with a stellar wind (solid). (From Gabler et al. 1989). Right: Energy distributions around the Lyman-jump for B-giant models with different mass-loss rates $\dot{M}$. Modifications in $\dot{M}$ lead to enormous changes in the ionizing flux. (From Najarro et al. 1996).

resonance transitions is reduced and the detailed balance is modified so that electrons are pumped to the excited level and the ground states are severely depopulated.

This has important consequences for the ionization of the ISM by hot stars. Gabler et al. 1991 demonstrated that the effect contributes to the solution of the longstanding problem of the "Zanstra-discrepancy" in Planetary Nebulae. It may also help to understand the ionizing population in Giant Extragalactic H II-regions of high excitation (see Gabler et al. 1992). In addition, it is crucial for the ground state continua of H and He I in B-stars, as has been shown recently by Najarro et al. 1996 and has important

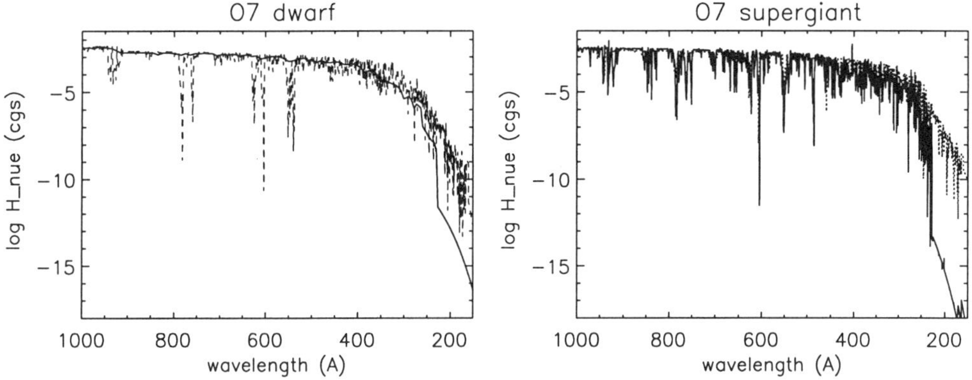

FIGURE 5. Left: Ionizing flux of a NLTE line-blanketed model with wind (Pauldrach et al. 1998, dashed) compared with a hydrostatic LTE model calculated with the Kurucz code (solid). Note that shortward of 400 A the difference is more than 1 dex. Right: Ionizing fluxes of two NLTE line-blanketed model with wind (Pauldrach et al. 1998) calculated for solar metallicity (solid) and for 0.2 solar (dotted).

consequences for the ionization of the ISM by a somewhat older population—such as the local ISM—when O-stars are already absent.

In Fig. 5 we compare ionizing UV fluxes of NLTE line-blanketed atmospheres with winds with those of their hydrostatic LTE counterparts. The differences are significant. The same is true for the comparison between NLTE models of different metallicity (Fig. 5). For the diagnostics of extragalactic emission line nebulae and galaxies these effects have important repercussions. They will also be crucial for the investigation of the re-ionization of the universe by a population of massive Population III stars (Haiman and Loeb 1997).

2.3. *Stellar winds and starforming galaxies at high redshift*

Stellar winds in hot stars produce broad emission or P Cygni-type line features which are easy to detect even in the noisy spectra of extremely remote objects. The most exciting detections of this type have been the starforming galaxies at redshifts of $z = 3$ (Steidel et al. 1996, Yee et al. 1996, see Fig. 6). All of the UV stellar wind features which we have been able to study quantitatively in the spectra of hot stars in Local Group galaxies are now recognized in the optical. In this sense, stellar winds are really a gift of nature, since they render the possibility of a quantitative interpretation of the starforming process at high redshifts. Of course, this requires the fundamental know-how of stellar spectroscopists in conjunction with population synthesis techniques. There are two possible ways to go in this direction: empirical population synthesis based on observed spectra of well studied stars, for instance in the Magellanic Clouds (see Fig. 7, extended HST observing campaigns in this direction are now under way) or population synthesis based completely on synthetic spectra computed with model atmospheres tested in the Local Group (see Fig. 8).

2.4. *The wind contamination of photospheric lines*

Almost every line in the optical spectra of luminous hot stars is affected by stellar wind emission which modifies the line profiles and leads to systematic errors, if model atmospheres which ignore the presence of winds are applied (see Fig. 9). However, using Unified Model Atmospheres that include the effects of winds one can turn the situation

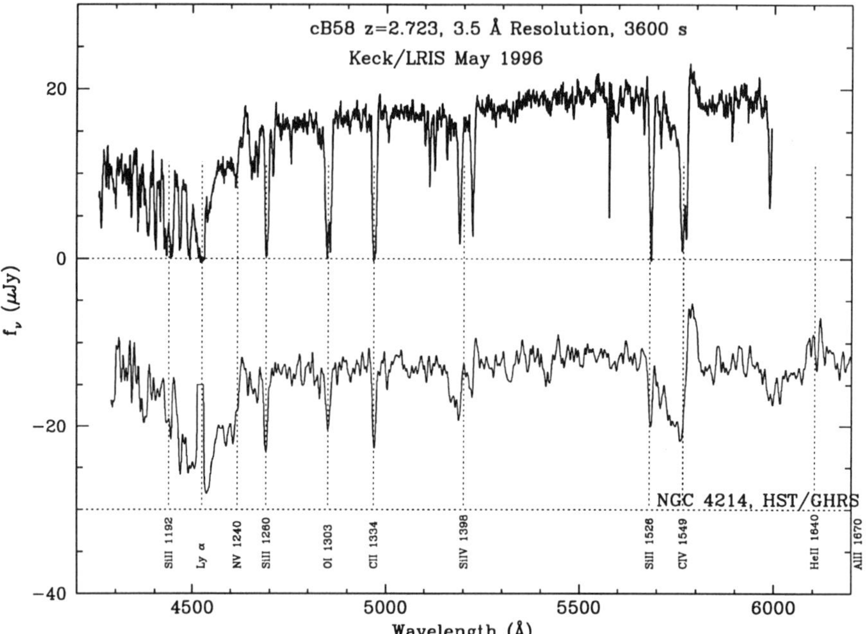

FIGURE 6. Optical spectrum of a galaxy at very high redshift ($z = 2.732$) compared with a "local" starburst. Figure from Steidel and Pettini (priv. comm.), for details see Steidel et al. 1996 and Yee et al. 1996.

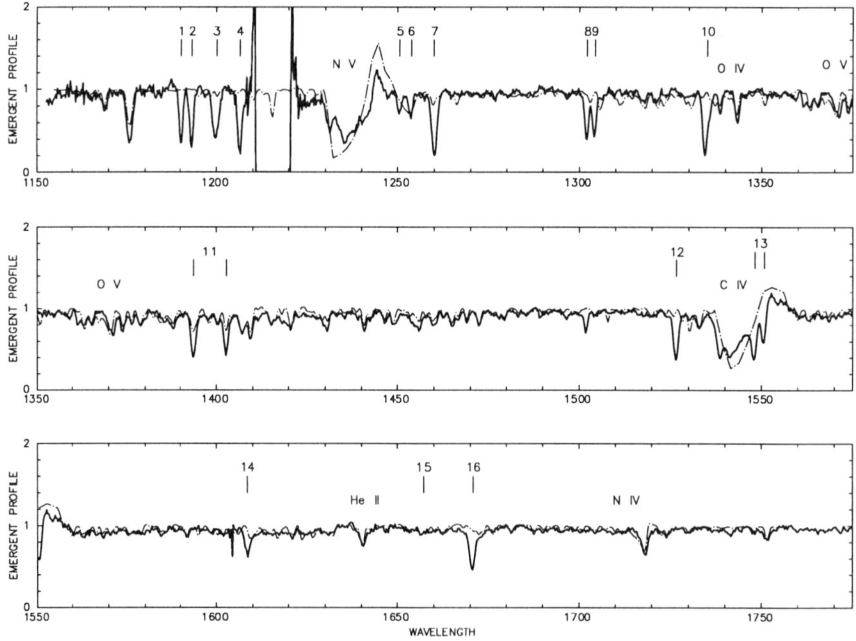

FIGURE 7. Final model atmosphere fit of the HST UV spectrum of the O6 V star AV243 in the SMC adopting a metallicity of $z = 0.2z_\odot$. Interstellar lines are indicated by vertical bars with numbers. From Haser et al. 1998.

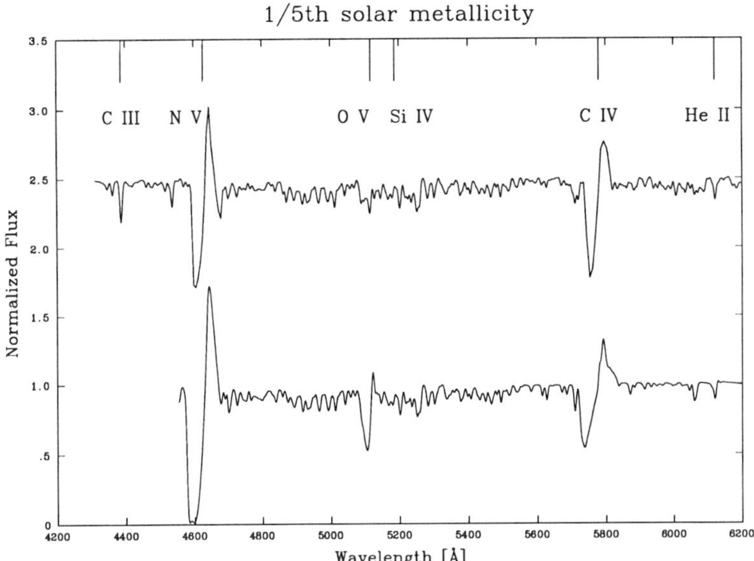

FIGURE 8. Calculated model atmosphere UV spectra of O-stars redshifted to $z = 2.732$ and degraded to 3 Å resolution. Top: Main sequence O6 object with T_{eff}=43000 K; bottom: O3-supergiant with T_{eff}=55000 K. 1/5 solar metallicity has been adopted. Note that the simulations do not include the interstellar lines.

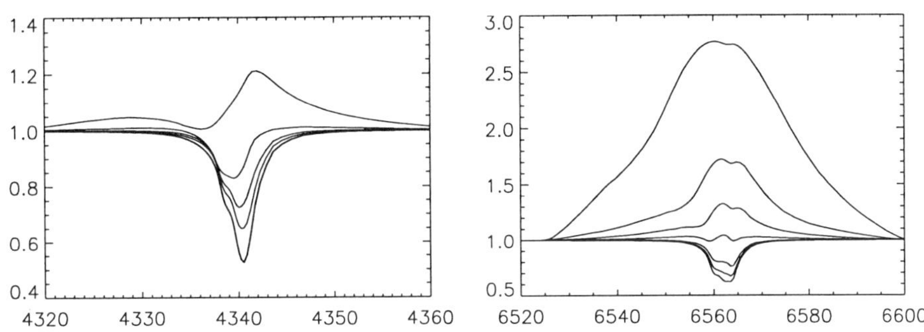

FIGURE 9. Balmer line profiles for a typical O5 Ia supergiant adopting mass-loss rates of $30, 10, 5, 2.5, 1.0, 0.05$ and 0.01×10^{-6} M$_\odot$/yr. Left: Hγ. Right: Hα.

around and determine not only all the necessary stellar parameters but also the stellar wind mass-loss rate from extreme stellar wind lines such as H$_\alpha$.

3. The concept of the Wind Momentum–Luminosity Relationship

The *Wind Momentum–Luminosity Relationship* is a new, straightforward way to look at winds in the context of stellar and galactic evolution. Moreover, and most importantly, it provides a new independent tool to derive extragalactic distances with an accuracy comparable to the Cepheid method. The basic concept is straightforward. Since the winds of the blue supergiants are a result of radiation pressure, we expect that the properties of the stellar winds must reflect the luminosities of the stars. With this

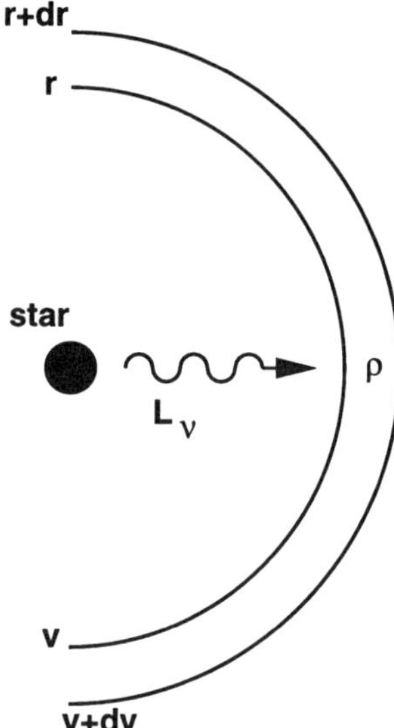

FIGURE 10. Sketch of a Blue Supergiant irradiating its own stellar wind envelope. L_ν is the spectral luminosity at frequency ν. v is the wind outflow velocity at radius r and ρ is the mass density.

in mind, we develop the following very simple idea. For radiation driven winds the mechanical momentum flow of a stellar wind $\dot{M}v_\infty$ should be a function of the photon momentum rate L/c provided by the stellar photosphere and interior

$$\dot{M}v_\infty = f(L/c) \ . \tag{3.1}$$

If this is true and if we are able to find this function f, this would enable us to determine directly stellar luminosities from the stellar wind by using the inverse relation

$$L = f^{-1}(\dot{M}v_\infty) \ . \tag{3.2}$$

In other words, by measuring the rate of mass-loss and the terminal velocity directly from the spectrum we would be able to determine the luminosity of a blue supergiant. This is an exciting perspective, because it would give us a completely new, purely spectroscopic tool to determine stellar distances. Quantitative spectroscopy would yield T_{eff}, gravity, abundances, intrinsic colours, reddening, extinction, $\dot{M}$ and v_∞. With the luminosity from the above relation one could then compare with the dereddened apparent magnitude to derive a distance.

3.1. *An analytical derivation of the WLR*

Knowing the functional dependence of the radiative line acceleration on velocity field and density, it is possible (see Kudritzki et al. 1989) to obtain rather precise analytical solutions for $\dot{M}$ and v_∞ as functions of the stellar parameters, which can then be used to calculate the stellar wind momentum (see Kudritzki et al. 1996a, Puls et al. 1996, Kudritzki 1998). Here we give a somewhat more simplified, but very instructive description of this approach.

We assume that the wind is stationary and spherical symmetric and obeys the equation of continuity

$$\dot{M} = 4\pi r^2 \rho(r)v(r) \ . \tag{3.3}$$

Then we consider the star as a point source of photons irradiating and accelerating its own stellar wind (see Fig. 10) and calculate the amount of photon momentum absorbed by one spectral line

$$\frac{L}{c}\frac{L_{\nu_i}(1-e^{-\tau_i})d\nu^{\text{width}}}{L} = \frac{L}{c^2}\frac{\nu_i L_{\nu_i}}{L}(1-e^{-\tau_i})dv \quad . \tag{3.4}$$

The first factor on the left side gives the total photon momentum rate provided by the star, the second describes the fraction absorbed by one spectral line in an outer shell of thickness dr. τ_i is the optical thickness of such an outer shell in the line transition i. $L_{\nu_i}d\nu^{\text{width}}$ is the stellar spectral luminosity at the frequency of line i multiplied by the spectral width of the line. This luminosity can in principle be absorbed by the line if it is entirely optically thick (i.e. $\tau_i \gg 1$). However, depending on the optical thickness only the fraction $(1 - e^{-\tau_i})$ is really absorbed. If we are in the supersonic part of the wind, then the spectral width $d\nu^{\text{width}}$ is not determined by the thermal motion of the ions but rather by the increment of the velocity outflow dv via the Doppler formula

$$d\nu^{\text{width}} = \nu_i\frac{dv}{c} \quad , \tag{3.5}$$

which leads to the right hand side of equation (3.4).

After calculation of the photon momentum absorbed by a single spectral line we can consider the momentum balance in the stellar wind. The photon momentum absorbed by all lines will just be a sum over all lines i of the expression shown on the right hand side of equation (3.4). Almost all of this absorbed momentum will be transformed into gain of mechanical stellar wind flow momentum $\dot{M}dv$ of the outer shell except the fraction $g(r)dM_r$, which is the momentum required to act against the gravitational force. ($g(r) = GM_*/r^2$ is the local gravitational acceleration and $dM_r = \rho 4\pi r^2 dr$ is the mass within the spherical stellar wind shell)

$$\dot{M}dv = \frac{L}{c^2}\sum_i \frac{\nu_i L_{\nu_i}}{L}(1-e^{-\tau_i})dv - G\frac{M*(1-\Gamma)}{r^2}\rho 4\pi r^2 dr \quad . \tag{3.6}$$

Note that this momentum balance also includes the photon momentum transfer by Thomson scattering, which leads to the correction factor $(1-\Gamma)$ in the local gravitational acceleration.

Now, the important next step is to deal with the sum over all lines in the above momentum balance. Here, the complication arises from the term in parentheses containing the local optical depth τ_i of each line. τ_i will not only be different for each of the thousands of lines driving the wind, it will also vary through the stellar wind as a function of radius. On the other hand, one of the enormous simplifications in supersonically expanding envelopes around stars is that the optical thickness is well described by (see for instance Castor 1970, Kudritzki 1988a)

$$\tau_i = \mathbf{k_i}\kappa_{\text{Thom}}\frac{v_{\text{therm}}}{dv/dr} \quad , \tag{3.7}$$

where v_{therm} is the thermal velocity of the ion and $\mathbf{k}_i$ is the (dimensionless) *line strength* defined as

$$\mathbf{k_i} = \frac{\kappa_i}{\kappa_{\text{Thom}}} \quad , \tag{3.8}$$

i.e. the opacity of line i

$$\kappa_i = \frac{1}{\Delta\nu_{\text{Dopp}}}\frac{\pi e^2}{m_e c}n_l f_{lu}\left(1 - \frac{n_u}{n_l}\frac{g_l}{g_u}\right) \quad , \tag{3.9}$$

in units of the continuous Thomson scattering opacity or in units of the local density

$$\kappa_{Thom} = n_e \sigma_e = \rho s_e \ .$$ (3.10)

σ_e is the cross section for Thomson scattering of photons on free electrons, n_e the local number density of free electrons. In a hot plasma with hydrogen as the main constituent mostly ionized and with $Y_{\mathrm{He}} = n_{\mathrm{He}}/n_{\mathrm{H}}$ and I_{He} the number of electrons provided per helium nucleus we have (m_{H} is the mass of the hydrogen atom)

$$s_e = \frac{1 + I_{\mathrm{He}} Y_{\mathrm{He}}}{1 + 4 Y_{\mathrm{He}}} \frac{\sigma_e}{m_{\mathrm{H}}} \ .$$ (3.11)

Thus, if the degree of ionization of helium is roughly constant as a function of radius r in the wind, s_e is also constant and we have

$$\tau_i = \mathbf{k_i} s_e \rho(r) \frac{v_{\mathrm{therm}}}{dv/dr} \ ,$$ (3.12)

with the line strength $\mathbf{k}_i$ proportional to oscillator strength f_{lu}, wavelength λ_i and the occupation number n_l of the lower level divided by the mass density ρ

$$\mathbf{k_i} \propto \frac{n_l}{\rho} f_{lu} \lambda_i \ .$$ (3.13)

Thus, the line strength is roughly independent of the depth in the atmosphere and is determined by atomic physics (f_{lu}, λ_i) and atmospheric thermodynamics (n_l/ρ). Since ρ and dv/dr vary strongly through the wind a line can be optically thick in deeper layers

$$\tau_i \gg 1 \Longrightarrow (1 - e^{-\tau_i}) dv \approx dv \ ,$$ (3.14)

and can become optically thin further out

$$\tau_i \ll 1 \Longrightarrow (1 - e^{-\tau_i}) dv \approx k_i \rho dr \ .$$ (3.15)

For the calculation of the momentum transfer in equation (8.26) this means that line contributions can have an entirely different functional form depending on the optical thickness of the lines.

This problem can be solved in a very elegant way by introducing a *line strength distribution function*

$$n(k, \nu) dv dk = \text{number of lines with } \nu_i \text{ from } (\nu, \nu + d\nu) \text{ and with k from (k, k+dk)}.$$

Since the hydrodynamic model atmosphere codes (see section 2) contain atomic data and occupation numbers for millions of lines, we can investigate the physics of the line strength distribution function. As it turns out (Castor, Abbott, & Klein 1975, Puls 1987, Kudritzki et al. 1988b, Puls 1993, Springmann 1997) the distribution in line strengths— to a very good approximation—obeys a power law

$$n(k, \nu) dv dk = g(\nu) d\nu k^{\alpha - 2} dk, 1 \leq k \leq \infty \ ,$$ (3.16)

independent (to first order) of the frequency. The exponent α depends weakly on T_{eff} and varies (in the temperature range of OB-stars) between

$$\alpha = 0.6 \ldots 0.7 \ .$$ (3.17)

α is mostly determined by the atomic physics and basically reflects the distribution function of the oscillator strengths. One can, for instance, show that for the hydrogen atom the distribution of the Lyman-series oscillator strengths is a power law with exponent $\alpha = 2/3$ (see Kudritzki 1998). It is important to realize that α is not a free parameter but, instead, well determined from the thousands of lines taken into account

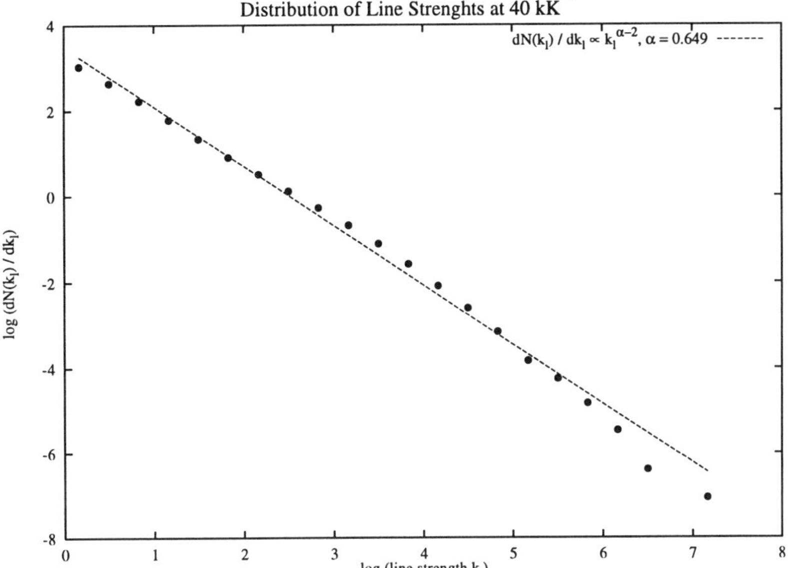

FIGURE 11. Logarithmic plot of the line strength distribution function for a stellar wind model with $T_{\text{eff}} = 40000$ K. Dots are numbers from the line list in the model atmosphere code. The dashed curve is the power law fit. For this fit the last three points at the largest k-values, which correspond to a total of only six lines, were ignored. From Springmann 1997.

in the model atmosphere calculations. An example for the power law dependence of line strengths is given in Fig. 11.

With the line strength distribution function of equation (3.16) we can replace the sum in the momentum balance of equation (3.6) by a double integral, which is then solved taking equation (3.12) into account (see also Kudritzki 1988a).

$$\sum_i \frac{\nu_i L_{\nu_i}}{L}(1 - e^{-\tau_i}) \longrightarrow \int_0^\infty \int_0^\infty (1 - e^{\tau(k)})\frac{\nu L_\nu}{L}n(k,\nu)d\nu dk \longrightarrow N_o \left\{ \frac{dv/dr}{\rho} \right\}^{\alpha-1} . \tag{3.18}$$

This means that the momentum transfer from photons to the stellar wind plasma depends non-linearly on the gradient of the velocity field. The degree of the non-linearity is determined by the steepness of the line strength distribution function α. N_o is proportional to the number of lines in the line strength interval $1 \le k \le \infty$.

With equation (3.12) we can return to the total momentum balance equation (3.6) to obtain

$$\dot{M}dv = \frac{L}{c^2}N_o \left\{ \frac{dv/dr}{\rho} \right\}^{\alpha-1} dv - G\frac{M_*(1 - \Gamma)}{r^2}4\pi r^2\rho dr . \tag{3.19}$$

Inserting eq. (3.3) we then arrive at the non-linear differential equation for the velocity field

$$r^2v\frac{dv}{dr} = \frac{L}{\dot{M}^\alpha}\frac{N_o}{c^2}(4\pi)^{\alpha-1}\left\{ r^2v\frac{dv}{dr} \right\}^\alpha - GM_*(1 - \Gamma) , \tag{3.20}$$

which looks much more complicated than it really is. The solution is easy (Kudritzki et al. 1989). We obtain $\dot{M}$ as the uniquely determined eigenvalue of the problem

$$\dot{M} \propto L^{1/\alpha}\left\{ M_*(1 - \Gamma) \right\}^{1-1/\alpha} , \tag{3.21}$$

and a velocity gradient

$$r^2 v \frac{dv}{dr} \propto GM_*(1 - \Gamma) \; , \tag{3.22}$$

leading to a velocity field with

$$v(r) \propto v_{\rm esc} \{1 - R_*/r\}^{0.5} \, , v_{\rm esc} \propto \left\{ \frac{GM_*(1 - \Gamma)}{R_*} \right\}^{0.5} \; , \tag{3.23}$$

and a terminal velocity proportional to the escape velocity $v_{\rm esc}$.

$$v_\infty \propto v_{\rm esc} \; . \tag{3.24}$$

Combining equations (3.21) and (3.24) we obtain the wind momentum rate

$$\dot{M}v_\infty \propto \frac{1}{R_*^{0.5}} L^{1/\alpha} \{M_*(1 - \Gamma)\}^{3/2 - 1/\alpha} \; , \tag{3.25}$$

which—as expected—depends strongly on the luminosity but also on the photospheric radius and the expression in the parentheses, which contains the stellar mass and distance from the Eddington limit. It is this expression which can vary significantly for different Blue Supergiants and , therefore, causes the large scatter in the observed correlations of mass-loss rates with luminosity and terminal velocity with escape velocity as discussed previously. However, for the product of mass-loss rate and terminal velocity, the stellar wind momentum rate, the exponent of the term in brackets should be—thanks to the laws of atomic physics—very close to zero, since $\alpha \approx 2/3$. This means that to first order the wind momentum rate should be determined by

$$\mathbf{\dot{M}v_\infty} \propto \frac{1}{R_*^{0.5}} \mathbf{L}^{1/\alpha} \; . \tag{3.26}$$

This is the Wind Momentum–Luminosity Relationship. It predicts a strong dependence of wind momentum rate on the stellar luminosity with an exponent determined by the statistics of the strengths of the tens of thousands of lines driving the wind. It also contains a weak dependence on stellar radius which comes from the fact that the stellar wind has to work against the gravitational potential when accelerated by photospheric photons.

4. The observed WLR of blue supergiants and Central Stars of PN in the Galaxy

It is, of course, challenging to compare the prediction of radiation driven wind theory with the observations to see whether an observed WLR does exist. This requires the determination of v_∞ and $\dot{M}$. While the measurement of the former is easy and straightforward by means of the blue edges of the P-Cygni profiles of UV resonance lines (for O-, B-, and A-supergiants) or H_α (only for A-supergiants with winds of significant strength; for examples and references, see for instance Kudritzki 1998), the determination of the latter with sufficient precision used to be problematic. However, the development of NLTE Unified Model Atmospheres as discussed in section 2 made it possible to use the strength of H_α as a wind line to determine $\dot{M}$ very precisely (see again Kudritzki 1998 for an overview). In this way, Puls et al. 1996 obtained accurate mass-loss rates for a large sample of galactic O-stars and provided a detailed discussion of the H_α line formation process and the possible error sources. Kudritzki et al. 1998b extended this work on the brightest galactic B- and A-supergiants with well determined distances. Fig. 12

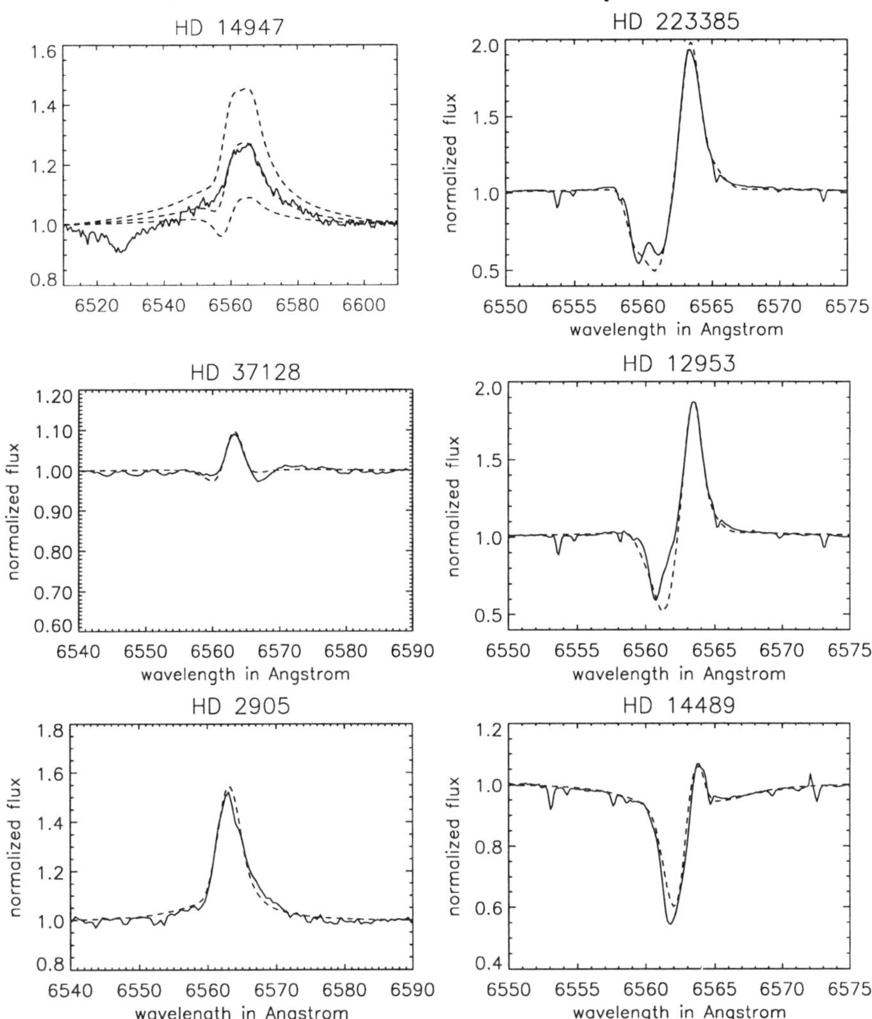

FIGURE 12. H_α line profile fits of the galactic supergiants HD 14947 (O5 Ia), HD 37128 (B0 Ia), HD 2905 (B0.7 Ia), HD 223385 (A3 Ia), HD 12953 (A1 Ia), HD 14489 (A2 Ia). The fits of HD 14947 are taken from Puls et al., 1996 and the three different curves indicate adopted changes in the model calculations of 25 percent in $\dot{M}$. The fits of all other objects are from Kudritzki et al., 1998, who used an improved version of the unified model atmosphere code developed by Santolaya-Ray, Puls and Herrero (1997) for the line profile fits.

indicates how accurately the observed H_α stellar wind line profiles can be reproduced by the theoretical models yielding mass-loss rates with a precision better than 0.1 dex. Fig. 13 combines the results from Puls et al. 1996 and Kudritzki et al. 1998b with regard to the observed wind momentum and reveals the clear existence of a WLR.

Fig. 13 may indicate a slight curvature of the observed WLR in the log wind momentum–log luminosity plane. Another interpretation could be the existence of three independent relations of slightly different height and slope for the three spectral types. The latter interpretation is supported by the investigation of the wind momentum of the B0.5 supergiant HD 38771 (κ Ori). This object is usually considered as a twin of HD 37128 (ϵ Ori, H_α profile displayed in Fig. 12), however, the recent astrometric results obtained with the Hipparcos satellite show that it is in the foreground of the Orion association

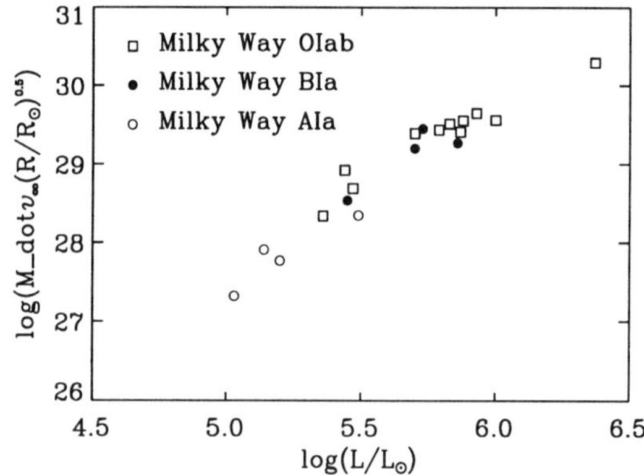

FIGURE 13. The observed WLR of the brightest galactic supergiants of spectral type O, early B and early A. Data from Puls et al., 1996 and Kudritzki et al. 1998. Wind momenta are given in cgs-units.

with a de-reddened absolute magnitude of only $M_V = -4.8^m$. This explains, why—contrary to HD 37128—H_α is in absorption and is fitted with a significantly smaller rate of mass-loss (see Fig. 14, upper left). The wind momentum of HD 38771—a low luminosity B-supergiant—is still larger than the wind momentum of the A-supergiants of comparable luminosity (see Fig. 14, upper right). We take this as an indication of a temperature dependence of the WLR and adopt three independent linear regression for the three spectral types (Fig. 14, bottom).

As mentioned in the introduction, not only hot stars of high mass but also low-mass objects in their post-AGB evolutionary phase such as Central Stars of PN exhibit strong observable winds. If the theory of radiation driven winds is conceptually correct, then these objects—although of completely different nature and in a completely different stage of evolution—should have wind momenta following the extrapolation of the WLR of massive stars towards lower luminosities. More exactly, since the Central Stars of PN are as hot as or even hotter than O-stars, we expect them to follow the WLR for O-stars.

Kudritzki et al. 1997 have used Keck HIRES spectra to investigate the wind momenta of CSPN. The result is striking (see Fig. 15). CSPN follow indeed the O-star extrapolation towards lower luminosities (although the scatter is admittedly large—for a discussion see their paper). Comparing with the wind momenta of A-supergiants, we conclude that the WLR must indeed be a function of effective temperature.

The basic idea of the existence of a WLR is thus very convincingly confirmed by the observations. The next step is to investigate, whether the theory of radiation driven winds is able to reproduce the observed relationship quantitatively—in particular its dependence on spectral type.

5. New wind models for supergiants of spectral type O, B and A

To be able to investigate the systematic behaviour of line driven winds across the whole HRD of hot stars (where "hot" means that hydrogen is ionized and can be as cool as $T_{eff} = 8000$ K for A-supergiants) in all different stages of evolution (including massive stars as well as post-AGB) we have developed a new approach to calculate the wind dynamics. This approach is based on the improvements achieved during the last decade with regard to atomic physics and line lists (see Pauldrach et al. 1998). We use the line list of $2.5\ 10^6$ lines of 150 ionic species and apply analytical formulae (see Lucy and Abbott 1993 , Springmann 1997, Springmann and Puls 1998) for a fast

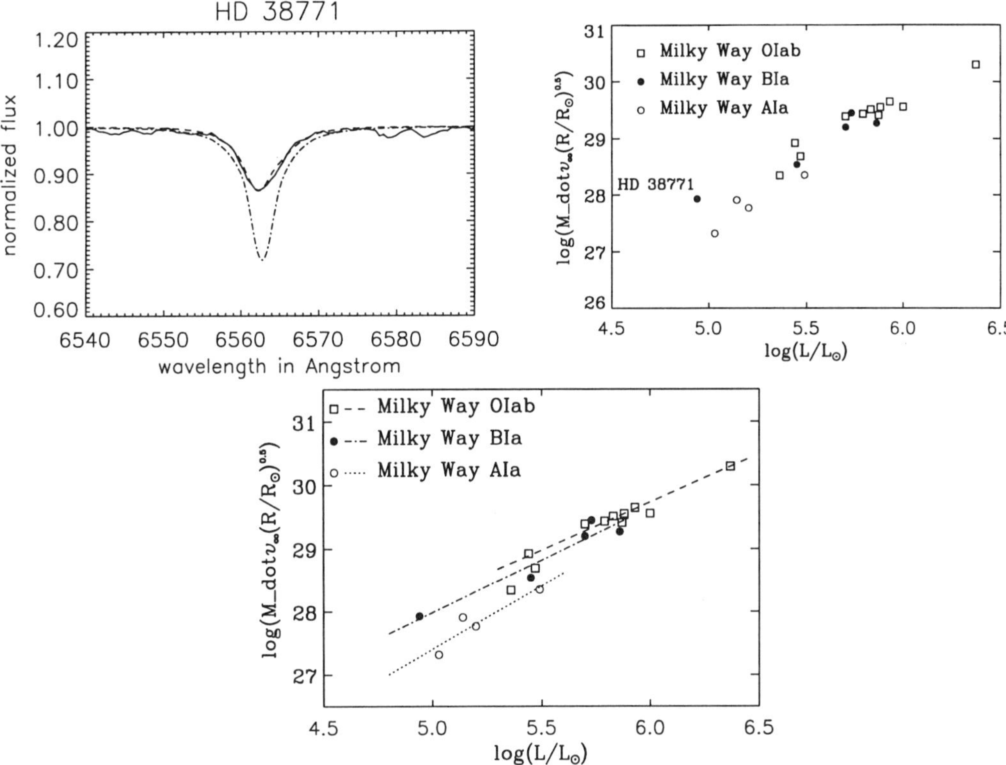

FIGURE 14. Upper left: H_α profile fit of the low luminosity supergiant HD 38771 (κ Ori, B0.5 Ia) with $\dot{M} = 2.8$ and $1.0\ 10^{-7}$ $M_\odot$/yr (dotted and dashed-dotted, respectively). The latter value is used to demonstrate that H_α even when in absorption is wind contaminated and allows a determination of $\dot{M}$. Upper right: The observed WLR including HD 38771. Bottom: Three independent linear regressions fitted to the three spectral types assuming a temperature dependence of the WLR.

approximation of NLTE occupation numbers to calculate the radiative line acceleration, which is then represented by a new parameterization using depth dependent force multiplier parameters (see Kudritzki et al. 1998a; note that now as an improvement real model atmosphere fluxes are used instead of Planck functions for the photospheric irradiation of the wind). Because of the depth dependent force multipliers a new formulation of the critical point equations is introduced (Kudritzki et al. 1998c), which includes the old algorithm (Pauldrach, Puls, & Kudritzki 1986) in the limit of neglected depth dependence. In this way, wind models can be calculated within a few seconds on a workstation for every hot star with specified effective temperature, mass, radius and abundances.

We have used this new approach to calculate wind models for typical parameters of A-, B- and O-supergiants. The results are compared with the observations in Fig. 16. Although much more work has yet to be invested in the future to compare theory and observation on a star by star basis, a first conclusion is that the theory is roughly able to reproduce the observed trends with regard to wind momentum and terminal velocity.

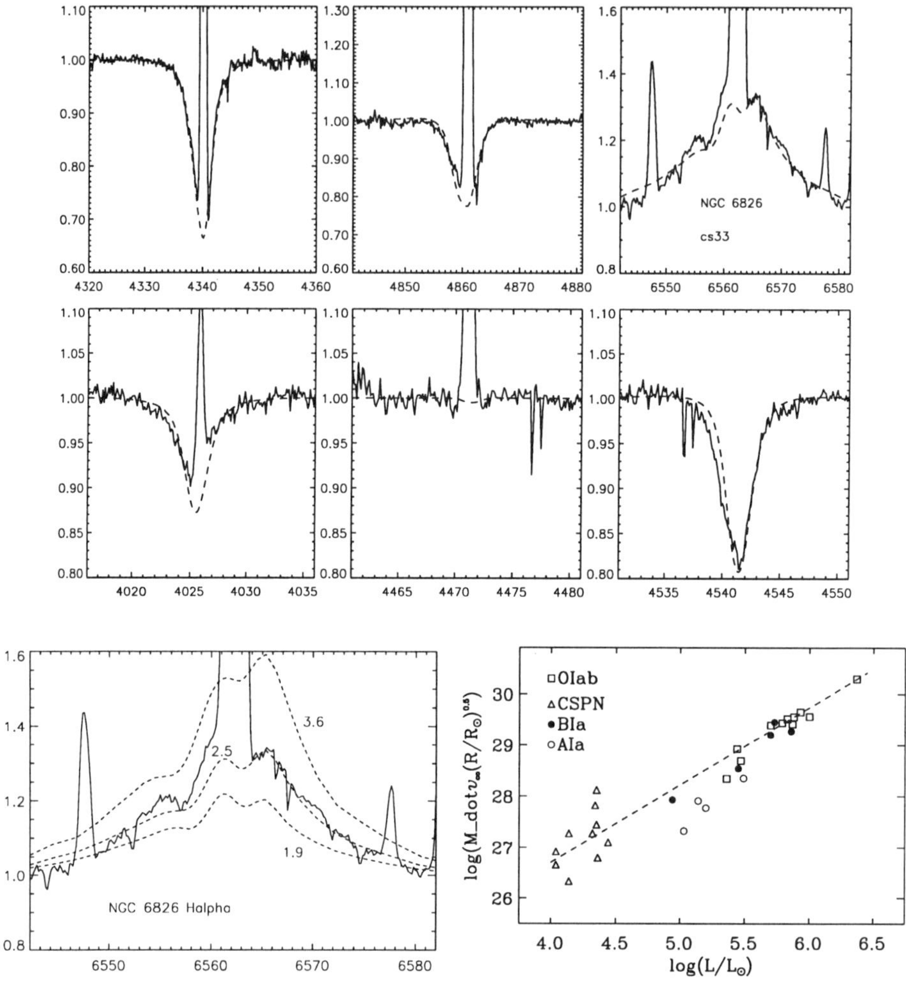

FIGURE 15. Top: Fit of the diagnostically important hydrogen and helium lines in the Keck HIRES spectrum of the PN Central Star NGC 6826 to obtain effective temperature, gravity and luminosity. Note that the presence of very strong nebular emission lines requires high resolution spectra. Lower left: H_α profile fits of NGC 6826 adopting changes of $\dot{M}$ by 20 percent. Again note the presence of nebular emission lines. Lower right: The observed galactic WLR including Central Stars of PN. the regression curve corresponds to the wind momentum of massive O-stars. (From Kudritzki et al. 1997).

6. Simple scaling relations for mass-loss, wind momentum and wind power

For many astrophysical applications, estimates of mass-loss, stellar wind momentum and energy are needed along sequences of evolutionary tracks. The WLR provides a simple way to obtain such estimates. From the observational and theoretical results obtained in the proceeding two sections, we suggest the use of two formulae

$$log\left\{\dot{M}v_\infty\left(\frac{R_*}{R_\odot}\right)^{0.5}\right\} = \mathbf{A} + \mathbf{x} * log(L/L_\odot) \ , \tag{6.27}$$

$$v_\infty = \mathbf{B} * v_{\text{esc}}^{\text{phot}} \ . \tag{6.28}$$

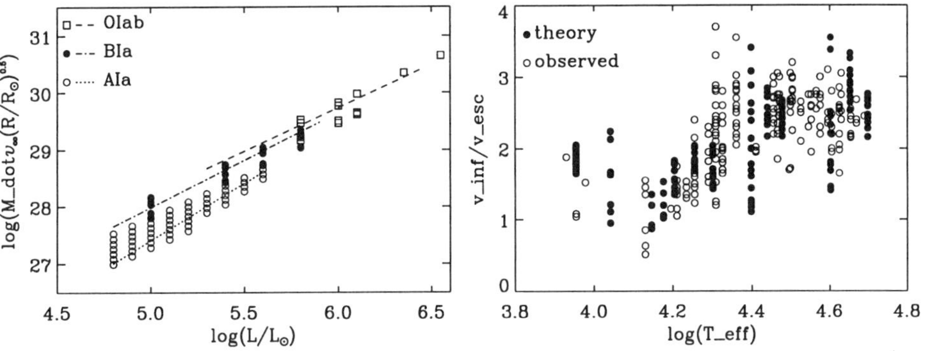

FIGURE 16. New wind calculations compared with the observations. Left: Wind momenta as function of luminosity. The dotted, dashed-dotted and dashed curves correspond to the observed regression curves of Fig. 14 for A-, B- and O-supergiants, respectively. The (open and filled) circles and squares correspond to the models. Their scatter at a given luminosity is a result of the different masses or effective temperatures adopted. Right: Ratio of terminal velocity to photospheric escape velocity as function of effective temperature. Open circles represent observations taken from Prinja and Massa (1998). Filled circles correspond to the calculations and the scatter at a given temperature is a result of the different masses and luminosities which were adopted.

spectral type	A	x	B
O Iab	20.65	1.51	2.5
B0 to B1 Ia	19.67	1.66	2.5
A0 to A3 Ia	17.37	2.00	1.5

TABLE 1. Fit formula coefficients for wind momentum and terminal velocity as function of spectral type. Work on the remaining spectral types is in progress.

The coefficients A, x, and B are given in Tab. 1. These formulae allow a quick computation of the dynamical properties of hot star winds. Note that the definition of the photospheric escape velocity includes a correction for the radiative acceleration due to Thomson scattering (see eq. (3.23)).

7. The wind momentum of the "Pistol Star"

As the result of the recent dramatic improvements in IR imaging and spectroscopy it became apparent that the Galactic Centre hosts a surprisingly large number of extremely luminous hot stars, many of them in advanced stages of stellar evolution (Krabbe et al. 1995, Najarro et al. 1994, Najarro et al. 1997). A spectacular example is the "Pistol Star", which has been studied by Figer, Najarro et al. 1998. According to their quantitative spectroscopic analysis using advanced Unified Model Atmospheres with stellar winds this object is very likely a mid-B-supergiant of $T_{eff} = 14000$ K and a huge radius of $R_* = 340$ $R_\odot$. The correspondingly high luminosity leads to the conclusion that the ZAMS-mass of the object must have been of the order of 250 $M_\odot$.

However, the wind momentum of the object is surprisingly low compared with its high luminosity. Figer, Najarro et al. 1998 obtain $\dot{M} = 3.8\ 10^{-5}$ $M_\odot$/yr and $v_\infty = 95$ km/sec

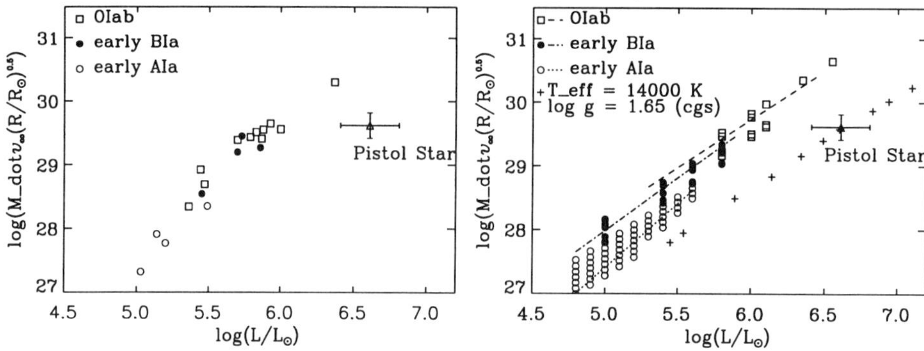

FIGURE 17. Left: The wind momentum of the very massive Galactic Centre "Pistol Star" compared with the wind momentum of the brightest blue supergiants in the solar neighbourhood. Right: A model sequence of wind momenta for $T_{eff} = 14000$ K and log g $= 1.65$ as function of luminosity reproducing the wind momentum of the Pistol Star. For comparison the theoretical models for early B-, and A-supergiants and O-stars are also shown together with the corresponding regression curves representing the observations.

for mass-loss rate and terminal velocity, respectively. Fig. 17 shows the wind momentum of the "Pistol Star" compared with the WLR of galactic supergiants.

What does this deviation from the observed WLR mean? Is the object in reality less massive and, therefore, less luminous? Or is it a cluster of unresolved stars rather than a single object? The careful work by Figer, Najarro et al. 1998 mostly rules out these possibilities.

Another explanation would be that at 14000 K radiation driven winds are much less effective and produce significantly less wind momentum. We have investigated this possibility by calculating wind models for B-supergiants with $T_{eff} = 14000$ K and log g $=$ 1.65 (cgs) as a function of luminosity. This sequence of models is also shown in Fig. 17. Indeed, the theory produces smaller wind momentum at this effective temperature and fits the observed value at the corresponding luminosity. (Note that log g = 1.65 corresponds to $M = 227$ M$_\odot$ at this luminosity, which would mean that the star has so far not lost much of its original mass).

8. The Wind Momentum–Luminosity Relationship and the determination of extragalactic distances

It has long been a dream of stellar astronomers to use the most luminous blue stars to determine the distances to other galaxies. For many decades, this dream has remained unfulfilled, mostly because the photospheres and spectral diagnostics are so enormously complicated by the presence of the strong stellar winds. However, now after many years of effort to include the effects of stellar winds in the quantitative spectroscopic diagnostics using hydrodynamic NLTE model atmospheres, the situation has changed dramatically and the presence of winds can be regarded as a gift of nature allowing the determination of stellar luminosities and, therefore, distances directly from the spectrum.

The clue is the existence of Wind Momentum–Luminosity Relationship. The basic technique for using the WLR is to derive the stellar parameters (temperature, gravity, metallicity) spectroscopically from optical absorption lines (see Kudritzki 1998 for references), and then to model the H_α profiles ($\rightarrow \dot{M}$ for O-, B- and A-supergiants, v_∞ for A-supergiants) and the UV P-Cygni line profiles ($\rightarrow v_\infty$ for O-, B- and A-supergiants) to obtain the wind momentum. Application of the empirically calibrated WLR appro-

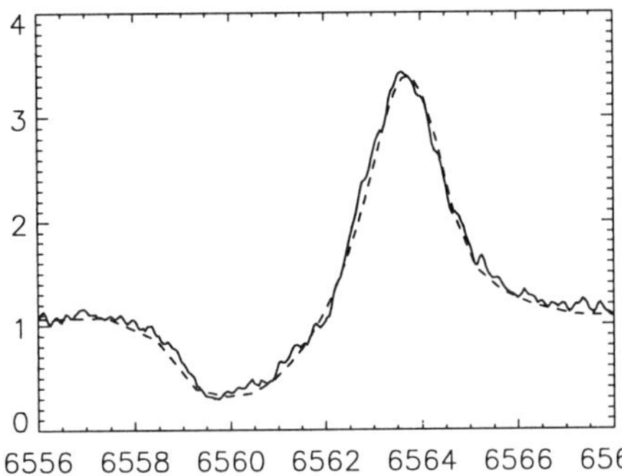
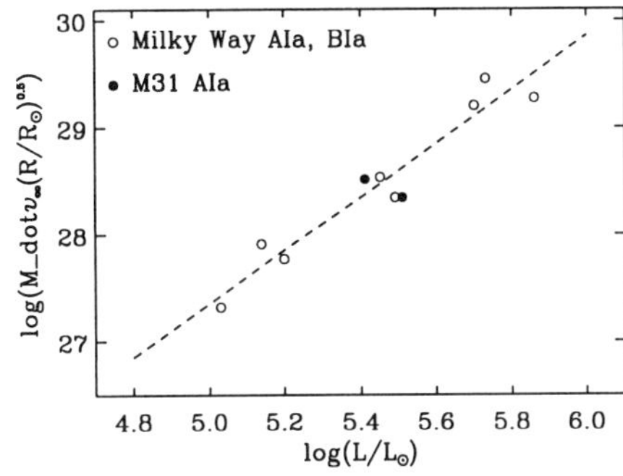

FIGURE 18. Left: H$_\alpha$ fit of the Keck HIRES spectrum of the M31 A2 Ia-O supergiant 41$-$3654 yielding a precise determination of $\dot{M}$ and v$_\infty$. Right: Wind momenta of two A-supergiants in M31 (obtained with Keck HIRES) compared with galactic A- and B-supergiants of the solar neighbourhood. Results from McCarthy et al. 1997.

priate to the spectroscopically determined metallicity then allows one to determine the intrinsic luminosity from which the distance follows once the stellar apparent magnitude and colours($\rightarrow$ reddening and extinction in conjunction with the stellar parameters determined spectroscopically) are known. In practice, one will adopt a distance for all stars in a galaxy to locate them in the momentum/luminosity plane and will then iterate the distance until the distribution coincides with the calibrated sample.

8.1. *A first investigation in M31*

A crucial test of the WLR method is the investigation of wind momenta of blue supergiants in M31. In an extensive collaboration (see acknowledgments), we have started a systematic study using the WHT and Keck telescopes for optical spectroscopy and the HST for multicolour photometry and UV spectroscopy. While most of the Cycle 6 and 7 HST observations have still to be carried out, we are able to present the first analysis of two M31 A-supergiants (see McCarthy et al. 1997) based solely on Keck HIRES optical spectroscopy in Fig. 18. The results are very encouraging. The H$_\alpha$ line profiles allow a precise determination of wind momenta. Adopting a distance modulus of 24.25 mag (Rozanski & Rowan-Robinson 1994) we find that both objects coincide very well with galactic WLR. McCarthy et al. 1997 estimate that with 10 to 20 objects it will be possible to obtain an independent M31 distance modules with an accuracy of 0.1 mag.

8.2. *Metallicity dependence. The Magellanic Clouds and the abundance gradient in M33*

It is very obvious that the WLR must depend on metallicity. Since the winds are driven by photon momentum transfer through metal line absorption, the wind momentum rate must be a function of stellar metallicity. As has been shown by Abbott 1982, Kudritzki et al. 1987 and Puls et al. 1996, this results in a metallicity dependence of $\dot{M}$ of the form

$$\dot{M} \propto \left[\frac{\epsilon}{\epsilon_\odot}\right]^{\frac{1-\alpha}{\alpha'}} , \tag{8.29}$$

where ϵ is the average metallicity and α is the exponent of the linestrength distribution function as defined in section 2. ($\alpha' = \alpha - 0.07$, see Puls et al. 1996).

The analysis of O-stars in the LMC and SMC by Puls et al. 1996 has indeed confirmed this metallicity dependence. Recent results for the WLR in the LMC and SMC are shown

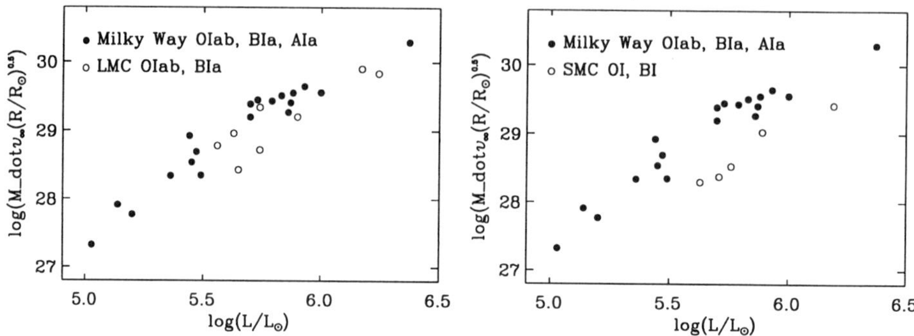

FIGURE 19. Wind momentum and luminosity of A-, B- and O-supergiants in the Galaxy compared with the LMC and SMC.

in Figs. 19. A small shift in wind momentum is found for the LMC, corresponding to an abundance difference of 0.3 dex. For the SMC the shift is more significant indicating an abundance lower by 0.8 dex. The differences in abundance inferred from shifts in wind momentum are in rough agreement with the abundances obtained directly from the quantitative analysis of HST and ground-based spectra (see Haser et al. 1998 and references in Kudritzki 1998). However, there are still not enough objects studied in the Magellanic Clouds to allow a firm conclusion. An open question—only to be answered by an investigation of a larger sample—is whether the exponent α, and therefore the slope of the WLR, change with abundance. This work is presently under way.

For the determination of extragalactic distances using either Cepheids or blue supergiants it is important to realize that abundances may vary not only from galaxy to galaxy but also within one galaxy because of the intrinsic abundance gradients. Blue supergiants in the outskirts of a spiral galaxy will, therefore, have weaker winds than objects in the inner regions. The pulsation properties of Cepheids might differ as well.

To investigate the influence of abundance gradients on the WLR method we have started a project on blue supergiants in M33, a Local Group galaxy with a well pronounced abundance gradient identified in stars as well as in H II-regions (Fig. 20). First results have been presented by McCarthy et al. 1995, where two A-supergiants have been studied, B 324 and M 117-A. B 324, located only 1.3 kpc away from the centre of M33, has roughly solar abundance, whereas M 117-A, at 4.5 kpc galactocentric distance, has a metallicity clearly below the SMC. The wind momenta of these two objects are strikingly different (see again Fig. 20). As to be expected from the abundance analysis, the solar composition object coincides roughly with the galactic WLR. On the other hand, the extremely metal poor object falls below the SMC relationship, again in agreement with expectations. *We conclude that with an empirical calibration of wind momentum as a function of metallicity, it will also be possible to apply the WLR method for distance determinations to spiral galaxies with a strong metallicity gradient.*

8.3. The potential of the WLR-method

The application of the WLR method does not necessarily require spectra of high S/N or high spectral resolution. As a very illustrative example we have used the H_α and H_β equivalent widths of A-supergiants published by Tully and Wolff 1984 to derive gravity and—most importantly—wind momenta (adopting $v_\infty = 150$ km/s for all objects). Fig. 21 shows the resulting correlation between wind momentum and absolute magnitude for those objects where a good simultaneous determination of gravity and mass-loss rate

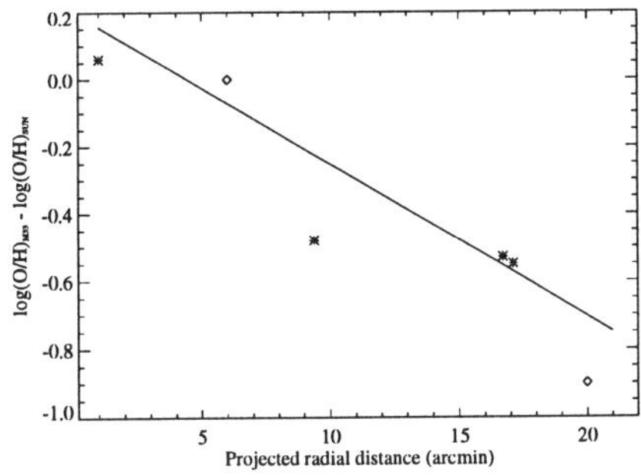
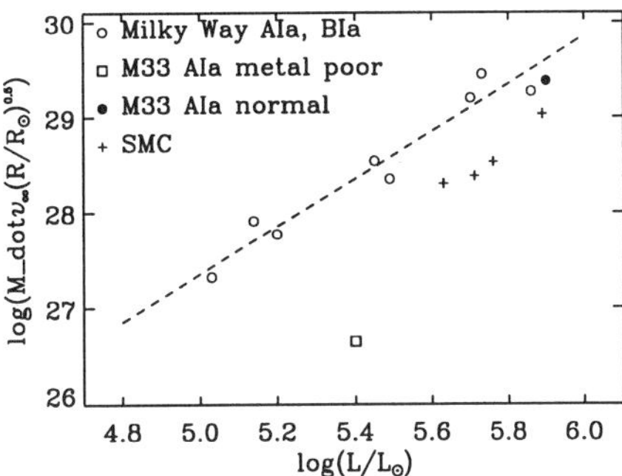

FIGURE 20. Left: The abundance gradient in M33. Asterisks refer to oxygen from B-supergiants, diamonds to iron from A-supergiants, The solid line represents the average result obtained for oxygen from H II-regions. From McCarthy et al. 1995 and Monteverde et al. 1997. Right: Wind momentum (in cgs units) and luminosity of the two A-supergiants in M33 compared with galactic and SMC supergiants. (From McCarthy et al. 1995).

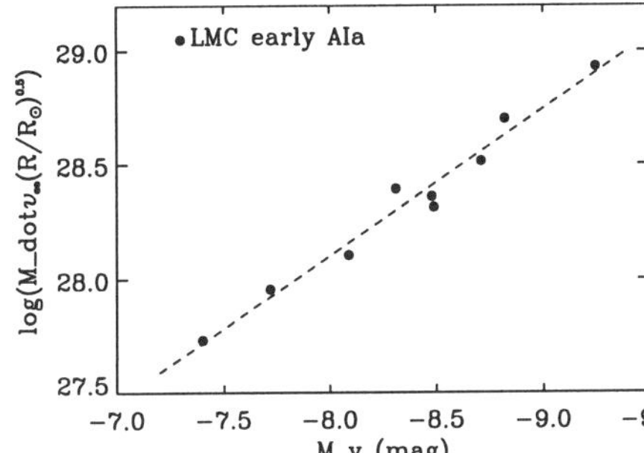
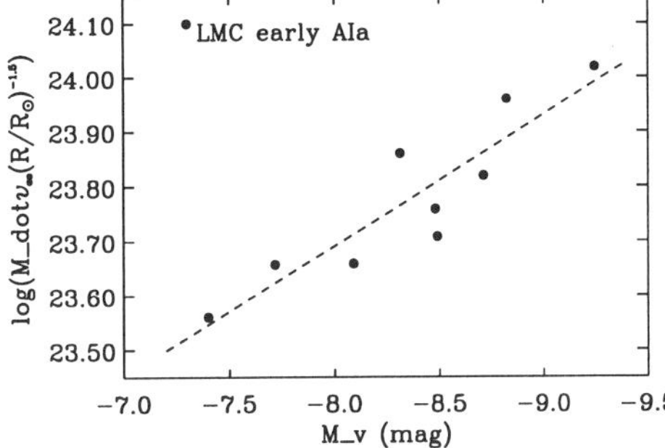

FIGURE 21. Left: Wind momentum determined from the Balmer line equivalent widths of LMC AO Ia—supergiants published by Tully and Wolff (1984) as function of absolute magnitude (adopting a distance modulus of 18.5 mag). From Knoerndel et al., 1998. Right: Same as the left figure, but instead of wind momentum the invariant Q of the H_α fitting (see text) is plotted.

was possible (for details, see Knoerndel et al. 1998). Despite the limited quality of the data, the tightness of the relation obtained is striking.

If one wants to use this observed relationship to estimate the possible accuracy of distance determinations, it is important to realize that determining a mass-loss rate from an H_α profile requires an assumption about the stellar radius and therefore the distance. Thus, both wind momentum and absolute magnitude depend on the radius adopted. However, as one can show analytically (Puls et al. 1996), the quantity

$$Q = \dot{M} v_\infty \left(\frac{R_*}{R_\odot} \right)^{-1.5} \tag{8.30}$$

is an invariant of the profile fitting, i.e. the strength of H_α as a wind line depends exactly on this quantity. As a consequence, we also plot Q as a function of M_V in Fig. 21 to quantify the distance independent scatter. The standard deviation in absolute magnitude from the mean relation $Q = f(M_V)$ is on the order of 0.3 mag. This confirms the estimate by McCarthy et al. 1997 that with ten to twenty objects per galaxy, distance moduli as accurate as 0.1 mag should be achievable.

The uncertainties in WLR distance moduli might, therefore, be comparable to those obtainable from Cepheids in galaxies. The advantage of the WLR-method, however, is that individual reddening (and therefore extinction) as well as metallicity can be derived directly from the spectrum of every object. Moreover, it is a new independent primary method for distance determination and can contribute to the investigation of systematic errors of extragalactic distances. However, the crucial question is to what distances will the method be applicable.

The best spectroscopic targets at large distances are A-supergiants. Since massive stars evolve at almost constant luminosity towards the red, A-supergiants are the optically brightest "normal" stellar objects because of the effects of Wien's law on the bolometric correction. In addition, for these objects we can determine the wind momentum solely by optical spectroscopy at H_α (without need of the UV). This means that we can use ground-based telescopes of the 8m class for spectroscopy rather than the tiny HST (which is then needed for accurate photometry only).

From Fig. 21 we see that the brightest A-supergiants have absolute magnitudes between -9 and -8 mag (objects like 41–3654 and 41–3712 in M31 are other examples). Such objects would be of apparent magnitude between 20 and 21 in galaxies 6 Mpc away, certainly not a problem for medium (2 Å) resolution spectroscopy with 8 m class telescopes. Even in a galaxy like M100 at a distance of 16 Mpc (Freedman et al. 1994b; Ferrarese et al. 1996) these objects would still be accessible at magnitudes around 22.5 and would yield wind momentum distances if medium resolution spectroscopy on a 8 m class telescope were combined with HST photometry. Indeed, the HST colour magnitude diagram published by Freedman et al. 1994a may show the presence of such objects in M100.

One might ask, of course, whether a medium resolution of 2 Å would still allow a sufficiently accurate determination of effective temperature, surface gravity, abundances, and wind momentum. We note that the rotational velocities of A-supergiants are typically on the order of $40 \, \mathrm{km \, s^{-1}}$, a broadening which is matched by a spectral resolution of roughly 1.4 Å at $H\alpha$. Therefore we expect that the medium resolution situation will not be dramatically worse, provided sufficient S/N can be achieved and accurate sky-, galaxy-, and H II region-background subtraction can be performed at very faint magnitudes. Experiments with observed spectra degraded with regard to S/N, resolution and sky emission indicate that such observations are challenging but feasible.

In summary, the WLR results obtained so far in the Galaxy and in the Local Group galaxies from LMC, SMC to M33 and M31 are very encouraging. We are optimistic that after further tests and calibration steps, the WLR method can make an independent contribution to the determination of extragalactic distances. The obvious next steps for future work are:

- the addition of more objects with reliable distances to the galactic WLR
- the investigation of the impact of spectral variability
- further calibration of the WLR metallicity dependence using LMC and SMC supergiants and other Local Group galaxies of even lower metallicity
- the addition of more targets in M31 and M33 including HST photometry and spectroscopy (the latter for B-supergiants to measure v_∞ from UV resonance lines) to obtain independent distances to these galaxies and to study their abundance gradients
- the investigation of blue supergiants beyond the Local Group out to the Virgo cluster of galaxies.

Work in all of these directions is now under way. *We are confident that, in the end, we will have a new distance determination method capable of reaching out to the Virgo and Fornax clusters of galaxies.*

Much of the work presented here is the result of an intense collaboration between many colleagues working on hot stars. It is a pleasure to thank Artemio Herrero and Ilu Monteverde from the IAC, Jim McCarthy (Caltech), Kim Venn (Minnesota), Stephen Smartt (La Palma) and the Munich crowd Rudi Gabler, Stefan Haser, Oliver Knoerndel, Margie and Danny Lennon (now at La Palma), Paco Najarro (now Madrid), Adi Pauldrach, Joachim Puls, Uwe Springmann and Gudrun Taresch.

In addition, I want to thank the staff and director of Steward Observatory, Tucson, Arizona, for their hospitality and support during the final phase of this work. Special thanks go to Dr. Tom Fleming for his careful reading of the final manuscript.

REFERENCES

ABBOTT, D. C. 1979 *Proc. IAU Symp. 83*, (eds. P. S. Conti & C. W. H. de Loore). p. 237.

ABBOTT, D. C. 1982 *ApJ* **259**, 282.

CASTOR, J. I. 1970 *MNRAS* **149**, 111.

CASTOR, J. I., ABBOTT, D. C., & KLEIN, R. I. 1975 *ApJ* **195**, 157.

FERRARESE, L. ET AL. 1996 *ApJ* **464**, 568.

FIGER, D. F., NAJARRO, F., MORRIS, M., MCLEAN, I. S., GEBALLE, T. R., GHEZ, A. M., LANGER, N. 1998 *ApJ*, in press.

FREEDMAN, W., ET AL. 1994a *ApJ* **435**, L31.

FREEDMAN, W. ET AL. 1994b *Nature* **371**, 757.

GABLER, R., GABLER, A., KUDRITZKI, R. P., PAULDRACH, A. W. A., & PULS, J. 1989 *A&A* **226**, 162.

GABLER, R., KUDRITZKI, R. P. & MENDEZ, R. H. 1991 *A&A* **245**, 587.

GABLER, R., GABLER, A., KUDRITZKI, R. P. & MENDEZ, R. H. 1992 *A&A* **265**, 656.

HAIMAN, Z., LOEB, A. 1997 *ApJ* **483**, 21.

HASER, S. M., PAULDRACH, A., LENNON, D. J., KUDRITZKI, R. P., LENNON, M., PULS, J., VOELS, S. A. 1998 *A&A* **330**, 285.

HILLIER, D. J., MILLER, D. L. 1998 *ApJ* **496**, 407.

KNOERNDEL, O., KUDRITZKI, R. P., PULS, J., LENNON, D. J. 1998, *A&A*, in preparation.

KRABBE, A., GENZEL, R., ECKART, A., NAJARRO, F., LUTZ, D., CAMERON, M., KROKER, H., TACCONI-GARMAN, L. E., THATTE, N., WEITZEL, L., DRAPATZ, S., GEBALLE, T., STERNBERG, A., KUDRITZKI, R. 1995 *ApJ* **447**, L95.

KUDRITZKI, R. P., PAULDRACH, A. W. A., PULS, J. 1987 *A&A* **173**, 293.

KUDRITZKI, R. P. 1988. In 18th Advanced Course of the Swiss Society of Astrophysics and Astronomy (Saas-Fee Courses) *Radiation in Moving Gaseous Media* (eds. Y. Chmielewski and T. Lanz). p. 1. Geneva Observatory.

KUDRITZKI, R. P., PAULDRACH, A., PULS, J. 1988 "Radiation driven winds of hot luminous stars." In *O-stars and Wolf-Rayet stars* (eds. P. S. Conti & A. B. Underhill), p. 173. NASA Sp-497.

KUDRITZKI, R. P., PAULDRACH, A. W. A., PULS, J., & ABBOTT, D. C. 1989, A&A **219**, 205.

KUDRITZKI, R. P., LENNON, D. J., HASER, S. M., PULS, J., PAULDRACH, A., VENN, K., & VOELS, S. A. 1996a. In *Science with the Hubble Space Telescope II* (eds. P. Benvenuti et al.). p. 285.

KUDRITZKI, R. P., MENDEZ, R. H., MCCARTHY, J. K., PULS, J., 1997. *Planetary Nebulae* (eds. H. J. Habing and H. J. G. L. Lamers). Proc. IAU Symp. 180. p. 64. Kluwer.

KUDRITZKI, R. P., SPRINGMANN, U., PULS, J., PAULDRACH, A. W. A., LENNON, M. 1998a, ASP Conf. Series Vol. 131, p. 299.

KUDRITZKI, R. P., PULS, J., LENNON, D. J., VENN, K. A., MCCARTHY, J. K., & HERRERO, A. 1998b *A&A*, in preparation.

KUDRITZKI, R. P., SPRINGMANN, U., PULS, J., PAULDRACH, A. W. A., LENNON, M. 1998c *A&A*, in preparation.

KUDRITZKI, R. P. 1998, Quantitative Spectroscopy of the Brightest Blue Supergiant Stars in Galaxies. In Proc. of *Stellar Astrophysics for the Local Group*, VIII Canary Island Winterschool for Astrophysics, (eds. A. Aparicio et al.). p 149. Cambridge University Press.

LUCY, L. B. & ABBOT, D. C. 1993 *ApJ* **405**, 738.

MCCARTHY, J. K., LENNON, D. J., VENN, K. A., KUDRITZKI, R. P., PULS, J., & NAJARRO, F. 1995 *ApJ* **455**, L35.

MCCARTHY, J. K., KUDRITZKI, R. P., LENNON, D. J., VENN, K. A., PULS, J. 1997 *ApJ* **482**, 757.

MONTEVERDE, I., HERRERO, A., LENNON, D. J., & KUDRITZKI, R. P. 1997 *ApJ* **474**, L107.

NAJARRO, F., HILLIER, D. J., KUDRITZKI, R. P. ET AL. 1994 *A&A* **285**, 573.

NAJARRO, F., KUDRITZKI, R. P., CASSINELLI, J. P., STAHL, O., HILLIER, D. J. 1996 *A&A* **306**, 892.

NAJARRO, F., KRABBE, A., GENZEL, R., KUDRITZKI, R. P., HILLIER, D. J. 1997 *A&A* **325**, 700.

PAULDRACH, A. W. A., PULS, J., & KUDRITZKI, R. P. 1986 *A&A* **164**, 86.

PAULDRACH, A. W. A., PULS, J., KUDRITZKI, R. P., MENDEZ, R. H. & HEAP, S. R. 1989 *A&A* **207**, 123.

PAULDRACH, A. W. A., KUDRITZKI, R. P., PULS, J., BUTLER, K., & HUNSINGER, J. 1994 *A&A* **283**, 525.

PAULDRACH, A. W. A., LENNON, M., HOFFMANN, T. L., SELLMAIER, F., KUDRITZKI, R. P., PULS, J. 1998 *ASP Conference Series Vol. 131*, p. 258.

PRINJA, R. K., MASSA, D. L. 1998, *ASP Conf. Series, Vol. 131*, p. 218.

PULS, J. 1987 *A&A* **184**, 227.

PULS, J. 1993 *Habilitationsschrift*, University of Munich

PULS, J., KUDRITZKI, R. P., HERRERO, A., PAULDRACH, A., HASER, S. M., LENNON, D. J., GABLER, R., VOELS, S. A., VILCHEZ, J. M., WACHTER, S., & FELDMEIER, A. 1996 *A&A* **305**, 171.

ROZANSKI, R., & ROWAN-ROBINSON, M. 1994 *MNRAS* **271**, 530.

SANTOLAYA-REY, E., PULS, J., & HERRERO, A. 1997 *A&A* **323**, 488.

SCHAERER, D. & SCHMUTZ, W. 1994 *A&A* **288**, 231.

SCHAERER, D. & DE KOTER, A. 1997 *A&A*, in press.

SPRINGMANN, U. 1997 *doctoral thesis*, University of Munich.

SPRINGMANN, U. & PULS, J. 1998 *ASP Conf. Series, Vol. 131*, p. 286.

STEIDEL, C. C., GIAVALISCO, M., PETTINI, M., DICKINSON, M., ADELBERGER, K. L. 1996 *ApJ* **462**, L17.

TARESCH, G., KUDRITZKI, R. P., HURWITZ, M., BOWYER, S., PAULDRACH, A. W. A., PULS, J., BUTLER, K., LENNON, D. J., HASER, S. M. 1997 *A&A* **321**, 531.

TULLY, R. B. & WOLFF, S. C. 1984 *ApJ* **281**, 67.

YEE ET AL. 1996 *AJ* **111**, 1783.

Mass loss from cool stars

By L. A. WILLSON

Department of Physics and Astronomy, Iowa State University, Ames, IA 50011

Physical processes important for mass loss from cool stars are reviewed in the context of heavily mass-losing asymptotic giant branch (AGB) stars. Hydrodynamical models that take into account departures from radiative equilibrium (RE) and local thermodynamic equilibrium (LTE) are quite different from otherwise similar models assuming LTE and/or RE. I stress that for the range of densities and temperatures that is relevant to theoretical studies of stellar winds, it is not possible to separate the radiative transfer from the hydrodynamic calculations, because when the dynamical timescale ($|1/\rho\; d\rho/dt|^{-1}$) is shorter than the thermal relaxation timescale, near-adiabatic compression and expansion lead to departures from radiative equilibrium. Observed mass loss relations such as Reimers' relation ($\dot{M} \propto LM/R$) are reproduced very accurately by the latest Bowen models but with a very different interpretation than is usually made of the significance of such relations: Such relations tell us the parameters of stars that are losing mass, rather than the mass loss rate associated with a given set of stellar parameters. I argue that such empirical relations are produced primarily by selection effects resulting from an extremely steep dependence of mass loss rates on stellar parameters. There are also at least two very different classes of models for mass loss from evolved cool stars currently in widespread use. These two kinds of models suggest very different interpretations of the radio emission ($\sim$ 8–12 GHz) observed and discussed by Reid and Menton (1997). I conclude that we are still in the "polytropic" era of interpreting mass loss observations for most stars, and that substantial improvements in our understanding of the nature and evolutionary role of mass loss from stars are likely to occur as the models improve.

1. The framework

Nearly all cool stars show evidence for winds. Evidence for stellar winds are found for stars along the main sequence, stars in the instability strip, G and K giants (redward of the "Linsky-Haisch" (1979) dividing line), red giants (on the first ascent Red Giant Branch (RGB), and the final ascent asymptotic giant branch (AGB) stars, supergiants (roughly, luminous red stars with $M > 5$ M$_\odot$), and post-AGB stars. For the highly evolved stars at the tip of the AGB—Miras, OH-IR stars, and other long period variable stars—mass loss determines their evolution and their fate.

In this review, I will emphasize AGB mass loss for several reasons. (1) The mass loss rates are high, so there is dramatic evidence for mass loss. (2) This mass loss stage plays a key role in stellar evolution. (3) Mass loss rates are probably different for low-metallicity stars, which has potentially profound implications for the appearance of stellar populations and for the chemical evolution of galaxies. (4) The majority of the papers brought up by searches on subjects related to mass loss from cool stars dealt with this mass loss epoch.

The theme for this review may be expressed succinctly in the words of Will Rogers:

"It ain't what you don't know that'll hurt you—it's what you do know that ain't true."

Most of what we "know" about cool star mass loss comes from empirical results interpreted via relatively simple models. Can we trust these results, or are there possibly big surprises in store? An example of what I mean by big surprises from recent astronomical history: Between 1935 and 1955, astronomers gradually became convinced that the main sequence is NOT an evolutionary track, and main sequence lifetimes are long. Detailed evolutionary calculations showed that the true evolutionary tracks wandered from the main sequence to the region of the giants, but with some of the evolutionary stages being short-lived so those regions of the HRD are lightly populated compared with the main sequence. The main sequence was recognized to be showing us which stars are converting H to He in their cores, rather than (as initially supposed) the evolutionary track followed by most stars.

In what follows, I will emphasize two kinds of tests of mass loss models. First, statistical comparisons with populations of stars: A new interpretation for the widely-used (or misused) Reimers' formula (Reimers 1979), and a perfect fit to the $P - L$ relation for Miras. Second, comparison with observations of individual stars: Radio continuum observations of cool Mira variables interpreted with two very different classes of models.

For most cool stars the energy and momentum going into mass loss are a very small "left-over" fraction of those going into structuring the inner atmosphere. For the luminosity going into removing mass from the star,

$$L_{\dot{\mathrm{M}}} = \frac{1}{2}\dot{\mathrm{M}}\left(v_\infty^2 + 2GM/R\right) \lesssim 10^{-4}L^* \text{ if } \dot{\mathrm{M}} \lesssim \dot{\mathrm{M}}_{\mathrm{Reimers}} \text{ and } v_\infty \approx v_{\mathrm{escape}}$$

Since the fraction of L going into mass loss is small, if we are even just slightly off in computing what happens to the rest of L we cannot hope to predict or understand the mass loss rates. Thus, we should be concerned about other ways that our models may be inadequate—for example, because the underlying atmospheric temperature distribution is not spatially homogeneous. Much published work on cool stars has used "empirical chromospheres" which are now suspect in light of new studies of the solar chromosphere. The reader is referred to papers by Ayres & Rabin 1996; Ayres 1981; and Carlsson & Stein 1997, 1995, and 1992.

2. Mass loss relations for Red Giants—A new interpretation

For the past 15 years, Bowen has been developing a code whose aim is to allow us to understand the mechanism and consequences of the mass loss that ends the red giant stages of intermediate mass stars (taken to be 0.7 to x solar masses with $5 < x < 8$ in Pop I). In general terms, the philosophy has been to identify all processes that affect the outcome and to find ways to incorporate those processes to a sufficient degree of accuracy that their effect on the results are modest and able to be estimated reliably. From the point of view of the evolutionary consequences, the "outcome" is the range of stellar parameters over which the mass loss rate becomes appreciable. This "cliff" location (see below) for oxygen-rich Miras is now reliable (within acceptable uncertainty in the terminal luminosity and final mass predicted) over the full range of uncertainty associated with the current approximations in all processes that have been found to be important. For details, see Bowen (1988 and 1999 (in prep)).

In tracing the evolution of AGB star mass loss, it turns out to be quite important to take into account that the stars are evolving towards increasing luminosity with decreasing effective temperature and, at the very end, also decreasing mass. Thus, for computing mass loss rates for stars evolving up the AGB, we have essentially three free parameters: L, M, and Z. From these, we can obtain all the other parameters of interest: Thus L and M together with evolutionary tracks give R and T_{eff}; M and R, with pulsation relation and mode, give pulsation period P. The six quantities L, M, Z, P, R and T_{eff} are all relevant parameters for the computation of mass loss rates, but the evolution can be traced with a subset of three of these because there are three constraints relating them to each other (evolutionary track, definition of T_{eff}, and *PMR* relation).

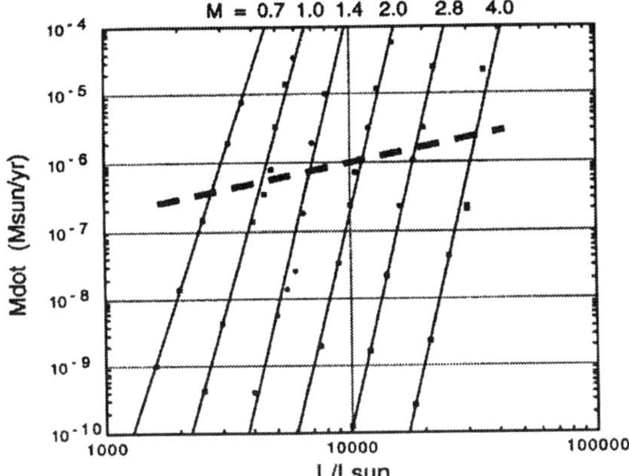

FIGURE 1. Mass loss rates for solar-metallicity stars evolving up the AGB, from Bowen's 1995 grid, displayed vs. $\log L$ for various values of stellar mass M. This M is not the progenitor mass but the "current" mass; each dot represents a calculated model. A star will evolve up the constant mass track until it gets close to the dashed line ($\dot{M}_{\mathrm{crit}} = 5 \times 10^{-7}\ M_* \ \mathrm{year}^{-1}$); then the stellar track will curve upwards to approach a vertical line (constant L) in this diagram. Note that the steep dependence of M on L shown here does *not* imply that the luminosity is the key factor governing the mass loss rate: rather, L is used to represent a point on the evolutionary track with T and R determined by L and the "LTM" relation (= the evolutionary track for mass M for a given metallicity Z).

Figure 1 shows $\log \dot{M}$ vs. $\log L$ for a grid of AGB mass loss models computed by Bowen in 1995. This set of models was constrained by evolutionary tracks from an analytic expression by Iben (1984) and a *PMR* relation by Ostlie and Cox (1986) for fundamental mode pulsation. From this grid of mass loss models, we obtain predicted mass loss rates $\dot{M}(L, M, Z)$. Then, analysis yields predictions for the terminal L and final M for a given M_{initial} and metallicity Z. Note that our choice of L as the "time" variable in no way implies that it is the physically dominant parameter determining mass loss rates: it is merely the most convenient choice. The lines drawn in Figure 1 are fits of the form $\dot{M} \propto L^{\alpha}\ M^{\beta}$, with the resulting α between 11 and 16, and β between 16 and 23, where L and M refer to the properties of a star at a particular moment in its evolution. *This steep dependence on M and L (or R and/or T_{eff}) is very different from most "empirical relations." Why?*

Further analysis shows that the "empirical relations" have been misleading us. Figure 2 shows the evolutionary tracks in a plot of $\log M$ vs. $\log L$ that correspond to the mass loss rates shown in Figure 1. Because the mass loss rates increase so precipitously in a narrow range of L for each M, each star evolves at nearly constant mass until it is close to the critical value of (L, M). There, $|d \log M/dt|$ crosses $|d \log L/dt|$ and the star loses the rest of its envelope at nearly constant L. We might say it "goes over the cliff" at this point. We sometimes refer to Figure 2 as the "lemming diagram," for obvious reasons.

Now we can look at the implications of this picture for empirical mass loss relations. Stars that are not yet near the cliff have low—unobservable—mass loss rates. Stars that have gone over the cliff have very short life expectancy, and mass loss rates high enough that they are detectable in the IR but not the visible. Thus, the majority of stars that are picked up in surveys for mass-losing red giants will be those that are close to the edge of the cliff. Figure 3 shows the result of taking the stars with $d \log L/dt = d \log M/dt$— the "cliff" stars—and those with $1/10$ and $10 \times$ the critical mass loss rate ("slope - 0.1" and "slope -10") and plotting their $\dot{M}$ vs. their LR/M. From the remarkable agreement

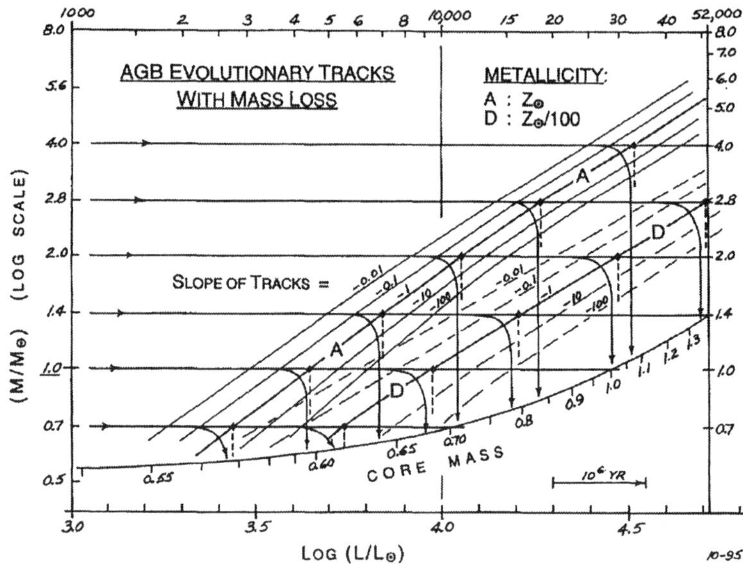

FIGURE 2. Evolution of AGB stars displayed as the evolution of M with increasing L. A star evolves to the right in this plot with roughly constant $d\log L/dt$. For this sample plot, Paczynski's core mass-luminosity relation was used to determine $d\log L/dt$ and to plot the core mass curve. Where $\dot{M} \simeq \dot{M}_{crit} = M\, d\log L/dt$ the star's evolutionary track is inclined by 45°; this is the edge of the "cliff" over which the stars fall to their deaths (to $M = M_{core}$). Evolutionary tracs and "cliff" lines are shown for two initial compositions: Solar and 0.01 times solar metallicity (lines A and D). The cliff shifts to higher L for lower Z because (a) the evolutionary tracks shift and (b) dust is only able to assist mass loss in the highest metallicity stars.

of the "cliff" locus with "Reimers' relation," we conclude that *Reimers' relation tells us the properties of stars that are losing mass and* **not** *the mass loss rate that arises from a certain set of stellar parameters.* It has been widely misunderstood and widely misused in stellar evolution and stellar population studies. Another way to say this: For a narrow range of stellar parameters there is a broad range of mass loss rates. The Reimers correlation is found because a relatively narrow range of mass loss rates is observable and the stellar parameters for those stars with observable mass loss rates are very narrowly constrained. Note also that the *maximum mass loss rates detected for any class of stars and for any mechanism of mass loss will always tend to be those for which* $\dot{M} = (M/L)dL/dt$ because stars do not spend much time where $\dot{M}$ is higher.

In Figures 4 and 5 further evidence is given that the Bowen models, with their precipitous mass loss rate increase with L or time, are in agreement with observational constraints. In Figure 4, a simple relation between $K - L$ and mass loss rate has been used to plot the Bowen model lines over the observations. It shows, among other things, that at a given P there is a wide range of $K - L$, therefore probably a wide range in mass loss rate. In contrast, at given P there is a narrow range of L for the Miras, identifying these as stars occupying a narrow range of L and R near the cliff, as is illustrated in Figure 5. Again, the remarkable fit to the observed $P - L$ relation was obtained without any "parameter-twiddling" in the grid of models.

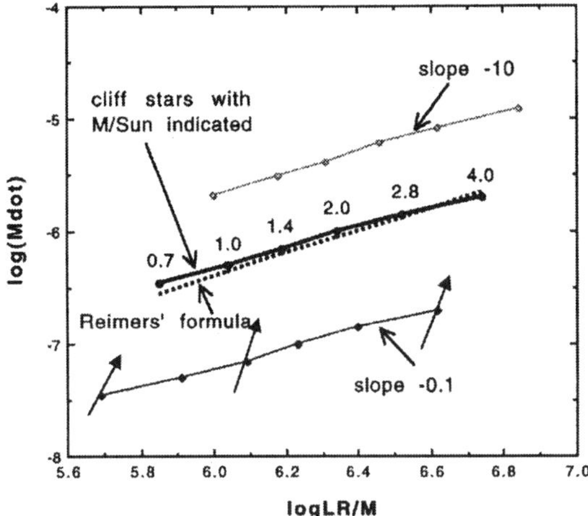

FIGURE 3. What is the significance of the "Reimers' relation," an observed correlation of the mass loss rate for red giants with LR/M? Here, the location of the "cliff" of Figure 2 ($\dot{M} = \dot{M}_{\mathrm{crit}}$) and of $\dot{M} = 0.1\ \dot{M}_{\mathrm{crit}}$ and $= 10\ \dot{M}_{\mathrm{crit}}$ (labeled "slope $-0.1''$ and "slope $-10''$) are plotted vs. LR/M. The empirical relation coincides neatly with the "cliff," thus suggesting that the observed relation results from a very steep dependence of mass loss rate on stellar parameters combined with a relatively narrow (± 1 dex) range of mass loss rates that are most likely to be detected. Evolution towards the cliff line occurs as indicated by the three arrows crossing the "slope $-0.1''$ line. Thus Reimers' relation and other observed correlations tell us the properties of stars that are losing mass at nearly the maximum rate, and not how a given star loses mass.

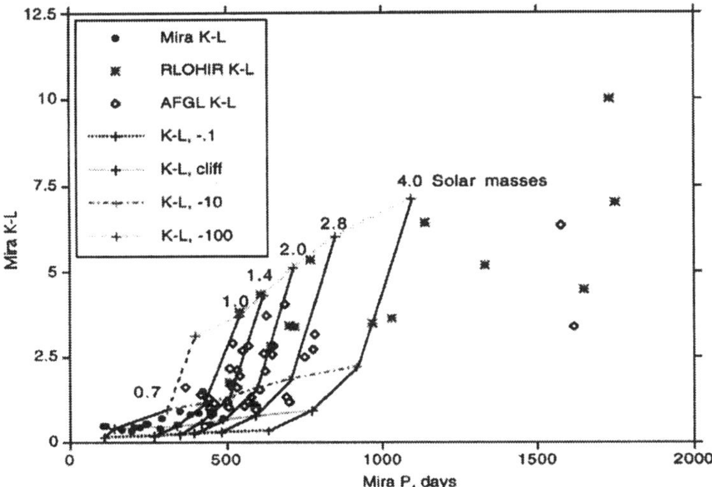

FIGURE 4. The IR color $K - L$ is a good indicator of mass loss rates when those are high enough (Lawrence et al. 1990). Using the mass loss rates and periods for the grid of models that produced Figures 3–6 with a formula related mass loss rate to the $K - L$ color ($K - L = 0.0275 + 624.5(\dot{M})^{1/2}$, from T. J. Jones (priv. comm.)) produces the grid of lines overlaying the observed $K - L$ vs. P for Miras and OH/IR stars. The Miras are located on or near the cliff line, while OH-IR stars and AFGL sources range up to much higher mass loss rates at the same periods. Thus, for a given P (implying a narrow range of stellar properties) the mass loss rate ranges over several orders of magnitude.

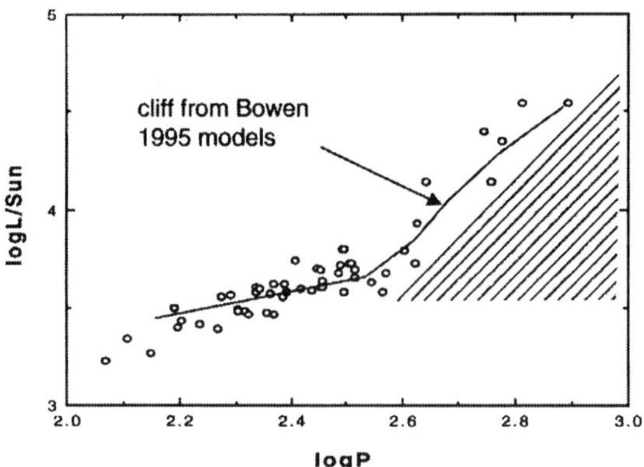

FIGURE 5. Luminosity vs. period for Miras. Data points are taken from Feast et al. 1989. In Figure 4, the cliff line followed the locus of the Miras, suggesting this identification of Miras as stars on the cliff edge. Here, again, the cliff line from Bowen's 1995 models fits the observations with no adjustment of parameters. The OH-IR stars occupy the shaded area; as a star loses mass at nearly constant L it moves to the right to occupy this region.

3. Two classes of Mira models

A stellar wind or mass loss from a star is the result of incomplete cancellation of energy and momentum flows into the atmosphere with energy lost to radiation and the restoring force of gravity. Usually, the energy going into mass loss is a very small fraction of the total luminosity. Usually, the velocity (at infinity) of the wind is within a factor of two of the escape velocity in the region where the wind receives most of its acceleration. Thus, to model $\dot{M}$ correctly we must know the conditions in the denser portions of the atmosphere.

There are two very different models for the atmospheres of pulsating AGB stars currently in the literature.

Class 1: Fundamental mode models (including Bowen's models). For these, radii are ~ 1 AU and effective temperatures ~ 2900–3300 K. Bowen's models include non-LTE coupling of the radiation field to the gas at low densities. This results in large departures from radiative equilibrium, both "refrigerated" regions where dust may form and "calorispheres" where the temperature stays above the radiative equilibrium temperature at all times.

Class 2: Overtone pulsation models or large static models. In these models, the assumed radii are $\gtrsim 2$ AU, with corresponding effective temperatures 2200–2800 K. Examples of such models are those computed by the Berlin (Sedlmayr, Fleischer, Gail, Gauger, Winters, Woitke) and the Vienna (Dorfi, Höfner) groups. Their models usually assume LTE coupling of the radiation field to the gas, and this results in much faster cooling or reheating of the low density gas, so that most of the atmosphere remains in radiative equilibrium.†

These two classes of models are contrasted in Figures 6, 7, and 8, showing $v(r)$, $T(r)$, and $\rho(r)$ for two cases that are identical except that in one case ("mostly non-LTE") the assumption of LTE is presumed to break down for densities below 10^{-10} gm cm^{-3} and in the other case this does not occur until the density goes below 10^{-16} gm cm^{-3} so the

† The Berlin group's first attempt to include non-LTE coupling suffered from a severe shortcoming: they essentially assumed that forbidden lines would dominate the cooling—as they do in interstellar conditions–even at the much higher densities relevant in the atmospheres of pulsating giant stars. See Willson and Bowen 1998 for further discussion of this point.

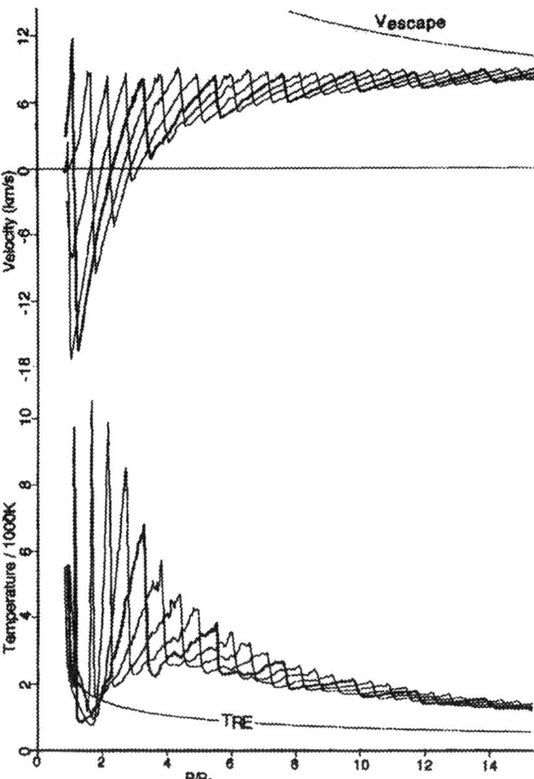

FIGURE 6. Velocity and temperature vs. radius at four phases for a Mira model by Bowen computed assuming a critical density of 10^{-10} gm/cm^{-3}. It includes dust formation near 2 R^* (where, thanks to the "refrigeration" caused by expansion between shocks, SiO is strongly supersaturated) and thus enhanced mass loss due to the transfer of momentum from the radiation field to the dust and from the dust to the gas.

model is in LTE throughout the region where the wind originates. In this "mostly-LTE" model, the temperature is equal to the radiative equilibrium temperature through most of the atmosphere. In this case, the variation in T_{RE} over the cycle that results from imposed variations in the photospheric temperature with phase is clearly seen. In the mostly non-LTE model, the temperature is usually quite far from the radiative equilibrium value outside of the deepest layers. The much weaker coupling of the radiation field with the gas in the non-LTE case produces both the deep "refrigerator" minimum in the temperature around 1.5–2 stellar radii and the elevated temperature (calorisphere) beyond about 3 stellar radii. Both of these models were told to form dust according to a smooth function of T_{RE} that rises abruptly near 2 stellar radii. For the non-LTE model, this is approximately where the refrigeration effect produces very low gas temperatures and high SiO supersaturation rations. For the mostly-LTE model, this assumption is probably not physically consistent; thus, the mass loss rate implied by Figures 7 and 8 is an overestimate in the LTE case.

Mass loss in Class 1 (compact, non-LTE) models occurs by a combination of pulsation and (for solar metallicity stars with $M > 1$ M$_\odot$) radiation pressure on dust. Non-LTE thermal physics plays an important role both in providing "refrigerated dust factory" regions between shocks and in providing thermal driving of the winds of low-Z stars. Class 2 (extended, mostly LTE) models can form dust at smaller R/R^* in a stationary

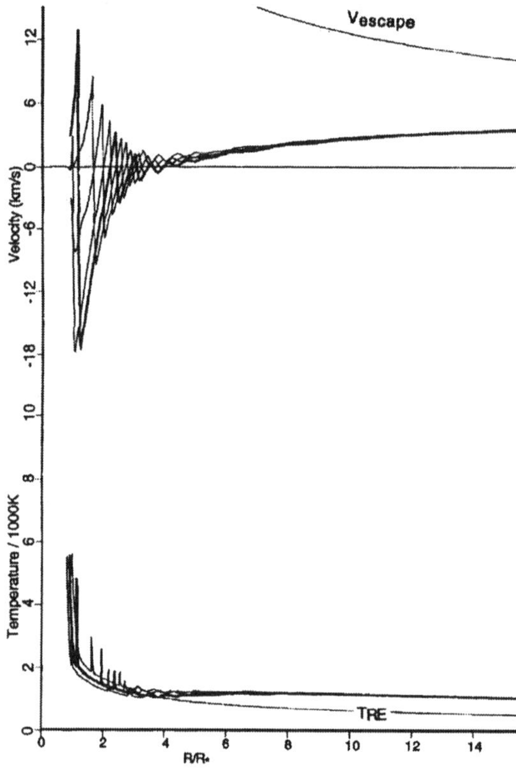

FIGURE 7. Velocity and temperature vs. radius at four phases for a model identical to that of Figure 8 except that the critical density is assumed to be 10^{-16} gm cm^{-3}. Here, also, dust formation is *assumed* to occur near 2 R^* even though there is no "refrigeration," and tests on the model show SiO is not supersaturated anywhere. This dust drives the outflow at large R/R^*.

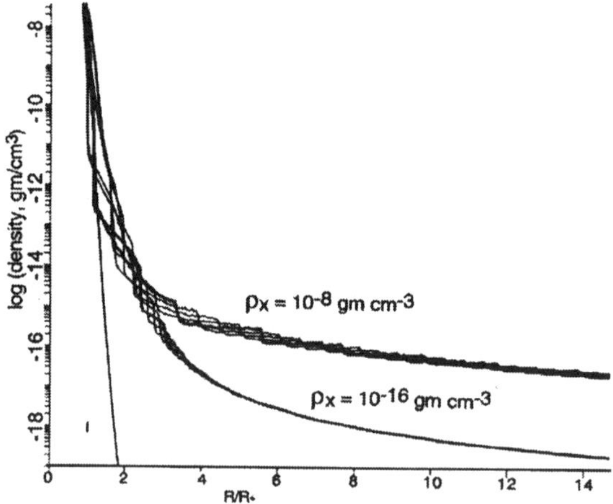

FIGURE 8. Density vs. radius for the two models of Figures 8 and 9. Note that not only is the temperature structure very different in these two cases, the velocity and density structure is also affected by the treatment of radiative relaxation.

flow (because they have lower T_{eff}), but heavy mass loss results only if the star is very cool, very luminous, and (for carbon dust) has very large C/O. As a result, a recently published mass loss formula from the Berlin work predicts a "cliff" at considerably higher L than the Bowen cliff shown in Figure 2, which leads to larger predicted final masses and higher terminal AGB luminosities than are observed.

What other kinds of observations can reveal which of these is the correct class of models to use? One possible "probe" is the observed radio continuum for Miras. Observations in the range 8–22 GHz (by Reid and Menton 1997, among others) showed that to a good approximation these had continua with:

$$S_\nu \propto \nu^2$$

consistent with a Rayleigh-Jeans blackbody spectrum with $R^2T \sim 10^4$ AU^2 K. (These are my units, translated from their results assuming the calibrations and distances that they used.) They also found relatively small amplitude time-variability (25%–100%). From these observations, Reid and Menton derived the following empirical model:

Starting with an extended stellar photosphere (= exponential density drop-off, with scale height $H \sim 2H_{\text{static}}$) around a large ($R^* \sim 2$ AU) star, they assumed that the gas would be in radiative equilibrium (thus at $T \sim 1560$ K in the region where the radio continuum forms). At such low temperatures there is very low radio opacity per particle, so they needed relatively high density ($\rho/m_{\text{H}} \sim 10^{12}$ cm^{-3}) at $R \sim 4$ AU to reproduce the observed signal level (corresponding to $R^2T \sim 10^4$ AU^2 K). Their model includes careful computation of the radio opacity at these low temperatures. It is a very fine example of an empirical model, built with care and (as far as I have been able to discover) completely internally consistent. However, it looks nothing like the theoretical dynamical models for Miras.

Approaching these observations from a very different perspective, let us consider what Bowen's dynamical models suggest about the origin of the radio signals from Miras. In Figure 9 we show the optical depth at 10 GHz as a function of radius in a Bowen model at four phases. The corresponding $T(r)$ and $\rho(r)$ are shown in Figure 10, and the resulting radio lightcurve in Figure 11. There is an abrupt increase in the optical depth at each shock front, and the dominant opacity is free-free scattering in post-shock ionized material. Because the opacity rises so abruptly, the star should appear as a blackbody with radius equal to the position of a rising shock at nearly all frequencies at a given phase, or nearly all phases at a given frequency. This picture suggests that the radio continuum observations are revealing a shock front rising from a relatively compact star, *not* a quiescent extended "radio photosphere."

In this interpretation, the signal comes from smaller R and higher T than in the Reid and Menton model. Some time-variation is expected. The details depend on the ionization (non-LTE beyond about 1.5 R^*) and the temperature and density structure. While R is much smaller than in the "radio photosphere" model, T is higher; R^2T is $\sim 10^4$ AU^2T as required. The models used to produce Figures 9–11 did not have enough spatial resolution at the shock front to accurately model the details of the (propagating and time-dependent) radio photosphere, which is formed in the material at the front of the shock during nearly all phases. Therefore, the appropriate T to use and hence the details of the predicted "light curve" are NOT reliable and do not establish whether this is or is not the correct picture. However, again without "parameter twiddling," the Bowen models give the right order of magnitude for the radio continuum observations, and predict that there should be some observable time-variability, possibly in the form of a short-lived spike of emission superimposed on a nearly-constant signal during the

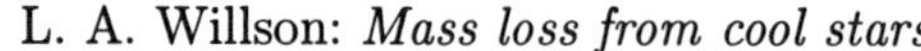
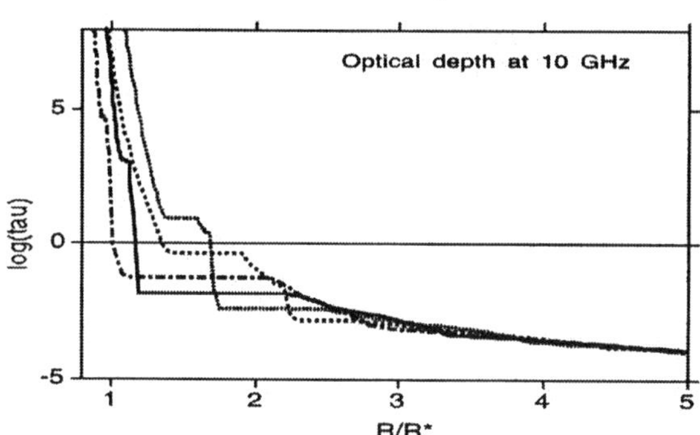

FIGURE 9. Optical depth at 10 GHz at four phases in a Mira model by Bowen. The same formulae for approximating the opacity as were used by Reid and Menton are used here. The sudden jump in the opacity occurs where there is an ionizing shock front. This suggests that the star should appear to have a very well-defined radius at 10 GHz at most phases.

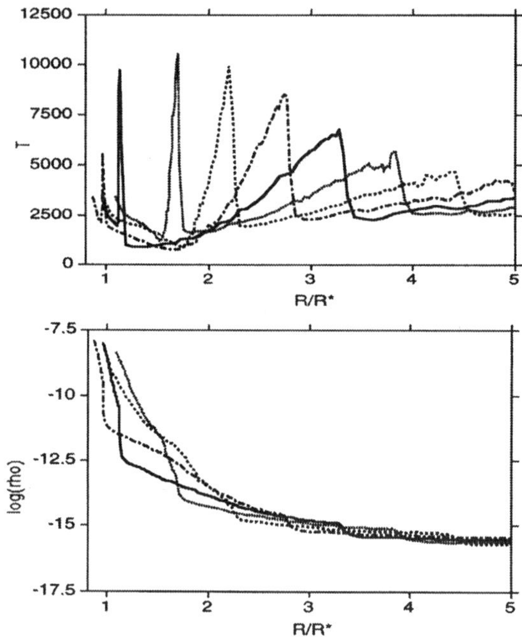

FIGURE 10. For the same model as i in Figure 1, $T(r)$ and $\rho(r)$ are displayed here. The location of the shock fronts is obvious. The expected radio signal is proportional to $R^2 T$ at the shock where the optical depth passes 1 in Figure 8.

rest of the cycle at each frequency, with a phase-shift in the time of the "spike" from one frequency to another.

It is very important, in comparing models with observations, to maintain consistent assumptions with respect to the micro-physics. For example, with Saha instead of non-LTE ionization used to predict a light curve, but using the same temperature and density structure from Bowen's (non-LTE) dynamical models, we would get a very different result (Figure 12). This illustrates an important, and more general, point: It has been quite common to try LTE radiative transfer in the context of a non-LTE dynamical model. This kind of thing has been done by Bessell et al. (1991), for example, in an attempt to

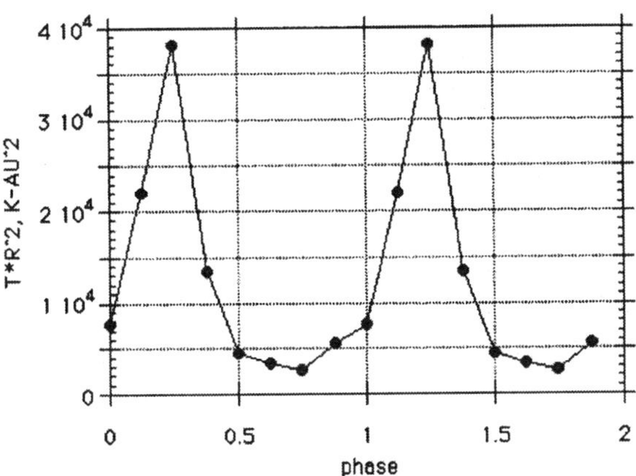

FIGURE 11. Radio lightcurves should be proportional to the quantity shown, R^2T in units of AU^2K, plotted here from the data of Figures 9, 10. The range of R^2T observed by Reid and Menton is roughly 1–3 AU^2K. The model phase (shown) has arbitrary zero-point, and is therefore not the visible or radio phase. While the extremely simple analysis of Figures 9, 10 clearly leads to an expected signal of the right order of magnitude, it does display greater variability than they detected. More detailed study of the likely appearance of the shock-photosphere may improve the agreement; it is also possible that the high-luminosity phase is shorter in duration than is pictured here, and hence does not show up in the observations yet.

model the spectroscopic features to be expected from a Bowen model, with spectacularly bad results. Here is why this is such a bad idea: When the radiation field and gas temperature are such that collisional transitions upward are immediately followed by radiative de-excitations, then the non-LTE cooling is slower than LTE cooling; a simple estimate is given by

$$S/S_{\mathrm{LTE}} = 1/(1 + \rho_x/\rho)$$

where ρ_x is the critical density at which collisional excitations = radiative excitations (with $\rho_x \lesssim 10^{-10}$ cm^{-3} for permitted transitions). Below this critical density the assumption of LTE breaks down. Thus, assuming LTE where non-LTE is appropriate will lead to an *overestimate* for the radiative losses from that region, or an underestimate for the thermal relaxation time. (Another way to see this: If collisional excitation is usually followed by radiative de-excitation, then the population of the excited level is usually $\ll$ the LTE value.) If the dynamical model is computed with non-LTE (slow cooling) and the radiative transfer with LTE (fast cooling) using the temperature structure from the slow-cooling model, the result will be that much more radiation will be computed to come out in the LTE model than was assumed to come out in the dynamical model. In essence, *it is a violation of conservation of energy to use LTE transfer or ionization in a structure computed with non-LTE, or vice versa.* For further discussion of this point, see Willson & Bowen (1998).

How bad can it be to oversimplify the energy equation? Consider a steady wind with constant outflow velocity and no mechanical energy input. If it is in radiative equilibrium, then $T_{\mathrm{RE}}(r) \sim (L/r^2)^{1/4} \sim r^{-0.5}$ at large distances. If it is expanding adiabatically, $T_{\mathrm{ad}}(r) \sim \rho^{\gamma-1} \sim \rho^{2/3} \sim r^{-4/3}$. Starting with $T = T_{\mathrm{RE}}$ at some point r_o gives $T_{\mathrm{ad}}/T_{\mathrm{RE}} = 0.56$ at $2\,r_o$ and 0.26 at $5\,r_o$. Clearly the difference can be significant. Note that such nearly-adiabatic expansion cooling will also affect the size of the Strömgren sphere in a stellar wind, because cooler gas has a higher recombination rate. A good rule-of-thumb for dynamical atmospheres: Where non-LTE is important, departures from RE are also

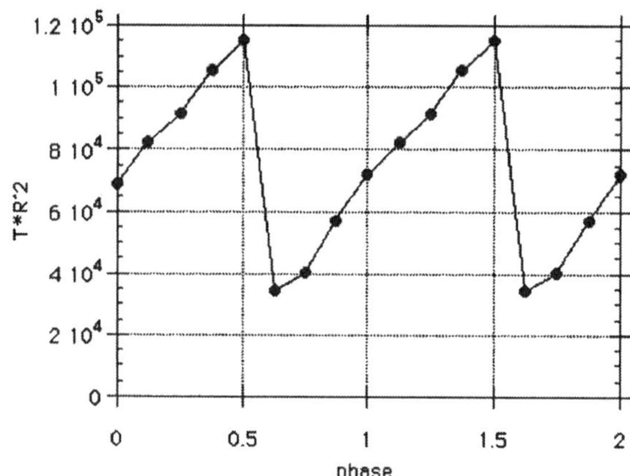

FIGURE 12. For contrast, the same Bowen dynamical model is used with Saha ionization to get a predicted radio lightcurve. Note the change of scale; the observed range falls entirely below this curve, which also has quite a different shape than the one in Figure 11. The dramatic difference between this figure and Figure 11 illustrates the importance of using correct and consistent physics in the modeling of the hydrodynamics and the derivation from the models of observable quantities.

likely. The reader is referred to discussions by Ruden, Glassgold & Shu (1990) and Goldreich & Scoville (1976).

Further note: It is often assumed in fitting radio observations that $v = $ constant and $n_e/n_H \approx$ constant. If so, the spectrum should have the appearance of $F_\nu \sim \nu^\alpha$ with $\alpha = 0.6$ (Wright and Barlow 1975; Panagia and Felli 1975). This should always be checked by having observations at two or more frequencies to get a reliable $\dot{M}$. In fact, in many cases, this is not the index that is found to provide the best fit. Many quoted mass loss rates or limits are based on inadequate data and/or oversimplified models (e.g. Brown et al. 1990); further observations need to be obtained, and better modeling of those observations needs to be done.

4. Conclusions and final comments

I believe that the following "radical statements" will prove to be correct:

• Correlations between observed mass loss rates and physical parameters of cool stars may be (and usually are) dominated by selection effects.

• Most observations have been interpreted using models that are relatively simple (stationary, polytropic, spherically symmetric, homogeneous) and thus "observed" mass loss rates or limits may be in error by orders of magnitude in some cases.

Correct modeling of the processes involved in driving mass loss from stars is hard, because these processes act in a parameter regime where we cannot make the usual simplifying assumptions. LTE no longer applies to at least some of the important transitions involved in cooling and heating the gas, complicating calculation of the thermal relaxation in the hydrodynamic code and requiring more complex radiative transfer routines. Relatively slow heating and cooling mean that thermal relaxation rates ($dlnT/dt$) will often be slower than dynamical relaxation rates ($dln\rho/dt$) so that the gas will not be in radiative equilibrium. Thus, *we cannot use even the most sophisticated radiative transfer codes to compute the expected spectral features and thermal relaxation rates without also taking into account the departures from radiative equilibrium produced by near-adiabatic compression and expansion of the gas.* The radiative transfer and the dynamics are inextricably linked because they are both produce important terms in the energy equation; separating them is analogous to tracking potential and kinetic energy separately in an elementary mechanics problem. In traditional stellar atmospheres, LTE and/or static conditions mean that the thermal relaxation time is short compared with the dynamical time scale. In the interstellar medium, dynamical time scales are very long, and the gas temperature is very weakly coupled to the radiation field; statistical and radiative equilibrium may be safely assumed for the transitions that need to be tracked in radiative transfer calculations. In modeling stellar winds, however, none of these is a safe assumption.

Simple models are inadequate and empirical relations may mislead us. What are we to do? We need to start making proper models for the computation of mass loss and the interpretation of mass loss data—to leave the "polytropic era" of mass loss studies. It will be a challenge, but modern computational techniques and facilities are just reaching the stage where they should be able to tackle the full complexity of modeling mass loss from cool stars.

REFERENCES

AYRES, T. R. 1981 *ApJ* **244**, 1064.

AYRES, T. R. & RABIN, D. 1996 *ApJ* **460**, 1042.

BESSELL, M. S., BRETT, J. M., SCHOLZ, M., & WOOD, P. R. 1989 *A&A* **213**, 209.

BOWEN, G. H. 1988 *ApJ* **329**, 299.

BOWEN, G. H. & WILLSON, L. A. 1991 *ApJ* **375**, L53.

BROWN, A. ET AL. 1990 *ApJ* **361**, 220.

CARLSSON, M. & STEIN, R. F. 1995 *ApJ* **440**, L29.

CARLSSON, M. & STEIN, R. F. 1997 *ApJ* **481**, 500.

FEAST, M. W., GLASS, I. S., WHITELOCK, P. A., & CATCHPOLE, R. M. 1989 *MNRAS* **241**, 375.

GOLDREICH, P. & SCOVILLE, N. 1976 *ApJ* **205**, 144.

HOLZER, T. E. 1987. In *Circumstellar matter; Proceedings of the IAU Symposium.* p. 289. D. Reidel Publishing Co.

IBEN, I., JR. 1984 *ApJ* **277**, 333.

LAWRENCE, G., JONES, T. J., GEHRZ, R. D. 1990 *AJ* **99**, 1232.

LINSKY, J. L. & HAISCH, B. M. 1979 *ApJ* **229**, L27.

OSTLIE, D. A. & COX, A. N. 1986 *ApJ* **311**, 864.

PANAGIA, N. & FELLI, M. 1975 *A&A* **39**, 1.

REID, M. J. & MENTEN, K. M. 1997 *ApJ* **476**, 327.

RUDEN, S. P., GLASSGOLD, A. E., & SHU, F. N. 1990 *ApJ* **361**, 546.

WILLSON, L. A. & BOWEN, G. H. 1988. In *Polarized Radiation of Circumstellar Origin* (eds. Coyne, G. V., Magalhães, A. M., Moffat, A. F. J., Schulte-Ladbeck, R. E., Tapia, S., & Wickramasinghe, D. T.). p. 485.

WILLSON, L. A. & BOWEN, G. H. 1998. In *Cyclical Variability in Stellar Winds* (eds. Kaper, L. & Fullerton, A. W.). p. 294. ESO Astrophysics Symposia.

WRIGHT, A. E. & BARLOW, M. J. 1975 *MNRAS* **170**, 41.

Selected problems in binary stellar-evolution theory

By PHILIPP PODSIADLOWSKI

Oxford University, Oxford OX1 3RH, U. K.

Since the majority of massive stars are members of binary systems, an understanding of the intricacies of binary interactions is essential for understanding the large variety of observed stellar systems. Here we review the main elements of binary stellar-evolution theory, particularly emphasizing the still remaining, sometimes very substantial gaps in our understanding and how future work may help to fill these.

1. Introduction

The application of stellar-evolution theory to the evolution of binary systems has been very successful in explaining to a large degree the reasons for the immense variety of observed types of binaries. I am particularly referring to the systematic studies of Iben and his collaborators (see Iben 1991 for a detailed review and references). In this contribution, I will summarize the main elements of binary evolution theory. However, its main purpose is not to emphasize the successes, but the outstanding problems. As I will show, in many cases there are still very substantial uncertainties in our understanding of fairly common types of binary interactions; in some cases I will even challenge some of the basic assumptions.

As is surprisingly little known in the astronomical community, most stars in the sky are actually members of binary (or multiple) systems. To zeroth order, *all* stars are members of binaries (see, e.g. the references in Podsiadlowski, Joss & Hsu 1991 [PJH] and Ghez 1996). Of course, in this context we will only be interested in interacting systems, i.e. systems where at least one binary component fills its Roche lobe during its evolution. The fraction of massive stars in interacting binaries can be estimated to be in the range of 30–50%. For example, Garmany, Conti & Massey (1980) found that 36% of massive stars are spectroscopic binaries with massive companions and orbital periods less than $\sim 1\,\mathrm{yr}$. This estimate would imply a *true* interacting-binary frequency of around 50%. It is worth noting that Roche-lobe overflow occurs more frequently in evolved phases simply because the radius of a star expands only by a factor of ~ 2 during the main sequence, while it expands by a factor of ~ 100 subsequently. This means that, for any plausible initial orbital-period distribution, a star is much more likely to encounter mass transfer after its main-sequence phase. Since stars spend the largest part of their lives on the main sequence, the majority of stars observed in the sky have not (yet) experienced a binary interaction, but almost half of them will do so in the future. This explains, at least in part, why stellar astronomers are often justified in neglecting the complications due to binary interactions.

2. The Roche Approximation

One of the key assumptions in modelling binary interactions is the Roche approximation (e.g. Kopal 1978). It assumes that the mass-losing star is centrally concentrated, rotates synchronously with the orbit and that the orbit has been circularized as a result of tidal dissipation in the tidally distorted envelope (e.g. Zahn 1975, 1977, 1989;

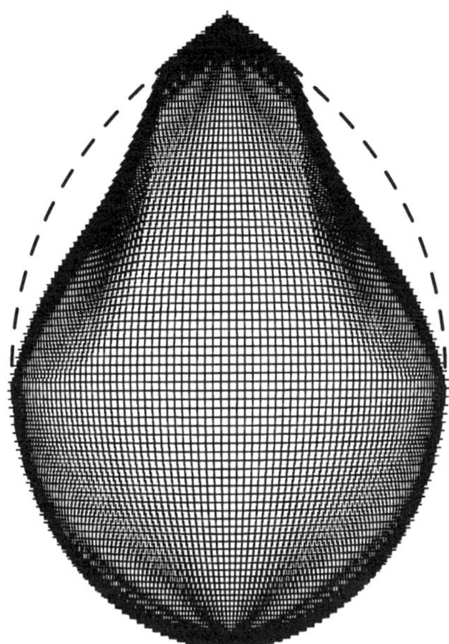

FIGURE 1. Irradiated, critical tidal lobe for a secondary in a low-mass X-ray binary (crosses) compared to the standard Roche-lobe surface (dashed curve). The irradiation model assumes a mass for the compact object of $1.4\,M_\odot$, companion mass of $0.5\,M_\odot$ and a X-ray luminosity of 10^{38} erg/s. The accretion disk, shadowing the secondary, has an opening angle of about $10°$ (after Phillips & Podsiadlowski 1999).

Lecar, Wheeler & McKee 1976; Krolik, Meiksin & Joss 1984; Tassoul 1995). However, observationally at least one class of binaries is known where mass-transfer occurs in very eccentric systems. These are the *VV Cephei systems* (e.g. VV Cep [Wright 1977], AZ Cas [Cowley, Hutchings, & Poppers 1977], KQ Pup [Cowley 1966]), where the mass-losing components appear to fill or be close to filling their respective critical tidal lobes at periastron; nevertheless the systems' eccentricities are typically larger than 0.5, suggesting that very little tidal circularization has taken place. The reason why these systems are not circularized despite their strong tidal interactions has not yet been studied in any detail. It may be a result of the continuous exchange of orbital and tidal energy back and forth between the stars and the orbit that, under certain conditions, can increase eccentricity, sometimes even in a chaotic manner (Mardling 1995a,b). This general issue may, however, be important for the survival of such systems, since it may help to prevent dynamical transfer which standard theory predicts for such wide systems with supergiant mass donors (see Sections 3 and 5).

Irradiation effects

When the surface of the mass-losing component is irradiated either by the companion's internal luminosity or because of accretion luminosity (in particular in the case of X-ray binaries), the external radiation pressure itself will modify the shape of the effective, critical potential (Drechsel et al. 1995; Phillips & Podsiadlowski 1999). Figure 1 illustrates this effect, showing what the surface of the secondary in a X-ray binary might look like if it is irradiated with a typical X-ray flux of 10^{38} erg/s and if the equatorial region of the secondary lies in the shadow of an extended accretion disk. The model assumes that the stellar surface can be represented by an effective equipotential that approximately takes the radiation force into account and ignores the effects of surface circulation currents. Despite of its approximate nature, it demonstrates that, for strongly irradiated stars, irradiation effects will modify ellipsoidal light curves (e.g. van Paradijs 1991) and affect the determination of radial-velocity curves. Since both of these are often employed in determining the masses of the components in X-ray binaries, care has to be taken that these effects, which can be quite substantial, are properly being modelled (see, e.g. Phillips, Shahbaz & Podsiadlowski 1999).

Irradiation effects can be even more dramatic in systems without an accretion disk where the illuminating flux directly impinges on the inner Lagrangian point, L_1 (e.g. in radio-pulsar binaries or stars orbiting supermassive black holes [e.g. Podsiadlowski & Rees 1994]). It is then possible that the Lagrangian point with the lowest potential is no longer the inner Lagrangian point, L_1, but one of the outer ones, typically L_2, which is shadowed from the external irradiation by the secondary itself. Then, the critical potential surface changes from an inner to an outer configuration. This means that, when the secondary fills its critical tidal lobe, mass transfer will take place through the outer Lagrangian point and lead to the formation of a *circumbinary excretion disk* rather than a standard accretion disk around the primary component. This may indeed be observed in the famous black-widow pulsar PSR 1957+20 (Fruchter, Stinebring & Taylor 1988), where the pulsar radiation is in the process of evaporating the secondary, apparently in a comet-tail-like outflow.

Finally, external irradiation may also affect the interior structure of the irradiated star, possibly causing significant expansion and thereby providing a means of driving mass transfer (see Podsiadlowski 1991 and Section 6).

3. Binary interactions

The main classification of mass transfer is based on the evolutionary stage of the mass donor at the beginning of the mass-transfer phase (Kippenhahn & Weigert 1967; Lauterborn 1970): case A (main-sequence), B (post-main-sequence, pre-helium-ignition), case C (post-helium-burning).

In addition, mass transfer can take place in two qualitatively very different modes: more-or-less conservative mass transfer and dynamically unstable mass transfer.

(Quasi-)conservative mass transfer usually occurs when the mass donor has a radiative envelope and the mass ratio of the mass-accreting to the mass-losing component is not too small (e.g. de Loore & de Grève 1992). Then, a large fraction ($\gtrsim 0.5$) of the mass lost from the primary is accreted by the secondary (e.g. Meurs & van den Heuvel 1989). Thus in this case, both components of the binary are affected, one by losing mass, the other by accreting it. During mainly conservative mass transfer, the orbital period tends to increase.

Dynamical mass transfer usually takes place when the secondary is a giant star with a deep convective envelope. In this case, mass transfer is dynamically unstable and leads to the formation of a common envelope surrounding the core of the giant and the secondary (Paczyński 1976). Due to friction between this immersed binary with the common envelope, the orbit of the binary starts to shrink. Dependent on how much energy is released in the orbital decay of the binary and is deposited in the envelope, two different outcomes are possible. If the deposited energy exceeds the binding energy of the envelope, the common envelope can be ejected, leaving a very close binary consisting of a white dwarf or helium star/Wolf-Rayet star and a normal companion star (which is only little affected by the common-envelope phase). If the energy is not sufficient to unbind the envelope, the less dense component of the immersed binary will ultimately be tidally destroyed and the two components merge completely to form a single, but rapidly rotating giant (Section 3.3).

In a typical binary scenario, a binary system may experience several different phases of mass transfer; this gives rise to an enormous variety of possible binary scenarios. In the following, I will only outline some of the main considerations.

3.1. *Mass loss*

Some 30–50% of all massive stars experience Roche-lobe overflow and mass loss at some point during their evolution. In most cases, they lose all of their hydrogen-rich envelopes and become white dwarfs or helium/Wolf-Rayet stars, depending on their mass. In some cases, however, it is possible for a *massive* primary to retain part of its hydrogen-rich envelope, in particular when mass transfer occurs very late during the evolution of the primary and when the initial mass ratio is very close to one. However, even in this case, at most a few solar masses can be retained in the envelope. The immediate supernova progenitor will be a *stripped supergiant* with a small envelope mass (Joss et al. 1988). It is interesting to note that the outer appearance of this star will be very similar to that of a star that has lost no mass at all, since the radius and the luminosity of the star are almost independent of the envelope mass (PJH), although the final supernova would be quite different from a standard Type II supernova (see Podsiadlowski et al. 1996 for a review).

3.2. *Mass accretion*

Just as the structure and appearance of the mass-losing star can be strongly affected, the structure and further evolution of the accreting companion can also be dramatically altered by mass transfer, in particular for massive systems which do not develop degenerate cores. This depends, however, strongly on the evolutionary stage of the accreting secondary at the beginning of the mass-transfer phase.

Secondary on the main sequence

If the secondary is still on the main sequence at the beginning of the mass-transfer phase, the secondary is usually *rejuvenated* and will behave subsequently (after the mass-transfer phase) like a single, but now more massive star (Hellings 1983; PJH). However, as was shown by Braun & Langer (1995), this need not be the case if accretion occurs very late on the main sequence and if semi-convection is very slow (combined with the Ledoux criterion for convective instability), since the convective core will then not grow significantly as a result of accretion. The subsequent evolution will resemble the evolution of a star that accreted after the main-sequence phase (see below); in particular, the star may never become a red supergiant and spend the whole post-main-sequence phase in the blue-supergiant region of the Hertzsprung-Russell (H-R) diagram.

Secondary has completed hydrogen core burning

If the secondary has already left the main sequence before the beginning of the mass-transfer phase, the subsequent evolution will generally differ quite substantially from the evolution of a normal single star. Since the mass of the helium core will not grow as a result of mass accretion, the main effect of post-main-sequence accretion is to increase the envelope mass relative to the core mass. As was first shown by Podsiadlowski & Joss (1989), this has the consequence that the star will now not become a red supergiant or, if it was a red supergiant at the time of the accretion phase, leave the red-supergiant region and spend its remaining lifetime as a blue supergiant. The final location of the star in the H-R diagram at the time of the supernova depends on how much mass has been accreted in the accretion phase. The star will be the bluer, the more mass it has accreted. The final supernova will be of the SN 1987A variety. Since a star has to accrete only a few percent of its own mass from a binary companion to be spun up to critical rotation, the progenitor is expected to pass through a phase where it is rapidly rotating. Rapid rotation may also induce large-scale mixing in the accreting star (even across chemical gradients). However, whether this happens depends critically on the angular-momentum transport inside and the structure of the accreting star.

3.3. *Binary mergers*

The most dramatic type of binary interaction is dynamical mass transfer leading to a common-envelope and spiral-in phase. If the orbital energy released during the spiral-in is sufficient to eject the common envelope, the end product is a short-period binary consisting of a white dwarf or Wolf-Rayet primary and a normal stellar companion. Unless the system becomes unbound in a first supernova, the secondary is also likely to experience mass loss in the future leading to a second mass-transfer phase, just as in the mass-accretion scenarios discussed in the previous subsection.

If the common envelope is not ejected, the two components will merge completely to produce a more massive and rapidly rotating single star. For massive systems, this end product resembles in many ways the outcome of the post-main-sequence accretion models in Section 3.2. In particular, it may again end its evolution as a blue supergiant rather than as a red supergiant either because of the added mass in the envelope (Podsiadlowski, Joss & Rappaport 1990) or because of the dredge-up of helium (Hillebrandt & Meyer 1989) or both. The resulting supernova may again be of the SN 1987A variety. PJH have estimated that the combined frequency for a blue-supergiant progenitor in either an accretion or a merger scenario is about 5% of all core-collapse supernovae with an uncertainty of about a factor of two.

The details of the merging process are, however, very poorly understood. Some of the main unsolved issues involve the amount of mixing and dredge-up of helium during the merging process and the question of unusual nucleosynthesis (hot CNO burning, s-processing) that may be expected in the intermediate mixing region (see Podsiadlowski & Spruit 1999 for a detailed discussion).

Complications

In this section, I only discussed the main types of interactions. As indicated earlier, many binaries experience more than one phase of mass transfer. For example, low-mass helium stars, produced in a first mass-transfer phase, will expand again after core helium burning to become helium giants and may fill their Roche lobes for a second time (so-called case BB mass transfer; de Grève & de Loore 1977; Delgado & Thomas 1981). This still does not exhaust the whole variety of multiple mass-transfer scenarios. In the systematic studies of Iben and his collaborators (see the references in Iben 1991), many examples of more than two mass-transfer phases can be found, leading to an almost limitless variety of binary scenarios (also see Bhattacharya & van den Heuvel 1991; Nomoto et al. 1994 for many further examples).

4. Early Case B mass transfer

One of the theoretically best studied cases of mass transfer is early case B mass transfer, in which the primary fills its Roche lobe after hydrogen-core burning in the Hertzsprung gap, i.e. before developing a deep convective envelope (e.g. de Grève & de Loore 1976; van den Heuvel 1976; van der Linden 1987; de Loore & de Grève 1992; Pols 1994; de Loore & Vanbeveren 1995). While in the earlier calculations mass transfer was usually assumed to be fully conservative, the later calculations often took significant mass loss from the system into account (typically of order 50%; Meurs & van den Heuvel 1989). In the more recent calculations (in particular Pols 1994) that try to follow the evolution of both components, the mass-losing and the mass-accreting one, self-consistently, it was often found that, in cases of early case B mass transfer, the secondary started to fill its own Roche lobe as a result of the accretion. These systems would therefore enter into a contact configuration where both binary components are surrounded by a common envelope. The parameter range in which a contact phase can be avoided was even smaller in the most up-to-date calculations by Braun (1997), mainly because of the use of the new, larger opacities (Iglesias, Rogers & Wilson 1992).

What does this mean for the survival of these systems? Do contact systems enter into a common-envelope phase and experience dramatic spiral-in as outlined in Section 3 and 5? While this is a possibility, it would have rather problematic consequences. One, some systems are known (e.g. v Sgr [Dudley & Jeffery 1993]) that almost certainly experienced early case B mass transfer without suffering significant spiral-in. Two, in most cases the common-envelope phase would lead to the complete merger of the two components. As already discussed in Section 3.3, the evolution of the resulting merged star would be quite different from a single star of the same mass and possibly end its evolution as a blue supergiant rather than as a red supergiant. This would almost certainly lead to an overproduction of supernovae of the SN 1987A variety (Langer 1997).

A possible solution to this conundrum is that, in the contact phase, these systems experience dramatic mass loss from the system but without the dramatic spiral-in usually associated with a common-envelope phase (also see de Loore & de Grève 1992); when the mass ratio has sufficiently been reversed, the systems become semi-detached again and subsequently follow the standard evolutionary picture of quasi-conservative mass transfer. This would explain the properties of the unusual helium-star binary v Sgr (Dudley & Jeffery 1993). Some recent theoretical modelling of the common-envelope phase (Podsiadlowski 1999a,b) also provides some support for this possibility, since it suggests that the physics of the common-envelope phase is fundamentally different when the common envelope is initially radiative.

5. Common-Envelope evolution

As is already clear from the previous discussion, common-envelope (CE) evolution is one of the least understood phases of binary evolution. This is unfortunate, since it is also one of the most important types of binary interactions: originally proposed by Paczyński (1976), it is the main mechanism by which an initially wide binary, often with periods of 100s of days to several decades, can be transformed into a very close binary with periods as short as 11 minutes. For recent reviews see Taam & Bodenheimer (1992); Webbink 1992; Iben & Livio (1993) and Livio (1996).

The theoretical problem already starts before the proper CE phase. In the standard picture, dynamical transfer occurs whenever the mass donor has a deep convective envelope and has a mass larger than $\sim 70\%$ of the mass of the accretor (Paczyński & Sienkiewicz 1972). This would predict that most red-giant donors should experience this fate. However, observationally that does not seem to be the case, since some late-type systems appear to be able to avoid the predicted spiral-in (see Webbink 1986; Eggleton & Tout 1989 and the VV Cep systems mentioned in Section 2). A possible solution may be mass loss by a stellar wind prior to the onset of mass transfer (possibly enhanced by the presence of the close binary companion [Tout & Eggleton 1988], which may help to stabilize mass transfer (also see Hjellming & Webbink 1987).

The second main problem is the question of when the common envelope can be ejected and when the spiral-in leads to the complete merging of the two components. This issue is usually addressed in terms of a simple energy argument, the so-called α-formalism, in particular in the context of binary population synthesis (e.g. de Kool 1996; Yungelson & Tutukov 1997). Here it is assumed that the envelope is ejected if the energy released in the orbital decay, $\Delta E_{\rm orb}$, times some unknown efficiency parameter, α, exceeds the binding energy of the envelope, $E_{\rm env}$, i.e.

$$\alpha \Delta E_{\rm orb} > |E_{\rm env}|.$$

Different authors use different representations for $E_{\rm env}$, and therefore the definitions of α vary substantially in the literature (see de Kool 1996). In addition, there is the more fundamental question whether it should include just the gravitational potential energy or also the thermal energy (see Han, Podsiadlowski & Eggleton 1995). While one might hope that population synthesis techniques could help to calibrate the efficiency factor α, there is no *a priori* reason to believe that α is the same for all systems. It must depend on the amount of friction in the spiral-in phase, the efficiency of energy transport in the common envelope and many other factors (see, e.g. the discussions in de Kool 1987 and Livio & Soker 1988). An obvious necessary condition for CE ejection is $\Delta E_{\rm orb} > |E_{\rm orb}|$. However, because energy is deposited in the envelope, the envelope will immediately start to expand. This in turn will reduce the friction and slow down the spiral-in. It is then possible that the spiral-in becomes self-regulated, where all the frictional energy deposited in the envelope is radiated away from surface without driving significant mass loss (Meyer & Meyer-Hofmeister 1979). *It should be noted that in this case the physical justification for the α-formalism becomes void.*

Theoretical progress in understanding common-envelope evolution can come from two different, but largely complementary approaches: studies of the static response of an envelope to the energy deposition caused by a spiralling-in binary (Meyer & Meyer-Hofmeister 1979; Podsiadlowski 1999a,b) and full hydrodynamical simulations (e.g. Taam & Bodenheimer 1992 and references therein). The first approach may help to clarify the various phases during a CE phase, since it is possible to follow the whole CE phase, while it is inadequate in describing distinctly dynamical aspects of the problem (e.g. the entrance into the common-envelope phase [Rasio & Livio 1996], dynamical envelope ejection, etc.). On the other hand, the hydrodynamical simulations have to use substantially simplified stellar physics and can only follow dynamical, but not the longer quasi-static phases of the CE phase. Fortunately, these different aspects are sufficiently decoupled from each other that it can be hoped that taken together they may ultimately provide a realistic description of this important phase of binary evolution.

6. Mass-transfer driving mechanisms

At the beginning of the mass-transfer phase, mass transfer is generally unstable, either thermally or adiabatically (Paczyński 1971), leading to a phase of "unstable" mass transfer. Once mass transfer has stabilized, an additional mechanism is required to continue driving mass transfer (see Bhattacharya & van den Heuvel 1991; Ritter 1996 for detailed reviews).

The simplest driving mechanism is the expansion of the mass donor itself, which is a consequence of its internal evolution. Algols where the mass donors are subgiants are perhaps the best known example. Generally, *evolution-driven mass transfer* is most important for stars in rapidly evolving phases, for example in the Hertzsprung gap (see

Section 4), or for stars evolving up the giant branch (see Webbink, Rappaport & Savonije 1993).

However, in many binaries (e.g. cataclysmic variables and low-mass X-ray binaries), the mass donors are evolving too slowly to account for the observed mass-transfer rates. In these cases, it is the *shrinking* of the orbit rather than the *expansion* of the donor star that drives mass transfer. This requires angular-momentum loss from the system. The two most important angular-momentum loss mechanisms are *gravitational radiation* and *magnetic breaking* (Verbunt & Zwaan 1981). The first process is well described by the Einstein formula for quadrupole radiation (e.g. Landau & Lifshitz 1962) and is only important for very close binaries, typically with periods less than $\sim 12\,\mathrm{hr}$. Magnetic breaking, however, is much less well understood. The basic idea is that the mass donor is a red dwarf that loses spin angular momentum in a magnetically coupled wind (similar to the Sun); but since the donor is tidally locked to the orbit, this decreases the orbital angular momentum of the system causing the orbit to shrink. An important unsolved question is what happens to magnetic breaking when the donor becomes fully convective, since at this point the dynamo process responsible for the star's magnetic field *may* cease to be operative. In the standard explanation for the observed period gap in cataclysmic variables, it has been postulated that magnetic breaking and hence mass transfer stops at this point (Rappaport, Verbunt & Joss 1983; Spruit & Ritter 1983). An important observational question therefore is whether the magnetic behavior of single stars changes significantly as one moves down the main sequence. At present, there appear to be no indications for a dramatic change at the spectral type where single stars are expected to become fully convective.

In X-ray binaries, irradiation effects become important and may enhance mass transfer. Two main mechanisms have been proposed (for a detailed review see D'Antona 1996): *irradiation-driven winds* (Ruderman, Shaham, & Tavani 1989; Ruderman, Shaham, Tavani & Eichler 1989) and *irradiation-driven expansion of the secondary* (Podsiadlowski 1991). In the first mechanism, irradiation drives a wind from the secondary, part of which will be accreted by the companion, while in the second, the external illumination changes the interior structure of the secondary causing it to expand and hence to transfer mass. Neither mechanism is very well understood. The problem with the first is the efficiency by which irradiation flux is converted into wind kinetic energy, while for the second it is presently not clear by how much the secondary expands, if at all. But even if there is no significant expansion, the calculations by Hameury et al. (1993) have shown that mass transfer will be cyclic, characterized by phases of enhanced mass transfer and separated by extended phases without mass transfer. This effect alone may help to explain why the observed mass-transfer rates in cataclysmic variables and low-mass X-ray binaries, systems with very similar properties in most other respects, are so different.

Finally, mass transfer can also be driven or initiated by *global fluid instabilities*, caused for example by the coupling and transfer of orbital angular momentum to spin angular momentum of one or both binary components (see Rasio 1996 for a detailed review). Such instabilities appear to be driving mass transfer in Cen X-3 (Kelley et al. 1983) and SMC X-1 (Levine et al. 1993). It most cases these instabilities will ultimately lead either to a common-envelope phase or the complete dynamical coalescence of the two components.

7. Binary interactions and peculiar stars

As I hope I have illustrated, binary interactions can affect the structure and evolution of stars in a multitude of different ways. It is therefore not surprising that many "peculiar" stars owe their peculiarities to previous binary interactions, even though the binary nature may not always be obvious. Two particularly good examples for the latter are barium stars (and the related CH stars) and symbiotic stars: for both classes of systems, the binary nature was only unequivocally proven in the early 80s (McClure 1983; Kenyon 1986).

There are cases where systems are generally assumed to be binaries, but where the binarity has not been established observationally. A prime example are low-mass X-ray binaries (e.g. van Paradijs 1995): only some 30% of these systems have known orbital periods. Indeed, some of these systems, classified as *binaries*, clearly are *single* objects. For example, the system 1E2259+586 (Davies, Wood & Coe 1990 and references therein) contains a young X-ray pulsar apparently accreting from an accretion disk, but the pulse timing shows no evidence for a companion star. Possibly, this object used to be in a binary, but the companion star was destroyed because of an asymmetric kick the newborn neutron star received as a result of the supernova explosion (e.g. Brandt & Podsiadlowski 1995). In this case, the X-ray pulsar would be accreting from the leftover remnant of the destroyed secondary.

Indeed, there are several channels of binary evolution that can *produce single objects*. The break-up of a binary in a supernova with or without a supernova kick (Thorsett 1999) can produce a single runaway star (Blaauw 1961). Dynamical interactions in dense stellar systems (e.g. globular clusters) can lead to the ejection of a star from a binary or higher-order multiple system (see Davies [1996] for a review). And the complete merging inside a common envelope (Section 3.3) leads to a single merged object probably with some highly unusual properties (with possible applications to sdO/B stars, RCrB stars, F Com, V Hyd, SN 1987A [Podsiadlowski & Spruit 1999]).

Thus, even though a star may be single now, it may well have been a member of a binary system in the past. Indeed, whenever one is confronted with a new stellar phenomenon, it is probably advisable to first thoroughly explore the possibility of a binary interaction as a cause of the phenomenon before starting to adjust the input physics in the stellar calculations.

REFERENCES

BHATTACHARYA, D. & VAN DEN HEUVEL, E. P. J. 1991 *Phys. Rep.* **203**, 1.

BLAAUW, A. 1961 *Bull. Astron. Inst. Netherlands* **15**, 265.

BRANDT, N. & PODSIADLOWSKI, PH. 1995 *MNRAS* **274**, 461.

BRAUN, H. 1997 *Ph.D. Thesis* LMU München, Chapter 6.

BRAUN, H. & LANGER, N. 1995 *A&A* **297**, 483.

COWLEY, A. P. 1966 *ApJ* **142**, 299.

COWLEY, A. P., HUTCHINGS, J. B., & POPPER, D. M. 1977 *PASP* **89**, 882.

D'ANTONA, F. 1996. In *Evolutionary Processes in Binary Stars* (eds. R. A. M. J. Wijers, M. B. Davies & C. A. Tout). p. 287. Kluwer.

DAVIES, M. B. 1996. In *Evolutionary Processes in Binary Stars* (eds. R. A. M. J. Wijers, M. B. Davies & C. A. Tout). p. 101. Kluwer.

DAVIES, S. R., WOOD, K. S., & COE, M. J. 1990 *MNRAS*, **245**, 268.

DE KOOL, M. 1987. *Ph.D. Thesis*, Amsterdam, Chapter 5.

DE KOOL, M. 1996. In *Evolutionary Processes in Binary Stars* (eds. R. A. M. J. Wijers, M. B. Davies & C. A. Tout). p 365. Kluwer.

DE GRÈVE, J.-P. & DE LOORE, C. 1976 *ApSS* **43**, 35.

DE GRÈVE, J.-P. & DE LOORE, C. 1977 *ApSS* **50**, 75.

Delgado, A. J. & Thomas, H.-C. 1981 *A&A* **96**, 142.

de Loore, C. & de Grève, J.-P. 1992 *A&AS* **94**, 453.

de Loore, C. & Vanbeveren, D. 1995 *A&A* **304**, 220.

Drechsel, H., Haas, S., Lorenz, R., & Gayler, S. 1995 *A&A* **294**, 723.

Dudley, R. E. & Jeffery, C. S. 1993 *MNRAS* **262**, 945.

Eggleton, P. P. & Tout, C. A. 1989. In IAU Colloq. 107, Algols (ed. A. H. Batten). p. 164. Kluwer.

Fruchter, A. S., Stinebring, D. R., & Taylor, J. H. 1988 *Nat* **333**, 237.

Garmany, C. D., Conti, P. S., & Massey, P. 1980 *ApJ* **242**, 1063.

Ghez, A. M. 1996. In *Evolutionary Processes in Binary Stars* (eds. R. A. M. J. Wijers, M. B. Davies & C. A. Tout). p. 1. Kluwer.

Han, Z., Podsiadlowski, Ph., & Eggleton, P. P. 1995 *MNRAS* **272**, 800.

Hameury, J.-M, King, A. R., Lasota, J.-P., & Raison, F. 1993 *A&A* **277**, 81.

Hellings, P. 1983 *ApSS* **96**, 37.

Hillebrandt, W., & Meyer, F. 1989 *A&A* **219**, L3.

Hjellming, M. S. & Webbink, R. F. 1987 *ApJ* **318**, 794.

Iben, I. Jr. 1991 *ApJS* **76**, 55.

Iben, I. Jr. & Livio, M. 1993 *PASP* **105**, 1373.

Iglesias, C. A., Rogers, F. J., & Wilson, B. G. 1992 *ApJ* **397**, 717.

Joss, P. C., Podsiadlowski, Ph., Hsu, J. J. L., & Rappaport, S. 1988 *Nat* **331**, 237.

Kelley, R. L., Rappaport, S., Clark, G. W., & Petro, L. D. 1983 *ApJ* **268**, 790.

Kenyon, S. J. 1986. *The Symbiotic Stars*. CUP.

Kippenhahn, R. & Weigert, A. 1967 *ZAp* **65**, 251.

Kopal, Z. 1978 *Dynamics of Close Binary Systems*. Reidel.

Krolik, J., Meiksin, A. & Joss, P. C. 1984 *ApJ* **282**, 466.

Landau, L. D. & Lifshitz, E. M. 1962 *The Classical Theory of Fields*. Pergamon.

Langer, N. 1997. In *Santa Barbara Workshop on Supernovae* (eds. A. Burrows, K. Nomoto, & F. Thielemann,
http://www.itp.ucsb.edu/online/supernova/snovaetrans.html.

Lauterborn, D. 1970 *A&A* **7**, 150.

Lecar, M., Wheeler, J. C., & McKee, C. F. 1976 *ApJ* **205**, 556.

Levine, A., Rappaport, S., Deeter, J. E., Boynton, P. E., & Nagase, F. 1993 *ApJ* **410**, 328.

Livio, M. 1996. In *Evolutionary Processes in Binary Stars*, (eds. R. A. M. J. Wijers, M. B. Davies & C. A. Tout). p. 141. Kluwer.

Livio, M. & Soker, N. 1988 *ApJ* **329**, 764.

Mardling, R. A. 1995a *ApJ* **450**, 722.

Mardling, R. A. 1995b *ApJ* **450**, 732.

McClure, R. D. 1983 *ApJ* **268**, 264.

Meurs, E. J. A. & van den Heuvel, E. P. J. 1989 *A&A* **226**, 88.

Meyer, F. & Meyer-Hofmeister, E. 1979 *A&A* **78**, 167.

Nomoto, K., et al. 1994 *Nat* **371**, 227.

Paczyński, B. 1971 *ARA&A* **9**, 183.

Paczyński, B. 1976. In *Structure and Evolution of Close Binary Systems* (eds. P. P. Eggleton, S. Mitton & J. Whelan). p. 75. Reidel.

Paczyński, B. & Sienkiewicz, R. 1972 *Acta Astron.* **22**, 73.

Phillips, S. N. & Podsiadlowski, Ph. 1999 *MNRAS*, submitted.

PHILLIPS, S. N., SHAHBAZ, T., & PODSIADLOWSKI, PH. 1999 *MNRAS*, in press.

PODSIADLOWSKI, PH. 1991 *Nat* **350**, 136.

PODSIADLOWSKI, PH. 1999a this symposium.

PODSIADLOWSKI, PH. 1999b in preparation.

PODSIADLOWSKI, PH., & JOSS, P. C. 1989 *Nat* **338**, 401.

PODSIADLOWSKI, PH., JOSS, P. C., & HSU, J. J. L. 1992 *ApJ* **391**, 246 (PJH).

PODSIADLOWSKI, PH., JOSS, P. C., POLS, O. R., & BRANDT, W. N. 1996. In *Evolutionary Processes in Binary Stars* (eds. R. A. M. J. Wijers, M. B. Davies & C. A. Tout). p. 181. Kluwer.

PODSIADLOWSKI, PH., JOSS, P. C., & RAPPAPORT, S. 1990 *A&A* **227**, L9.

PODSIADLOWSKI, PH. & REES, M. J. 1994. In *The Evolution of X-Ray Binaries* (eds. S. S. Holt & C. S. Day). p. 71. AIP.

PODSIADLOWSKI, PH. & SPRUIT, H. 1999 in preparation.

POLS, O. R. 1994 *A&A* **290**, 119.

RAPPAPORT, S., VERBUNT, F., & JOSS, P. C. 1983 *ApJ* **275**, 713.

RASIO, F. A. 1996. In *Evolutionary Processes in Binary Stars* (eds. R. A. M. J. Wijers, M. B. Davies & C. A. Tout). p. 121. Kluwer.

RASIO, F. A. & LIVIO, M. 1996 *ApJ* **471**, 366.

RITTER, H. 1996. In *Evolutionary Processes in Binary Stars* (eds. R. A. M. J. Wijers, M. B. Davies & C. A. Tout) p. 223. Kluwer.

RUDERMAN, M., SHAHAM, J., & TAVANI, M. 1989 *ApJ* **336**, 507.

RUDERMAN, M., SHAHAM, J., TAVANI, M., & EICHLER, D. 1989 *ApJ* **343**, 292.

SPRUIT, H. C. & RITTER, H. 1983 *A&A* **124**, 267.

TAAM, R. E. & BODENHEIMER, P. 1992. In *X-Ray Binaries and Recycled Pulsars* (eds. E. P. J. van den Heuvel & S. A. Rappaport). p. 281. Kluwer.

TASSOUL, J. L. 1995 *ApJ* **444**, 338.

THORSETT, S. 1999, these proceedings.

TOUT, C. A. & EGGLETON, P. P. 1988 *ApJ* **334**, 357.

VAN DEN HEUVEL, E. P. J. 1976. In *Structure and Evolution of Close Binary Systems* (eds. P. P. Eggleton, S. Mitton & J. Whelan). p. 35. Reidel.

VAN DER LINDEN, T. J. 1987 *A&A* **178**, 170.

VAN PARADIJS, J. 1991. In *Neutron Stars: Theory and Observation* (eds. J. Ventura & D. Pines). p. 289. Kluwer.

VAN PARADIJS, J. 1995. In *X-Ray Binaries* (eds. W. H. G. Lewin, J. van Paradijs, & E. P. J.van den Heuvel). p. 536. CUP.

VERBUNT, F. & ZWAAN, C. 1981 *A&A* **100**, L7.

WEBBINK, R. F. 1986. In *Critical Observations versus Physical Models for Close Binary Systems* (eds. K.-C. Leung & D.-S. Zhai). p. 403. Gordon & Breach.

WEBBINK, R. F. 1992. In *X-Ray Binaries and Recycled Pulsars* (eds. E. P. J. van den Heuvel & S. A. Rappaport). p. 269. Kluwer.

WEBBINK, R. F., RAPPAPORT, S., & SAVONIJE, G. J. 1983 *ApJ* **270**, 678.

WRIGHT, K. O. 1977 *JRASC* **71**, 152.

YUNGELSON, L. & TUTUKOV, A. 1997 In *Advances in Stellar Evolution* (eds. R. T. Rood, & A. Renzini). p. 237. CUP.

ZAHN, J.-P. 1975 *A&A* **41**, 329.

ZAHN, J.-P. 1977 *A&A* **57**, 383.

ZAHN, J.-P. 1989 —it A&A **220**, 220.

White dwarf stars

By D. E. WINGET

McDonald Observatory and Department of Astronomy, University of Texas, Austin, TX 78712

In this review I will take the reader on a brief guided tour of the white dwarf stars. We will examine the context making this field of growing significance to large portions of the physics and astrophysics communities. We will look at galactic cosmochronology as an important application requiring a detailed knowledge of the basic parameters of the white dwarf stars. We will visit the frontier of our understanding of these parameters and look at the problems we face in their determination We end our tour with a summary and a glimpse of the future as seen by your guide.

---

1. Motivations and context

The white dwarf stars are historically well established as cosmic laboratories for testing physics under extreme conditions. We will briefly examine their role in this context in the past, present, and immediate future. They are the only stable evolutionary state for the overwhelming majority of all stars, and so present us with a very important boundary-condition for prior stellar evolution. They have become important in several cosmological contexts: as candidates for dark matter, micro-lensing objects, and as the progenitors of supernovae of type Ia (SNIa). SNIa apparently exhibit significant peak-magnitude variations with redshift. If not due to differences in internal properties of the SNIa progenitors, the white dwarf stars, these magnitude variations imply SNIa could be used to measure curvature in the Hubble diagram. Finally, the white dwarf stars have been recognized for more than forty years to be accurate clocks, owing to the simplicity of their evolution and so have become important in determining the cosmic chronology. We examine all of these aspects in what follows.

1.1. *Physics and the white dwarf stars*

Large-scale experiments in physics and astronomy must sail the stormy seas of politics on a course plotted by legislators. Scientific leadership, and ultimately it's derivative technological leadership, are slipping away in the United States as the long-term view of the national good is fogged by regional self-interest and short-term gain. In this environment astronomy, with its broad popular appeal, has often fared better than physics. This has enabled some noble projects like the Hubble Space Telescope to navigate these troubled waters successfully, while others such as the Super Conducting Super Collider survive many of the hardships only to be dashed upon the rocks of budget cuts at the last possible moment.

Fortunately, nature is conducting large-scale physics experiments in space, and we have only to pay the cost of observing these experiments, not constructing them. We find an example of this in the white dwarf stars. Here we can explore the physics of matter under a wide range of extreme conditions, well outside the domain accessible in even the best-funded terrestrial laboratory. An illustration of the large range of temperatures and pressures we can expect inside the white dwarf stars appears in Figure 1.

Understanding the existence of white dwarf stars provided an early test of the quantum theory of matter (see Van Horn 1977 for an elegant discussion). Solving the mystery of the source of internal pressure in an object the mass of the sun and the size of the Earth gave us a novel demonstration of the combined action of the Pauli exclusion principle, and

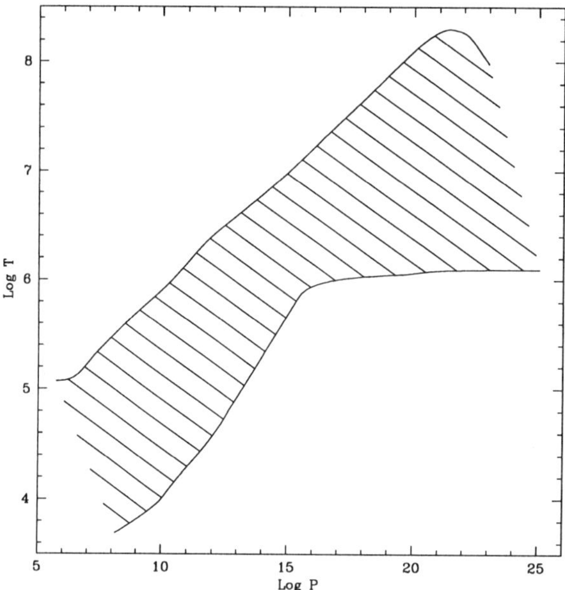

FIGURE 1. The region of the Pressure-Temperature plane spanned by the white dwarf star models. All quantities are in cgs units. The extreme boundaries are approximately marked at the upper left by a 0.6 $M_\odot$ model appropriate to the hot pre-white dwarf stars and the lower right boundary by a 1.2 $M_\odot$ model representing the coolest observed white dwarf stars.

the Heisenberg uncertainty principle (Fowler 1926). Chandrasekhar (1935, and references therein) showed that the speed of light forced a limit to the pressure support possible from degenerate electrons, and thus a maximum mass possible for a white dwarf star. Later we would see that the white dwarf stars could provide a test not only of special relativity and quantum mechanics, but also general relativity. The strong gravity implies a detectable gravitational redshift. This was first reliably measured by Greenstein & Trimble (1967) and has become an important way to measure the masses of white dwarf stars Reid (1996, and references therein).

A glance at Figure 1 makes evident the range of physical processes dominating the structure and evolution of the white dwarf stars. A 0.6 $M_\odot$ model sequence representing the hot, pre-white dwarf stars defines the upper boundary of the region (from Kawaler 1986) and a 1.2 $M_\odot$ model sequence representing the massive coolest white dwarf stars defines the bottom boundary (from Montgomery 1998).

At the extreme temperatures expected in the hot pre-white dwarf stars, we find the energy loss by photons completely swamped by neutrino losses a factor of ten larger. Witness the temperature inversion evident in the hot model. The temperature maximum is not at the central density maximum because the neutrinos refrigerate the deep central regions. This unusual temperature gradient implies a negative photon-luminosity in central regions of these models.

Many types of neutrinos are expected in theoretical models of these stars, but plasmon neutrinos dominate the energy loss in most of the regions where neutrinos are important (see Itoh et al. 1996 for the most current theoretical rates). This is the *only* environment known where we have an opportunity to study the production of plasmon neutrinos. These rates drive the evolution of the star, so a measurement of evolutionary timescales in these stars will provide a direct measure of plasmon neutrino rates, thereby testing the theory of electro-weak interactions in a new domain.

As the models evolve beyond the hot, pre-white dwarf stage, photon cooling dominates, gravitational contraction is dramatically reduced as the interior equation of state hardens into that of a strongly degenerate electron gas. Mechanical and thermal properties separate (see Van Horn 1971 for a nice discussion) at these stages and below. The degenerate electrons provide the dominant pressure, while the thermal motions of the ions make a negligible and ever-dwindling contribution to the mechanical support. The roles of electron and ion are reversed in their contribution to the energy. The heat capacity of the degenerate electrons is negligible (Mestel 1965) and the only significant source of energy is the reservoir of thermal energy in the nearly classical ideal gas of the ions. Energy transport in the interior is dominated first by neutrino energy transport, and then by conduction, and in the outer layers by both radiation, and in some cooler temperature regions by the convection associated with partial ionization of the most abundant element at the surface.

In the central regions where neutrinos once reigned supreme, the Coulomb interactions become increasingly important. First the ratio of Coulomb energy exceeds the thermal ion energy and the Coulomb liquid region is reached. Later, as temperatures drop further the ratio climbs to about 180 and we expect a change in the state of matter as the ions to crystallize into a lattice. Nearly forty years have passed since Salpeter (1961), and independently Kirzhnits (1960) and Abrikosov (1961), showed that the ions should crystallize, and still there is no empirical proof that this effect occurs.

Subsequent work has shown that the transition to a solid state may leave a signature not just in the structure of the white dwarf stars, but in their evolution. Van Horn (1968) showed that a first-order phase transition should release latent heat, thereby providing an additional source of energy and prolonging the white dwarf cooling times. A further effect was suggested by Stevenson (1977, 1980). He investigated the possibility that Carbon and Oxygen interiors might not crystallize as an alloy but undergo phase separation upon crystallization. Early work by Mochkovitch (1983); Barrat, Hansen & Mochkovitch 1988; Garcia-Berro et al. (1988) indicated that the phase separation, and trace element separation as well, may provide significant additional energy sources through the re-distribution of the ions of differing chemical potential energy.

After crystallization spreads from the center towards the surface layers, Mestel & Ruderman (1967) pointed out that the heat capacity in the crystallized regions will no longer be due to thermal particle motions in a gas, but to lattice vibrations in the form of phonons. When the interior temperature drops below the Debye temperature in the solid, higher energy levels become increasingly inaccessible and the heat capacity drops. Ostriker & Axel (1969) showed that the change in heat capacity associated with the solid phase should result in a much more rapid evolution after crystallization, ultimately ending in a Debye-cooling phase where the interior heat capacity falls off with T^3.

Theoretical modeling indicates that further cooling has dramatic effects on the surface regions as well. As the surface temperatures drop, gravitational settling dramatically influences the distribution of chemical elements in white dwarf stars, and the chemical stratification produced by prior evolution. The heavier elements tend to diffuse towards the centers. The surface layers become partially-ionized. In this way radiative energy transport gives way to turbulent transport in the comparatively high-pressures of the white dwarf envelope. The turbulent energy transport moves inward to ever-higher pressures as the star cools. The cooling causes the degeneracy boundary in the envelope to move towards the surface. Eventually the base of the partial-ionization zone and the degeneracy boundary collide and the turbulence is halted by the efficient energy transport associated with the degenerate electrons. Depending on the relative thickness of the various layers, as the turbulent convection reaches it's deepest extent, some of the effects of gravitational settling and chemical diffusion may be reversed. This latter effect may well alter the observed surface abundances as a function of the star's effective temperature.

Even the outermost layers become exotic by terrestrial standards, as the temperatures lower to the range of the coolest observed white dwarf stars. We begin to observe molecular H and now even a form of molecular He is suspected in the high-pressure atmospheres of cool He-white dwarf stars (see Bergeron, Ruiz, & Leggett 1997 and references therein, hereafter BRL).

We glance again at Figure 1, and realize that we must understand the entire shaded region, if we are to model the white dwarf stars. Or phrased more positively, this defines what the white dwarf stars can teach us about the behavior of matter under extreme conditions.

1.2. *White dwarf stars as a boundary condition for stellar evolution*

Cluster observations give us our best current estimate for the upper mass limit for main sequence stars becoming white dwarf stars. Depending on the theoretical models used the values range from 5 to 9 $M_\odot$ (Romanishin & Angel 1980, Weidemann & Koester 1983, Reimers & Koester 1994, Jeffries 1997). This implies that more than 98% of all stars will eventually become white dwarf stars, and from a more personal perspective, the Sun will become one. The connection with prior evolution is summarized in a quote from Brian Warner, "Inside every red giant is a white dwarf waiting to get out."

An interesting consequence of the relatively large upper limit for the initial mass is that for all stars more massive than 2 $M_\odot$ they ultimately recycle into the interstellar medium more mass than they remove in the form of the remnant white dwarf. While most of the re-cycled material is unlikely to have undergone significant nucleosynthesis, the abundances of the relatively fragile species may be dramatically affected by recycling.

The extreme difference in the initial main-sequence masses and the final white dwarf masses underscores the large role mass-loss must play in the post-main sequence, pre-white dwarf evolutionary phases. The qualitative stages by which the star moves off the main sequence we believe to be well understood. Core H, then He, burning followed by H- and He-burning shells surrounding a C/O core. But just how the mass is lost, what are the relative masses of the remnant H-layer and He-layer, and the relative distribution of C and O in the core, are still only very poorly understood. Measurements of the residual layer masses and the distribution of C/O in the core are therefore of great interest as boundary conditions for these complex evolutionary stages.

1.3. *White dwarf stars and cosmology*

The white dwarf stars are also believed to be the progenitors of SuperNovae of type Ia. An understanding of their structure and dynamical properties may improve our understanding of the resultant supernova event. Effects such as crystallization, phase separation, isotopic fractionation, and dynamical/pulsational modes of energy transport, must be understood. Only then can the SNIa be reliably interpreted, and effects such as the recently discovered variations in the brightness of these events with redshift be properly interpreted.

White dwarf stars in clusters have also been used to derive cluster distances, thereby affecting age estimates based on main-sequence turn-off or post main-sequence evolution. The idea is to use the sequence of white dwarf stars observed in the cluster and a sequence of local white dwarfs as a calibration to derive the cluster distance. This has been done for NGC6752 by Renzini et al. (1996) using the method suggested by Renzini (1991, 1998); also see the paper on 47 Tuc in these proceedings by Zoccali et al.

Recently, cool white dwarf stars have been invoked as a possibly large source of halo dark matter (see Isern et al. 1998, and Kawaler 1996 and references therein). In a similar vein, they have been examined as the lensing object to explain the observed microlensing events following the report of Alcock et al. (1997) that the microlensing events observed towards the Large Magellanic Cloud are produced by halo objects with an average mass of ~ 0.5 M$_\odot$. Numerous theoretical investigations of the halo white dwarf stars have consistently shown that halo white dwarf stars in these numbers are unlikely (Tamanaha et al. 1990, Adams & Laughlin 1996, Fields, Mathews & Schramm 1997, Chabrier, Segretain, & Mera 1996, and also see Kawler 1996 for constraints from the Hubble Deep Field).

Our final illustration of the significance of the white dwarf stars is their use as cosmic chronometers. We discuss this by way of illustration of an important application of white dwarf star studies in the next section.

2. White dwarf stars as galactic chronometers: The age of the disk

Four reasons combine to make the white dwarf stars excellent chronometers for measuring the age of the various stellar populations of the galaxy. The first reason is that the white dwarf stars are representative of the general population of stars; as we have seen, most stars are, or will become, white dwarf stars. Second they are a homogenous class of objects with a very narrow distribution of total stellar masses between $0.4 M_{wd}/M_\odot 1.1$ and a mean mass of $< M/M_\odot >= 0.590 \pm 0.134$ (Bergeron et al. 1995). Further, on theoretical grounds (see Iben 1991 for an excellent summary), we expect all of the objects with masses in the range $0.5\text{--}1.1 M/M_\odot$ to have C/O cores, with thin overlying He-layers, and most an additional surface H-layer. Third, high surface gravity ($\log g \sim 8$), typically slow rotation rates and low magnetic fields, nuclear energy and gravitational energy generation rates near zero, argue that these objects are physically simple. Fourth, the unavailability of energy sources other than the residual thermal energy of the ions implies that the evolution of white dwarf stars is primarily a simple cooling problem; there is a straightforward relation between the age of a white dwarf and its luminosity. These factors combine to imply that the white dwarf stars provide an archaeological record of the history of star formation in the galaxy.

Two excellent recent reviews of this subject are Chabrier (1998) and Isern (1999). I send you to these references for a different perspective, and more complete treatment of the problems of chemical fractionation in the cool white dwarf stars.

2.1. *White dwarf evolution as a cooling problem*

Many years ago, Mestel (1952) demonstrated the simplicity of the evolution of a star supported by electron degeneracy pressure and without significant nuclear energy sources. He constructed an analytical model for white dwarf evolution by treating the structure as if it consisted of only two layers. The inner layer contains most of the mass of the star, and is assumed to be isothermal because of strong e^--degeneracy. For the same reason, the electrons don't contribute significantly to the heat capacity; it comes almost entirely from the ions, which are assumed to behave as a classical ideal gas. The thin non-degenerate outer layer forms an insulating blanket and controls the rate at which the energy from the ion reservoir is leaked out into space. The specific rate is controlled by the radiative opacity at the boundary between these two layers, and is assumed to obey Kramer's law, $\kappa = \kappa_o \rho T^{-3.5}$. The Mestel theory has been nicely summarized by Van Horn (1971), Wood (1990) and most recently by Kepler and Bradley (1995). As expected it gives a simple form for the age-luminosity relation,

$$\log(\tau_{\mathrm{cool}}) \approx Const. - (5/7)\log(L/L_\odot). \tag{2.1}$$

This analytical model for white dwarf cooling is both a good picture to have in your head, and as shown by Iben and Tutukov (1984), gives surprisingly good agreement with the predictions of detailed numerical models. We can think of the key physical effects not included in the Mestel theory as perturbations to Mestel theory to a good approximation. Or, we can even construct a modified Mestel theory including them (see for example Wood 1990).

As discussed by Lamb & Van Horn (1975), we can expect five key physical effects to produce deviations from the predictions of classical Mestel theory. These are neutrino energy loss, possible residual nuclear burning in the H-layer (dismissed by Lamb & Van Horn and previous authors, but see Iben & Tutukov 1984, and Iben & MacDonald 1986), gravitational contraction, surface convection, and crystallization. In the latter we include the closely related phase separation as well as the subsequent Debye cooling ($C_v ion$ is proportional to T^3).

2.2. *The white dwarf luminosity function and the age and history of the galactic disk*

The idea of using white dwarf stars as chronometers has a long history beginning with Mestel (1952). Schwarzshild (1958) wrote, "We see that the cooling times we have derived in this manner are of just the right order of magnitude. The cooling times are sufficiently short so that many stars have had time enough to reach those values of the internal temperature which... give the observed low-luminosities of the white dwarfs. At the same time, the cooling times are sufficiently long so that a cooling star will not become unobservably faint too quickly." This made clear that if we could observationally determine the population of the coolest white dwarf stars, we would have the oldest white dwarf stars (allowing for mass effects), and thereby a constraint on the ages of the oldest stars. This idea was further developed by Schmidt (1959) in a galactic context, and later by D'Antona & Mazitelli (1978).

This concept was challenged by Ostriker and Axel (1969), which we recall were the first to consider the effects of Debye cooling on white dwarf evolution. They pointed out that the onset of Debye cooling would produce a sharp decrease in the number of cool white dwarf stars observed below $M_v \approx 13$–14. It is important to note that this value was based on calculations for white dwarfs of 1 $M_\odot$, which was believed at the time to be a typical white dwarf mass (Greenstein & Trimble 1967). We thought that Greenstein (1969) had identified the coolest population of white dwarf stars. This latter turned out to be an observational selection effect, but it was an extremely important result for the field in that it stimulated the future search by Liebert and collaborators (Liebert 1979, Liebert, Dahn & Monet 1988), which did find a downturn in the white dwarf luminosity function.

At the 1979 meeting in Rochester, N.Y. (I.A.U. Colloquium No. 53 entitled "White Dwarfs and Variable Degenerate Stars") Jim Liebert (Liebert 1979) announced the new evidence for the downturn in the white dwarf luminosity function as a paucity of cool degenerates. He pointed out that this could be the result of accelerated cooling, or because of the finite age of the disk, but it was not an observational selection effect. Recently completed evolutionary calculations by Van Horn, Lamb, and myself indicated a cooling age in the vicinity of this downturn to be 9–10 Gyrs. During the course of the meeting I drew this to the attention of Jim Liebert, Gilles Fontaine, and my thesis advisor, Hugh Van Horn. We realized that this could reflect the finite age of the disk, and give us a meaningful measure of its age. Other investigators, however, were finding much younger ages for white dwarf stars at the luminosity where Liebert observed the downturn (D'Antona & Mazzitelli 1978 Shaviv & Kovetz 1976) and so the prevailing opinion at the meeting was the paucity was not due to the age of the disk, but due instead to Debye cooling (D'Antona & Mazzitelli 1978, and the later work of Mazzitelli & D'Antona 1986).

Improved observational results (Liebert, Dahn, and Monet 1988, 1989 hereafter both papers are referred to as LDM), and further calculations of theoretical white dwarf luminosity functions incorporating a distribution of masses and an estimate of the average main-sequence evolutionary time, convinced us that the turndown Liebert had found was indeed a measure of the age of the galactic disk. Later, at the urging of Gerry Wasserberg, we published this method of cosmochronometry along with a preliminary result of $\tau_{disk} \approx 9.3 \pm 2.0$ (Winget et al. 1987). This work also demonstrated the possibility of using the shape of the luminosity function prior to the turndown to constrain constitutive physics in the models and the star formation history of our galaxy. Icko Iben refereed this paper, and after carefully examining the methods and results, he concluded that this was correct. He published (Iben & Laughlin 1989) a careful re-examination of this method of obtaining a similar age for the galactic disk using the evolutionary models of Winget et al. as well as Mestel theory. This work was more sophisticated than Winget et al. in that they modeled the initial-final mass relation and main-sequence lifetimes as a function of mass.

There was still a problem in this method of obtaining an age for the galactic disk, in that there is a broad dispersion in ages for the coolest white dwarf stars depending on which group carried out the evolutionary calculations. Winget and Van Horn (1987) demonstrated that this variation was due to very different treatments of the constitutive physics (for example inclusion of crystallization, treatment of convection), or different choices of model parameters (stellar mass, or choice of surface layer masses). When these effects were taken into account they demonstrated remarkable agreement between the results of various investigators. Some still believed, however, that the downturn could still be the result of short cooling times and subsequent Debye cooling, as indicated by the comments of D'Antona following the paper by Winget and Van Horn (1987, p. 376).

Further theoretical calculations employing more sophisticated treatments were carried out by Weidemann & Yuan (1989), Noh & Scalo (1990) and culminated in the work of Wood (1990, 1992, and 1995). By this time there was a consensus that the white dwarf luminosity function gave one of the most accurate measures for the age of the galactic disk.

With all this effort, and enormous improvements in the treatment of the evolutionary models and the constitutive physics it is important to see how far we have come since the beginning of this field more than 40 years ago. To establish that it is amusing to compare the current results with those published for the age of the lowest luminosity white dwarfs by Schwarzschild. For the most current results, we quote the results of Leggett, Ruiz, & Bergeron (1998; hereafter LRB). They used the LDM sample supplemented with new optical and infrared data, and derived temperatures from the dramatically improved cool model atmospheres of Bergeron, Saumon, & Wesemael (1995). Using Wood's theoretical models they obtain

$$\tau_{\text{disk}}(1988) = 8 \pm 1.5 \text{ Gyr}, \tag{2.2}$$

while Schwarzschild, armed with Mestel theory, obtained,

$$\tau_{\text{disk}}(1958) = 8 \times 10^9 \text{ yr}. \tag{2.3}$$

With tongue in cheek, we note that in 40 years we have gained both error bars, and a new expression for units. In a more serious vein we must point out that with the same LRB data, and the phase-separation models of Montgomery (1998) or Salaris et al. (1997) the value for the age becomes 9 Gyr. Also, an entirely independent sample of cool white dwarfs in wide binaries reported by Oswalt et al. (1996; hereafter OSWH) gives an age of 9.5 Gyr with the Wood (1995) models, or 10.5 Gyr with the Montgomery (1998) or Salaris et al. (1997) phase separation models.

Clearly there is much more to the story but this remarkable consistency underscores why the white dwarf stars are such useful chronometers: they are simple!

2.3. *Uncertainties in white dwarf cosmochronometry*

Wood (1990, 1992, and 1995) systematically explored the effects of uncertainties in the various surface layer masses, and core composition distributions of C and O. Most importantly, he demonstrated that uncertainties in star-formation rates, initial final mass-relations and our description of disk-inflation did not contribute significantly to the error budget.

If we look a bit more closely, and with a bit more skepticism, at the theoretical evolutionary models and the observations, both, we see there is much work to be done.

2.3.1. *Some theoretical evolutionary considerations*

For example, Wood (1995) gave a somewhat smaller result for the age of the disk from the LDM sample: 7.5 ± 1 Gyr. The difference between Wood (1990.1992) and Wood (1995) results for the age of the disk is about 2 Gyr. The reason is the incorporation of a H-layer, a thicker He-layer, mixed C/O core compositions, and a new generation of radiative and conductive opacities (Iglesias, Rogers & Wilson 1993, Itoh, Hayashi, & Kohyama 1993). To underscore the significance of the layer mass, an order of magnitude change in the He-layer mass gives ~ 0.75 Gyr change in the age of the turndown in the luminosity function (Wood 1990, 1992). Interestingly, uncertainties in the much thinner H-layer have less effect. Wood (1990) gives approximately 0.5 Gyr for the total effect due to variation of the fractional H-layer mass from 0 to 10^{-4}. Since his models do not include residual nuclear burning effects, it is reasonable to use the value from Winget & Van Horn (1987), 0.7 Gyr, even though this latter estimate is from comparison of inhomogeneous evolutionary calculations by different investigators.

Realizing the importance of accurate ages for the cool white dwarf stars, a series of investigators have examined the possible effects of phase separation of C and O at crystallization, as well as the separation of trace elements (as described above). Their estimates of the additional cooling time caused by the energy supplied in these processes ranged from 0.5–6 Gyr (Mochkovitch 1983; Barrat, Hansen, & Mochkovitch 1988; Garcia-Berro et al. 1988; Chabrier 1993; Hernanz et al. 1994; Segretain et al. 1994; Isern et al. 1997; and Salaris et al. 1997). Much of the range in these numbers comes from differences in the phase diagram assumed. The most recent estimates of C/O phase separation from Salaris et al. (1997), and the first fully self-consistent evolutionary calculations by Montgomery (1998), give an age increase of ~ 1.0 Gyr. This result depends of course on the distribution of the C and O in the white dwarf prior to crystallization as well.

Additionally, it is possible that settling of trace elements such as ^{22}Ne or ^{56}Fe may cause a further increase in the cooling times for white dwarf stars (see for example Isern 1999, Xu & Van Horn 1992, Segratain et al. 1994). Isern et al. (1997) point out, however, that the effect on the ages of the coolest white dwarf stars is likely to be small because of their presumed low abundances of heavy elements. For a full discussion of this see Isern et al. (1997) and especially the recent review by Isern (1999) and references therein.

2.3.2. *Some observational considerations*

Uncertainties in the age of the disk from the cool white dwarf stars arise not only from the theoretical modelling, but also from the observations themselves. Currently there are two independent samples of cool disk white dwarf stars: the LDM sample of 43 field white dwarf stars taken from the Luyten Half-second catalog (Luyten 1979), as modified by LRB, and the OSWH sample of 50 white dwarf stars obtained from observations of common proper-motion binary stars. We might naively take the difference of 1.5 Gyr in the turndown for the two samples as an estimate of the observational error. This is certainly too simplistic, but as we will see it gives a reasonable estimate of the errors.

Both sets of authors in various combinations have objectively criticized the results of both surveys. The work is sufficiently difficult that the best I can do is summarize the main problems faced by each determination of the white dwarf luminosity function (hereafter LF) based on the discussion in LRB, and Wood & Oswalt (1998).

A difficulty with the LRB result (and its predecessor LDM) rests in the problem of completeness. The method for computing a luminosity function from a proper-motion selected sample with known parallaxes and magnitudes is due to Schmidt (1975), and is commonly known as the $1/v_{\mathrm{max}}$ method. Here v_{max} refers to the maximum volume over which a star could contribute to the sample, given the upper and lower limits on proper motion and apparent magnitude of the sample. The completeness of the sample can then be checked by calculating the mean value of the ratio of the actual volume containing each star to the maximum volume, v/v_{max}. For a complete sample the mean value of the ratio should be near 0.5, for the LDM/LRB sample it is 0.37. LRB point this out, but note that the agreement of their cool LF with the hot LF (LDM and references therein), argues that the effect on the space densities is no greater than their quoted uncertainties of $\sim 50\%$.

The major difficulties with the OSWH sample are most important in the errors of the last bin, the critical bin for establishing the turndown. OSWH use $V - I$ color and optical spectra, interpreted assuming the mass is 0.6 $M_\odot$ to determine their luminosities. This makes the determination of the white dwarf spectral type difficult, which ultimately give luminosities differing by 40% (LRB). LRB also point out the problems with contamination of the photometry by the companions; they point out that one of the three stars, *LHS 290*, in the last bin of OSWH was observed by LRB, and the I data was too contaminated to be used. Their *BVRJHK* data, however, enabled them to determine that the star is more luminous than determined by OSWH by a factor of 4. Further, since the other two stars in the OSWH sample are fainter, and have similar companions at similar distances LRB argue that the luminosities of these objects may be similarly compromised.

Note that the related problem of possible low-luminosity companions, undetected in some fraction of the objects, may make the problem a common one for both LFs. This is particularly sobering in view of the results of Saffer, Livio, & Yungelson (1998) that roughly one sixth of so-called field white dwarf and hot subdwarf stars turn out to be binaries.

Ultimately in both current samples we are faced with small-number statistics in the most critical luminosity bins. We must ask the question, "Can we believe the turn-down in either sample?" For some ideas and answers we can turn to the very enlightening statistical study of the LF by Wood & Oswalt (1998). They carried out Monte Carlo simulations Of proper motion and magnitude limited samples using the $1/v_{\mathrm{max}}$ method. Their results are extremely encouraging for the field. They show that the $1/v_{\mathrm{max}}$ method is reliable, and that the uncertainties are $\sim 50\%$ for a sample size of 50—in excellent agreement with the estimates of LRB for a sample of 43. If the sample size is increased to 200 the limits drop to $\sim 25\%$. Further, they point out that the uncertainties in the age of the disk quoted in the observational LF papers are consistent with their Monte Carlo calculations. There is one sobering conclusion of this study: even in the limit of large sample size, statistical variations in the hot LF are too large for these populations to be useful in determining fluctuations in the recent star formation rate as a function of time. Time will tell if this conclusion is valid.

2.4. *White dwarf ages as a measure of the evolution of our galaxy*

The above work has been confined to the disk in the solar neighborhood. Clearly it is important to extend this work to the other components of the galaxy as well. In particular the relative age of the disk and halo of our galaxy will give as information about how the galaxy formed. This will be our template for understanding the formation and development of spiral galaxies in general. The importance of the halo luminosity function was well understood by LDM, who made a first pass at the problem, but the faint halo and globular cluster white dwarf stars were not within reach. Claver (1995) has shown that digital surveys in progress should be able to reach the halo downturn and hence give an age. He also points out the importance of finding the coolest white dwarf stars in open and globular clusters. All of this work must await the future, but as Richer et al. (1995) note, the refurbishment of the HST with the advanced camera will bring the faint end of the white dwarf luminosity function for M44 within reach in 1999, so by the time you read these words we may already know the answer!

3. Parameters and problems: Skeletons in the closet

For the application just described, white dwarf cosmochronometry, as well as most of the previous applications discussed in the first section, we must identify all the parameters necessary to describe the structure and evolution of the white dwarf stars, and measure their distributions. Space considerations prohibit completeness and we point the reader to the extensive review of Kepler & Bradley (1995,KB) for a more thorough treatment of the material in this and following sections. Here we will examine a selected set of white dwarf parameters, and briefly discuss what problems arise, placing emphasis on advances since 1995.

3.1. *The number of known white dwarf stars*

To get better statistics on all of the parameters of the white dwarf stars, it is clear that we must have more stars. This number should increase dramatically soon based on the results of surveys such as the Sloan survey (Szalay 1998, and references therein), the Texas Deep Sky Survey (Claver 1995), and others planned or in progress (see Smith 1997). With more stars we can think about constructing and comparing LFs separately for stars of various spectral types, to study their origins and evolution, or for stars of a given mass to isolate star-formation effects from constitutive physics. The applications to studies of small or rare classes of white dwarf stars are many and obvious. For example, our understanding of the magnetic white dwarf stars, the DAB and DBA stars, or any of the classes of pulsators, to name only a few.

3.2. *The fraction of binaries*

We rely heavily upon our observations of (presumed) single field stars to draw conclusions about single-star evolution. The recent work of Saffer, Livio, & Yungelson (1998) demonstrates how dangerous this can be. In radial velocity observations of 153 field white dwarf stars and sdB stars presumed single. They found 23 were radial velocity variables. As we have seen, undetected binaries in LF samples may produce problems in determinations of luminosity, especially when they occur in bins with relatively few objects.

3.3. *The distribution of white dwarf masses*

The most basic parameter of a white dwarf star is total stellar mass. This continues to be of great interest from the time of the discovery of Sirius B, to the present. Masses can be determined from orbital solutions for binary stars, asteroseismology, or measurements of surface gravity. The first two techniques yield by far the most accurate measures. Only the latter technique has yielded a statistically significant sample of objects. Once the surface gravity is measured, we can infer a mass only with the help of a Mass-Radius relation, currently the best available relation come from the evolutionary calculations of Wood (1990, 1992 and 1995). These have been used in all of the recent calculations mentioned below.

There are two ways to determine surface gravities. The first depends on model atmosphere calculations combined with photometry, parallax information or low-resolution spectroscopy. For this reason analyses of objects of different surface abundances (spectral types, see discussion in the next subsection) are carried out independently. The largest sample of stars to-date is 129 DA white dwarfs analyzed by Bergeron, Saffer, & Liebert (1992). They initially obtained $< M_{DA} >= 0.562$, using evolutionary models with inappropriate surface He-layers. Bergeron et al. (1995) revised this determination by applying the mass-radius relation appropriate for DA white dwarfs (Wood 1995) and obtained

$$< M_{DA} >= 0.590 \pm 0.13 \ \mathrm{M_\odot}, \tag{3.4}$$

for the sample. This difference based on the theoretical models illustrates the sensitivity of the mean mass determination to the mass-radius relation used.

Similar analyses for the two large classes of non-DA stars were recently carried out. For the DB white dwarf stars, Beauchamp et al. (1996) report

$$< M_{DB} >= 0.585 \pm 0.063 \ \mathrm{M_\odot}. \tag{3.5}$$

Dreizler & Werner (1996) have measured the mean mass of the hot He-dominated DO white dwarfs. The find

$$< M_{DO} >= 0.59 \pm 0.08 \ \mathrm{M_\odot}. \tag{3.6}$$

The second method of determining surface gravities is from the measurement of gravitational redshifts. The redshift is sufficiently small that only the DA (H-spectra) white dwarf stars have sharp enough line cores for this method to work (Greenstein & Trimble 1967). In addition the method can only be applied to stars with well-determined radial velocities, so it is further restricted to white dwarf stars in binaries or open clusters. This method does have the advantage of avoiding any systematic errors that may be inherent in techniques involving model atmospheres. In this sense it gives an independent calibration of the first method. The largest and most recent sample is 53 DA white dwarf stars observed by Reid (1996). For the field stars in his sample, he obtains excellent agreement with the results of the photometric and spectroscopic work. He finds his distribution peaks at 0.58 $\mathrm{M_\odot}$, with a median of 0.581 $M_\odot$, and 70% of the stars give $M > 0.55 \ \mathrm{M_\odot}$.

The consistency of the gravitational redshift measurements of Reid (1996) using common proper-motion binaries in the field with the single-star results is encouraging. It supports the idea that there are no large systematic errors in the mass determinations.

It also argues that the assumption of a mean mass of 0.6 $M_\odot$ by OSWH in constructing their luminosity function from a sample of white dwarf stars in common proper-motion pairs is valid.

We can also be encouraged that the mean-mass determinations have been consistently in the vicinity of 0.6 $M_\odot$, with a narrowly peaked distribution, for the past 20 years since Koester, Schulz, & Weidemann (1979). As we have seen above, this result is independent of spectroscopic type, and illustrates just how homogeneous the class of white dwarf stars really is.

3.4. *Origin and evolution of the surface compositions*

The white dwarf stars can be divided into two distinct categories according to their surface abundances: those that are hydrogen rich, and those that are not. For the most part, we can claim that the nearly pure H and He compositions exhibited by most white dwarfs make sense. They are the result of chemical diffusion in the presence of a strong gravitational field as pointed out by Schatzman (1958). As the field matured, however, we came to understand the importance of other processes such as radiative levitation, winds and mass loss in the hot stars, convective mixing and accretion in the cool ones. Many attempts have been made to incorporate these complicating physical processes with a mixture of success and failure (see Fontaine & Wesemael 1987, 1991, 1997 and BRL and references therein).

There is a very useful, if necessarily somewhat complicated, taxonomy based on the observed spectral type and described in detail in Sion et al. (1983). Those that display the signature of H-rich photospheres we call DA stars; everything else we call non-DA. The DAs comprise roughly 80% of all white dwarf stars, although this is a function of temperature (Sion 1984). The dominant non-DA population are the He-rich, DO and DB stars. Until we get to the very cool stars this way thinking of the surface layers seems to work well, with one exception, the DB gap.

This concept (but not the taxonomy) breaks down at low temperatures where He and later H become increasingly transparent and difficult to detect. This gives rise to one of the outstanding problems in understanding the atmospheres of the cool white dwarf stars: what is the surface composition of the cool white dwarf stars? The composition, as we have seen, is important in determining the luminosities from the observational data. The relation between temperature and surface composition in the coolest white dwarf stars is complicated and poorly understood. For a detailed discussion of this important frontier of white dwarf research see Bergeron, Saumon, & Wesemael (1995), BRL and references therein.

We find the DA spectrum all the way from the hot white dwarf stars to the cool, defining the DA spectroscopic sequence. The He-rich sequence starts with the hottest stars of spectral type DO, with $T_{eff} > 45,000$ K. DOs show HeII at the higher temperatures, and a mixture of HeI and HeII spectra at the lower temperatures. Next, in order of temperature, we find the DB stars with $30,000$ K $> T_{\rm eff} > 11,000$ K, displaying HeI lines only. Below $T_{\rm eff} \sim 11,000K$ HeI lines cannot be excited and the spectra are featureless, DC, spectra. The problem with the non-DA spectroscopic sequence is the gap, discovered by Liebert et al. (1986): there are no known He-rich objects between 45,000 K and 30,000 K. This gap poses a serious challenge to ideas about white dwarf evolution and the origins of the spectral types. For an excellent recent discussion of this area the reader is referred to Dreizler & Werner (1996) and Fontaine & Wesemael (1997) and references therein.

The existence of the gap seems to destroy the simple picture of two independent channels into the white dwarf stars, coming, for example, from the He-rich PNN and

H-rich PNN stars. Any such scheme would require an accelerated evolution of the non-DA stars through the gap; this seems quite implausible at present. There is also a change in the ratio of DAs to non-DAs around $T_e = 10,000$ K and cooler (Sion 1984; Bergeron, Ruiz, & Leggett 1997). This scheme is dealt a further serious blow by the result that the mean masses of the DA and non-DA populations are the same.

This led Fontaine & Wesemael (1987,1991, 1997) to propose a self-consistent theory of spectral evolution of white dwarf stars. Their attempt to explain the evolution of surface chemical composition in the context of the mechanisms competing with gravitational settling. Though this theory has some problems, it remains the most viable one. It has been described in detail in the references above and summarized in a historical context by KB. We mention only a few of the main features of the model here.

The essence of the original idea was that all white dwarf stars have a common origin. Whether descended from H- or He-rich PNN, some H remains in the star, in total amount ranging from an upper limit of roughly 10^{-4} set by nuclear burning down to 10^{-15}. This H diffuses upwards as the star cools through the DO temperature range, and accumulates at the surface. Even the relatively small amounts assumed present in DO stars ($M_{He} < 10^{-13}M_*$, only as little as 10^{-15} is required to be optically thick; Provencal & Shipman 1998) becomes optically thick by the blue edge of the DB gap, $T_e \approx 45,000$ K, at which point all the white dwarf stars are DAs, including the once-and-future non-DA stars. At the red-edge of the gap, $T_e \approx 30,000$ K, the underlying He-layer in the former DO stars becomes convective as the result of the development of a He partial-ionization zone, the small surface H-layer is mixed in to the underlying deep He-convection zone. The DA stars with $M_H > 10^{-13}M_*$ have too much H to form a sufficient convection zone to mix the H down, and it remains, as always, a DA.

This model at once explains the DB gap, and because the DA and non-DA are similar except for trace amounts of H, it is quite consistent with the result that all the spectral types have essentially the same mean mass. Also, because the majority of DA stars might be expected to have comparatively thin H-layer masses distributed in some fashion between the 10^{-13} and 10^{-4} fractional mass limits, we can understand the increase in the number of non-DA stars observed below $T_e \approx 10,000$ K. Here again convective mixing is the logical explanation as the surface H-convection zone deepens, and reaches the He-layer below. Koester (1976) and Vauclair & Reisse (1977) had already used this result to argue for relatively thin surface H-layers in many DA stars. So we seem to be building a compelling case, but unfortunately it is only an illusion.

The spectral evolution model, in the above form, suggests that a wide range of relatively thin H-layers should be found in DAs, and yet the bulk of the asteroseismological evidence suggests that most of the DA stars have relatively thick surface H-layers (Clemens 1993,1994, 1995; Fontaine et al. 1994; Bradley 1996 and references therein). If typical DA H-layers are substantially thicker than about $10^{-10}M_*$ we lose not only the explanation of the gap, but the change in the DA to non-DA ratio below 10,000 K. This along with the disappearance of many of the other arguments suggesting thin H-layers led Fontaine et al. (1994) to remark, at the end of their paper, "Thus although there is no question that spectral evolution *does occur* in white dwarfs (the most obvious manifestation of which is the existence of the so-called DB gap), it is clear from these considerations that current models calling on thin hydrogen layers in a majority of DA stars will require substantial revisions."

Another telling blow to this appealing model is independent of asteroseismology and comes from the recent HST spectroscopic observations of Provencal & collaborators. The spectral evolution explanation of the DB gap implies trace amounts of H should remain in the hottest DB stars. H was found by Provencal et al. (1996) and Provencal & Shipman

(1998), but in abundances orders of magnitude too small to be consistent with spectral evolution models.

To add to the confusion, we know there are additional channels into the white dwarf stars besides the PNN stars (see KB), there are the subdwarf stars, and the Interacting-Binary White Dwarf (IBWD) stars. The subdwarf stars tend to have lower masses than the mean mass of the white dwarf stars. The narrower mass dispersion of the DB stars then argues that these stars must feed the DA channel only, but then this implies that a He-rich sdO star becomes a DA. With He-abundances between 50% and 99% (Thjell et al. 1994) there may be sufficient H left to make a thick H-layer DA.

The IBWDs are a bigger problem, or a solution. Which they are isn't clear yet. These mass-transfer binaries exhibit nearly pure-He in their spectra. The low-mass donor ($M_D/M_\odot \approx 0.05$) will almost certainly be either completely consumed, or remain as an invisible detached companion. Thus the end product will be a single DB white dwarf, and thus should contribute to this population of DB stars Nather, Robinson & Stover (1981).

As pointed out by Provencal (1994) and Provencal et al. (1997), there may be enough of these interacting, nearly pure-He systems to make a significant fraction of the single DB white dwarf stars observed, or possibly all of them. The binary origin of a significant fraction of the DB stars isn't eliminated in this case by the similarity of the mean masses. Recall that the mass of the donor is extremely small, and that the binary nature may have prevented the core mass of the primary from reaching the average white dwarf mass. It is then only a small stretch to imagine that the mean mass of stars produced in this way is the same as the mean mass of the DAs. Further, the accretion temperature is about 30,000 K, so these objects wouldn't enter the white dwarf cooling track until just below the DB gap. Then we could have all of the PNN stars with sufficient H to make DAs, so the DOs still disappear, and the DAs all have thick hydrogen. The DBs might then be presumed to come only from the IBWDs, and we have solved the problem of the gap.

Now on to another problem. If all DB stars come from IBWD, then *GD 358* is a descendent of an IBWD, and so should have a cooler core than a white dwarf on the cooling track at the same surface temperature, having been heated by prior accretion. This possibility was investigated by Nitta(1996), Nitta & Winget (1998). by examining the pulsation spectrum of GD 358 for the signature of a hot core. They find no signature of a cool core. So either GD 358 went through a very long accretion phase (10^8 yr or more) which brought the core into thermal equilibrium, or it is not the descendent of an IBWD. Such a long accretion phase pushes the theoretical limits, and so we conclude that it is most likely not a descendant of the IBWD stars. As Ed Nather would say, "Another perfectly good theory shot down by a damned eye-witness."

If by now you are thinking that we are giving with one hand and taking with the other, you are right! It is clear that we have no completely satisfactory model for the DB gap. We are far from a complete understanding of the origin and evolution of the surface abundances in white dwarf stars, arguably the simplest of all stars. This is sobering indeed for those working on photospheric abundances in much more complicated post-main sequence stars. Who said nature is not malicious? I guess he never worked on white dwarf stars.

3.5. *White dwarf rotation rates and magnetic fields*

Rotation and magnetic fields both break the spherical symmetry of the white dwarf stars. Not surprisingly, there is considerable overlap in the techniques to measure both (see for example the discussions in Heber, Napiwotzki, & Reid 1997, hereafter HNR), so we will include both in this section.

Rotation rates in white dwarf stars can be measured in three ways. From high-resolution spectroscopic measurements of the shapes of line cores, from observations of rotational variation in polarization in magnetic white dwarf stars, and from high-resolution temporal spectroscopy: asteroseismology.

The first method gives the most sensitive limits for DAs because of their narrow line cores (Greenstein & Trimble 1972, and see Greenstein & Peterson 1973), but less stringent limits can be obtained for DBs as well. HNR point out that the observations of DBs by Beauchamp et al. (1996), therefore set limits because they see no broadenings beyond their models which do not include rotation or magnetic fields. These methods have been applied to a sample of about 25 stars (Greenstein & Petersen 1973; Pilachowski & Milkey 1984, 1987; Koester & Herrero 1988) until the observations of 13 additional DAs with the Keck telescope by HNR brought the total to 38. The first group of 25 stars give limits of $v \sin i < 60$ km/s, and HNR drive limits of $v \sin i < 8$–40 km/s. Together, these non-detections give an impressive demonstration that white dwarf stars are surprisingly slow rotators.

Measurements of rotation rates can be made by observing periodic changes in the polarization or field strength (see below) in some magnetic white dwarf stars. Rotation rates and limits can be found in the recent review of Schmidt & Smith (1995,hereafter SS), Putney (1997) and Jordan (1997). As pointed out by Schmidt & Norsworthy (1991) these seem to fall into two groups: those with $1 hr P_{\rm rot}$ several days, and those with lower limits of $P_{\rm rot} \gtrsim 100$ yr. From the tables of SS and employing the limits from Putney (1997), Jordan (1997) gets 12–15 stars in the first group and 4–5 in the second. Interestingly, there seems to be no simple correlation of fields strength with rotation period (SS); however, note that all with $P_{\rm rot} > 100$ yrs have fields of $B > 100$ MG. There also seems to be no correlation with T_e and rotation period, or field strength (although see the discussion of magnetic field measurement.

I have excluded the most rapidly rotating magnetic white dwarf: the fascinating hot DA *REJ 0317–853* (Barstow et al. 1995) which has a period of 725 s, and a mass apparently close to the Chandrasekhar mass ($M_* \approx 1.35$ M$_\odot$). This is because it has both a higher mass and temperature than its nearby (600 AU) companion, and so is likely the product of an interaction (Jordan 1997).

Asteroseismology gives us rotation rates for a handful of pulsating stars. This comes from the breaking of the degeneracy of the azimuthal quantum number by rotation. This produces fine-structure splitting of the pulsation frequencies (see BK, and Winget 1999 for a discussion). In the limit of slow-rotation this is directly proportional to the rotation frequency, with a proportionality constant which is a function of the degree, ℓ, of the mode, and only weakly dependent on the eigenfunction (see Winget et al. 1991 for a discussion). Currently there are seven rotation rates measured in this way from asteroseismology. There are 3 DOV stars, 1 DBV, and 3 DAV stars. The shortest period comes from 5 hr in the DOV star *PG 2131+066* (Kawaler et al. 1995); and the longest from the DAV *GD 165* with a period of 2.4 d (Bergeron et al. 1993). Again, and this time somewhat surprisingly, there is no clear dependence of rotation rates with spectral type or temperature.

Certainly these rotation rates are not what one would expect by scaling from main sequence rotation rates. This seems to suggest that substantial angular momentum is lost from the cores during pre-white dwarf evolution, perhaps transferred to the outer layers and lost during the process of mass-loss.

3.5.1. *Magnetic fields*

Magnetic fields in white dwarf stars have recently been reviewed by SS, KB, Putney (1997), Jordan (1997), and Kanaan (1996). We will follow closely the discussions in Putney (1997) and I recommend this clearly written and extensive review to the interested reader.

Searches for magnetic white dwarf stars can be done in two ways. Historically the first way is to look for broadband circular polarization in the continuum, but it only picks up stars with very strong fields, $B > 50$ MG. Recently, much lower field strengths have been accessible using the method of circular spectropolarimetry. This measures the two oppositely polarized σ-components of Zeeman split absorption features. In this way SS was able to achieve limits as low as 86 kG, Putney reached limits of 30 kG, and Schmidt & Grauer achieved a startling limits ~ 1 kG in observations of a few selected pulsating white dwarf stars.

Detected magnetic field strengths range from 3×10^4 G to 10^9 G, with roughly constant detection rates per decade in field strength (SS). It is very interesting that as yet there is no evidence for a rapid increase in the number of magnetic fields at low field strengths. Either this doesn't occur, or we have not reached low enough field-strengths to observe it.

We are just beginning to reach field strengths suggested by asteroseismological observations, however (Schmidt & Grauer 1997). Deviations from uniform frequency splitting can occur for small magnetic fields. This deviation is proportional to m^2, and so moves the outlying m-components to higher frequency, removing the symmetry of the multiplet fine-structure with respect to the central component. Observations of such asymmetries in the fine-structure of *GD 358* indicated the possible presence of a small magnetic field of order 1300 G (Winget et al. 1994).

The magnetic white dwarfs do have a curious temperature distribution as pointed out by Kanaan (1995) and Putney (1997). We find magnetic fields at all representative temperatures, with an increased concentration near low temperatures consistent with white dwarf cooling theory. There does exist a surprisingly large number of white dwarfs with temperatures in the range from 14,000 K to 16,000 K, just hotter than the blue edge of the DAV instability strip, and hence just at the onset of surface convection. Putney (1997) speculates that there may be a dynamo which starts up at these temperatures. Interestingly, there are no DAV stars with large magnetic fields (see above), and we currently know of no stars in the DAV temperature range that are magnetic. I would like to point out that it may also be an evolutionary effect if the magnetic fields prevents convection—as expected in a MG strength field. Then the high radiative opacities may cause the cooling to slow and increase the discovery probability, until energy transport becomes more efficient at lower temperatures. We are currently looking into this curious distribution with the calculation of the evolution of forced radiative models. This effect almost certainly must occur, but it may not be enough to produce the observed spike. It also may be a combination of both effects, we shall see.

A final curiosity, consistent with Putney's ideas and pointed out by her, is that the strong fields (greater than 50 MG) are found below 16,000 K while weaker fields are found at all temperature ranges. This too needs to be understood.

4. Speculations for the future

To build our understanding of the most common and most stable of all stellar objects, the white dwarf stars, is to explore new physics, stellar evolution, and the formation and evolution of spiral galaxies.

How shall we do this? We must improve our knowledge of the basic physical parameters of the white dwarf stars. We must understand the origin of the white dwarf stars. How many channels feed the white dwarf stars? What is the origin of their spectral types and how and why does this change with temperature? What role do crystallization and possible phase separation play in the different white dwarf populations. Are there differences in the evolution of white dwarf stars with progenitors of wildly differing chemical compositions.

There are many ways to attack these problems using both theoretical and observational tools, as outlined above. I offer here my speculations on which ways will be the most effective.

As we have seen, one of the most accurate ways to measure the internal parameters of the white dwarf stars is to use the analysis of normal modes of oscillating white dwarf stars: asteroseismology. I have just completed an extensive review of this field (Winget 1999), and I refer the interested readers to that paper for progress and problems in this exciting field. Here I will mention only three of the many exciting developments.

The first relates to the hot subdwarf stars. Their position in the H-R diagram implies that they must contract and join onto the white dwarf cooling sequence. They are clearly a significant class of white dwarf progenitors, but their evolutionary status has remained relatively mysterious. The discovery of a new class of subdwarf B star variables (SdBV), pulsating in nonradial p-modes (Kilkenny et al. 1997), offers the possibility of seismological exploration of their interiors and evolution. Theory led the way in the investigations of these stars, as their existence was anticipated on theoretical grounds in a dramatic and important paper by Charpinet et al. (1996). Asteroseismology of these multi-periodic pulsators should soon yield up many of their secrets.

The second relates to potential breakthrough in understanding the cool white dwarfs. The massive pulsating DA white dwarf (DAV) *BPM 37093* is expected to be roughly 80% crystallized by mass. This gives us the opportunity to test our 40 year old ideas about crystallization in white dwarf interiors for the first time (Montgomery 1998, Winget et al. 1997). Mode identifications with the HST may soon make this possible. If successful, this will help eliminate one of the uncertainties in our estimates of the ages of the cool white dwarf stars, thereby aiding significantly in the accuracy of white dwarf cosmochronology.

Potentially the most significant development for exploring the physics, including crystallization and phase separation will come from a new theoretical approach to asteroseismology being developed by Travis Metcalfe for his PhD thesis. The idea is to vary the two critical parameters determining the pulsations frequencies in a model, the Brunt-Väisälä frequency and the acoustic frequency. The variations are adjusted using the genetic algorithm approach, independent of the input physics from the starting model, and a best fit to observed frequencies is obtained. This will allow us to explore the ultimate limits of white dwarf asteroseismology, in a way we not restricted by the shortcomings of our current input physics. This is especially important now as the recent work testing the white dwarf mass-radius relation (Provencal et al. 1998) indicates that there may be some serious flaws in our understanding of the interior EOS, or composition, or both. This technique offers to revolutionize our under

There are very few, if any, aspects or our studies not directly benefited by the dramatic increase in numbers of white dwarfs of all ages, colors, and kinds. The new digital sky surveys currently in progress, most notably the Sloan survey, will increase the number of your particular favorite kind of white dwarf by one to two orders of magnitude. These surveys will produce a sample of halo white dwarf stars, allowing comparison of the shape of the disk and halo luminosity function. The shape of the disk luminosity function will be known to great precision, opening the possibility of searching for evidence of bursts of star formation (Noh and Scalo 1990, Wood 1990, 1992, and especially Wood & Oswalt 1998).

Statistically significant samples of relatively rare objects will push the envelope of our understanding of the physics of the white dwarf stars. Imagine having between 10 and 100 crystallizing pulsators to study instead of just one! This overwhelming increase in data is coming at a time when we have the technology to handle it, rather than just be swamped by it. Further, large telescopes are coming on line all over the world. These telescopes will allow us to peer much deeper into space; the bulk of the new objects will almost by definition be fainter than the objects we currently observe. Finally, and just as exciting, is the near certainty that we will discover new types of stars, the like of which we have not yet dreamed!

I acknowledge support from the NSF through grant AST-9315461 and by NASA through grant number NAG5-2818 of the Astrophysics Theory Program, and through grant number G0-07401.01-96A from the Space Telescope Science Institute, which is operated by the Association of Universities for Research in Astronomy, Inc., under NASA contract NAS5-2655. I also thank Mike Montgomery, Ed Nather, Naoki Itoh, Dave Stevenson, Ed Salpeter, Jerry Ostriker, Icko Iben, Jr., Atsuko Nitta, Travis Metcalfe, Judi Provencal, Jim Liebert, Rex Saffer, Mario Livio, Gilles Chabrier, and Jordi Isern for stimulating discussions on subjects related to this paper. I thank Mike Montgomery, Betty Friedrich, Steve Kawaler, Paul Bradley, Atsuko Nitta, Travis Metcalfe and especially Karen Winget, for significant help in preparing this manuscript.

REFERENCES

ABRIKOSOV, A. A. 1960 *Zh. Eksp. i. Teor. Fiz.* **39**, 1798.

ADAMS, F., & LAUGHLIN, G. 1996 *ApJ* **468**, 586.

ALCOCK, C. ET AL. 1997 *ApJ* bf 486, 697.

BARRAT, J. L., HANSEN, J. P., & MOCHKOVITCH, R. 1988 *A&A* **199**, L15.

BARSTOW, M. A., BURLEIGH, M. R., FLEMING, T. A., HOLBERG, J. B., KOESTER, D., MARSH, M. C., ROSEN, S. R., RUTTEN, R. G. M., SAKAI, S., TWEEDY, R. W., & WEGNER, G. 1995 *MNRAS* **272**, 531.

BEAUCHAMP, A., WESEMAEL, F., BERGERON, P., LIEBERT, J. & SAFFER, R. A. 1996. In *Hydrogen-Deficient Stars* (eds. S. Jeffery & U. Heber). vol. 96, p. 295. ASP.

BERGERON, P., ET AL. 1993 *AJ* **106**, 1987.

BERGERON, P., LIEBERT, J., & FULBRIGHT, M. S. 1995 *ApJ* **444**, 810.

BERGERON, P., RUIZ, M.-T., & LEGGETT, S. K. 1997 *ApJS*, **108**, 339.

BERGERON, P., RUIZ, M.-T., LEGGETT, S. K., SAUMON, D., & WESEMAEL, F. 1994 *ApJ*, **423**, 456.

BERGERON, P., SAFFER, R. A., & LIEBERT, J. 1992 *ApJ*, **394**, 228.

BERGERON, P., SAUMON, D., & WESEMAEL, F. 1995 *ApJ*, **443**, 764.

BERGERON, P., WESEMAEL, F., LAMONTAGNE, R., FONTAINE, G., SAFFER, R. A., ALLARD, N. F. 1995 *ApJ* **449**, 258.

BRADLEY, P. A. 1996 *ApJ*, **468**, 350.

CHABRIER, G. 1993 *ApJ* **414**, 695.

CHABRIER, G. 1998. In *Proc. IAU Symp. 189* (ed. Bedding et al.). p. 381. Kluwer.

CHABRIER, G., SÉGRETAIN, L. & MÉRA, D. 1996 *ApJ* **468**, L21.

CHANDRASEKHAR, S. 1935 *MNRAS* **95**, 207.

CHARPINET, S., FONTIANE, G., BRASSARD, P., & DORMAN, B. 1996 *ApJ* **471**, L103.

CLAVER, C. F. 1995 *Ph.D. Thesis* Univ. Texas.

CLEMENS, J. C. 1993 *Baltic Astron.* **2**, 407.

CLEMENS, J. C. 1994 *Ph.D. Thesis* Univ. Texas.

CLEMENS, J. C. 1995. IN *Proc. 9th European Workshop on White Dwarf Stars* (EDS. D. KOESTER & K. WERNER). P. 294. SPRINGER.

D'ANTONA, F., & MAZZITELLI, I. 1978 *A&A* **66**, 453.

DREIZLER, S., & WERNER, K. 1996 *A&A* **314**, 217.

FIELDS, B. D., MATHEWS, G. J., & SCHRAMM, D. N. 1997 *ApJ* **483**, 625.

FONTAINE, G., BRASSARD, P., WESEMAEL, F., & TASSOUL, M. 1994 *ApJ* **428**, L61.

FONTAINE, G., & WESEMAEL, F. 1987. IN *The Second Conference on Faint Blue Stars* (EDS. A. G. D. PHILIP, D. S. HAYES, & J. LIEBERT). IAU COLLOQ. 95, P. 319. DAVIS.

FONTAINE, G. & WESEMAEL, F. 1991. IN *The Photospheric Abundance Connection* (EDS. G. MICHAUD & A. TUTUKOV). IAU SYMP. 145, P. 421. REIDEL.

FONTAINE, G., & WESEMAEL, F. 1997. IN *White Dwarfs* (EDS. J. ISERN, M. HERNANZ, E. GARCÍA-BERRO). P. 397. KLUWER.

FOWLER, R. H. 1926 *MNRAS* **87**, 114.

GARCÍA-BERRO, E., HERNANZ, M., MOCHKOVITCH, R. & ISERN, J. 1988 *A&A* **193**, 141.

GREENSTEIN, J. L. 1969 *Com. Astroph. & Space Sci.* **1**, 62.

GREENSTEIN, J. L. & PETERSON, D. M. 1973 *A&A* **25**, 29.

GREENSTEIN, J. L. & TRIMBLE, V. L. 1967 *ApJ* **149**, 283.

HEBER, U., NAPIWOTZKI, R., & REID, I. N. 1997 *A&A* **323**, 819.

HERNANZ, M., GARCÍA-BERRO, E., ISERN, J., MOCHKOVITCH, R., SEGRETAIN, L., & CHARBRIER, G. 1994 *ApJ* **434**, 652.

IBEN, I., JR. 1991 *ApJS* **76**, 55.

IBEN, I., JR. & LAUGHLIN, G. 1989 *ApJ* **341**, 312.

IBEN, I., JR. & MACDONALD, J. 1986 *ApJ* **301**, 164.

IBEN, I., JR. & TUTUKOV, A. V. 1984 *ApJ* **282**, 615.

IGLESIAS, C. A., ROGERS, F. J., & WILSON, B. G. 1992 *ApJ* **397**, 717.

IGLESIAS, C. A., WILSON, B. G., ROGERS, F. J., GOLDSTEIN, W. H., BAR-SHALOM, A., & OREG, J. 1995 *ApJ* **445**, 855.

ISERN, J. 1999 *J. Phys: Cond. Matter*, in press.

ISERN, J., GARCIÁ-BERRO, E., HERNANZ, M., MOCHKOVITCH, R., TORRES, S. 1998 *ApJ* **503**, 239.

ISERN, J., MOCHKOVITCH, R., GARCIÁ-BERRO, E., HERNANZ, M. 1997 *ApJ* **485**, 308.

ITOH, N., HAYASHI, H., & KOHYAMA, Y. 1993 *ApJ* **418**, 405; erratum: 1994 *ApJ* **436**, 418.

ITOH, N., HAYASHI, H., NISHIKAWA, A., & KOHYAMA, Y. 1996 *ApJS* **102**, 411.

JEFFRIES, R. D. 1997 *MNRAS*, **288**, 585.

JORDAN, S. 1997. In *White Dwarfs* (eds. J. Isern, M. Hernanz, E. García-Berro). p. 397. Kluwer.

KANAAN, A. 1996 *Ph.D. Thesis*, Univ. Texas.

KAWALER, S. D. 1986 *Ph.D. Thesis*, Univ. Texas.

KAWALER, S. D. 1996 *ApJ* **467**, L61.

KAWALER, S. D., ET AL. 1995 *ApJ* **450**, 350.

KEPLER, S. O. & BRADLEY, P. A. 1995 *Baltic Astron.* **4**, 166.

KILKENNY, D., KOEN, C., O'DOHOGHUE, D., & STOBIE, R. S. 1997 *MNRAS* **285**, 640.

KIRZHNITS, D. A. 1960 *Soviet Phys.–JETP* **11**, 365.

KOESTER, D. 1976 *A&A* **52**, 415.

KOESTER, D. & HERRERO, A. 1988 *A&A* **332**, 910.

KOESTER, D., SCHULZ, H., & WEIDEMANN, V. 1979 *A&A* **76**, 262.

LAMB, D. Q. & VAN HORN, H. M. 1975 *ApJ* **200**, 306.

LEGGETT, S. K., RUIZ, M. T., & BERGERON, P. 1998 *ApJ* **497**, 294.

LIEBERT, J. 1979. In *IAU Coll. No. 58* (eds. Van Horn, H. M. & Weidemannn, V.). p. 146. University of Rochester Press.

LIEBERT, J., DAHN, C. C., & MONET, D. G. 1988 *ApJ* **332**, 891.

LIEBERT, J., DAHN, C. C., & MONET, D. G. 1989. In *White Dwarfs* (ed. G. Wegner). IAU Colloq. 114, p. 15. Springer.

LIEBERT, J., WESEMAEL, F., HANSEN, C. J., FONTAINE, G., SHIPMAN, H. L., SION, E. M., WINGET, D. E., & GREEN, R. F. 1986 *ApJ* **309**, 241.

LUYTEN, W. J. 1979. *The LHS Catalogue, Second Edition*. University of Minnesota.

MAZZITELLI, I. & D'ANTONA, F. 1986 *ApJ* **308**, 706.

MESTEL, L. 1952 *MNRAS* **112**, 583.

MESTEL, L. 1965. In *Stars and Stellar Systems* (eds. L. H. Aller & D. B. McLaughlin). Vol. 8, chapt. 5. University of Chicago Press.

MESTEL, L., & RUDERMAN, M. A. 1967 *MNRAS* **136**, 27.

MOCHKOVITCH, R. 1983 *A&A* **122**, 212.

MONTGOMERY, M. H. 1998. *Ph.D. Thesis*, Univ. Texas.

NATHER, R. E., ROBINSON, E. L., & STOVER, R. J. 1981 *ApJ* **244**, 269.

NITTA, A. 1996. *Master's Thesis*, Univ. Texas.

NITTA, A. & WINGET, D. E. 1998 *BaltA* **7**, 141.

NOH, H.-R. & SCALO, J. 1990 *ApJ* **352**, 605.

OSTRIKER, J. & AXEL, L. 1969. In *Low Luminosity Stars* (ed. S. Kumar). p. 357. Gordon & Breach.

OSWALT, T. D., SMITH, J. A., WOOD, M. A., & HINTZEN, P. 1996 *Nature*, **382**, 692.

PILACHOWSKI, C.A. & MILKEY, R. W. 1984 *PASP* **96**, 281.

PILACHOWSKI, C. A. & MILKEY, R. W. 1987 *PASP* **99**, 836.

PROVENCAL, J. L. 1994. *Ph.D. Thesis*, Univ. Texas.

PROVENCAL, J. L. & SHIPMAN, H. L. 1998 *BaltA* **7**, 149.

PROVENCAL, J. L., SHIPMAN, H. L., HOG, E., & THEJLL, P. 1998 *ApJ* **494**, 759.

PROVENCAL, J. L., SHIPMAN, H. L., THEJLL, P., VENNES, S., & BRADLEY, P. A. 1996 *ApJ* **466**, 1011.

PROVENCAL, J. L., ET AL. 1997 *ApJ*, **480**, 383.

PUTNEY, A. 1997 *ApJS* **112**, 527.

REID, I. N. 1996 *AJ* **111**, 2000.

REIMERS, D. & KOESTER, D. 1994 *A&A* **285**, 451.

RENZINI, A. 1991. In *Observational Tests of Cosmological Inflation* (eds. T. Shanks et al.). p. 131. Kluwer.

RENZINI, A. 1999, *these proceedings*.

RENZINI, A. ET AL. 1996 *ApJ* **465**, L23.

RICHER, H. B. ET AL. 1995 *ApJ* **451**, L17.

ROMANISHIN, W. & ANGEL, J. R. P. 1980 *ApJ* **235**, 992.

SAFFER, R. A., LIVIO, M. & YUNGELSON, L. R. 1998 *ApJ*, in press.

SALARIS, M., DOMINGUEZ, I., GARCÍA-BERRO, E., ISERN, J. & MOCHKOVITCH, R. 1997 *ApJ* **486**, 413.

SALPETER, E. E. 1961 *ApJ* **134**, 669.

SCHATZMAN, E. 1958. *White Dwarfs*. North Holland Publishing Co.

SCHMIDT, M. 1959 *ApJ* **129**, 243.

SCHMIDT, M. 1975 *ApJ* **202**, 22.

SCHMIDT, G. D. & GRAUER, A. D. 1997 *ApJ* **488**, 827.

SCHMIDT, G. D. & NORSWORTHY, J. E. 1991 *ApJ* **366**, 270.

SCHMIDT, G. D. & SMITH, P. A. 1995 *ApJ* **448**, 305.

SCHWARZSCHILD, M. 1958. *Structure and Evolution of the Stars*. Dover.

SEGRETAIN, L., CHABRIER, G., HERNANZ, M., GARCÍA-BERRO, E., ISERN, J., MOCHKOVITCH, R. 1994 *ApJ* **434**, 641.

SHAVIV, G. & KOVETZ, A. 1976 *AA* **51**, 383.

SION, E. M. 1984 *ApJ* **282**, 612.

SION, E. M., GREENSTEIN, J. L., LANDSTREET, J. D., LIEBERT, J., SHIPMAN, H. L. & WEGNER, G. 1983 *ApJ*, **269**, 253.

SMITH, J. A. 1997. *Ph.D. Thesis*, The Florida Institute of Technology.

STEVENSON, D. J. 1977 *Proc. Ast. Soc. Australia* **3**, 167.

STEVENSON, D. J. 1980 *J. Physiqué* **41**, C2-61.

SZALAY, A. 1998. In *Astrophysics and Algorithms: a DIMACS Workshop on Massive Astronomical Data Sets*, in press.

TAMANAHA, C. M., SILK, J., WOOD, M. A., & WINGET, D. E. 1990 *ApJ* **358**, 164.

THEJLL, P., BAUER, F., SAFFER, R., LIEBERT, J., KUNZE, D., & SHIPMAN, H. 1994 *ApJ*, **433**, 819.

VAN HORN, H. M. 1968 *ApJ* **151**, 227.

VAN HORN, H. M. 1971. In *White Dwarfs* (ed. W. Luyten). IAU Symp. 42, p. 97. Reidel.

VAN HORN, H. M. 1977. In *The Uncertainty Principle and the Foundations of Quantum Mechanics* (eds. W. C. Price and S. S. Chissick) p. 441. Wiley.

VAUCLAIR, G. & REISSE, C. 1977 *A&A* **61**, 415.

WEIDEMANN, V. & KOESTER, D. 1983 *A&A* **121**, 77.

WEIDEMANN, V. & YUAN, J. W. 1989. In *White Dwarfs* (ed. G. Wegner). IAU Colloq. 114, p. 1. Springer.

WINGET, D. E. 1999 J. Phys: Cond. Matter, in press.

WINGET, D. E., HANSEN, C. J., LIEBERT, J., VAN HORN, H. M., FONTAINE, G., NATHER, R. E., KEPLER, S. O., & LAMB, D. Q. 1987 *ApJ*, **315**, L77.

WINGET, D. E., KEPLER, S. O., KANANN, A., MONTGOMERY, M. H., & GIOVANNINI, O. 1997 *ApJ*, **487**, L191.

WINGET, D. E. ET AL. 1994 *ApJ* **430**, 839.

WINGET, D. E. & VAN HORN, H. M. 1987. In *The Second Conference on Faint Blue Stars* (ed. A. G. D. Philip, D. S. Hayes, & J. Liebert). IAU Colloq. 95, p. 363. Davis.

WOOD, M. A. 1990 *Ph.D. thesis*, Univ. Texas.

WOOD, M. A. 1992 *ApJ* **386**, 539.

WOOD, M. A. 1995. In *White Dwarfs* (ed. G. Wegner). IAU Colloq. 114, p. 1. Springer.

WOOD, M. A. & OSWALT, T. D. 1998 *ApJ* **497**, 870.

XU, Z. W. & VAN HORN, H. M. 1992 *ApJ* **387**, 662.

The Pistol Star and massive stars in the Galactic Center

By DONALD F. FIGER,[1] FRANCISCO NAJARRO[2]
AND NORBERT LANGER[3]

[1]University of California, Los Angeles, Division of Astronomy, Department of Physics &
Astronomy, Los Angeles, CA 90095-1562

[2]CSIC, Instituto de Estructura de la Materia, Dpto. Fisica Molecular Serrano 121, 28006
Madrid, Spain

[3]Institut für Theoretische Physik und Astrophysik, Am Neuen Palais 10, D-14469 Potsdam,
Germany

A significant portion of the massive stars in the Galaxy are located within 50 pc of the Galactic
Center. The stars have a variety of ages between 1 and 10 Myr, giving rise to Wolf-Rayet and
red supergiant types. Most of these stars reside in three massive clusters: the Central Cluster,
Quintuplet Cluster, and Arches Cluster. The clusters have similar masses, $\approx 10^4$ M$_\odot$, but the
first two are substantially older (3–5 Myr) than the last (1–2 Myr). This spread in age, with
most other parameters being equal, gives us a unique opportunity to test models which predict
massive star evolution. We discuss the content in these clusters, with a particular emphasis on
how well they fit into our current understanding of stellar evolution. In particular, we discuss
the Pistol Star, one of the most luminous, and, therefore, initially most massive, stars in the
Galactic Center. Its evolutionary status places the star in the unstable Luminous Blue Variable
stage, providing an extraordinary opportunity to test stellar evolution models for massive stars
near their Eddington limits. We find a lower luminosity limit of $10^{6.6}$ L$_\odot$, $T_{\rm eff} \approx 10^{4.15}$ K,
and a helium-enriched surface, consistent with the star's advanced evolutionary status. Our
evolutionary tracks suggest an initial mass of ~ 200 M$_\odot$ and age of 1.7–2.1 Myrs. We interpret
the star and its surrounding nebula as an LBV which has recently ejected large amounts of
material. We expect that many of the massive stars in the Galactic Center will soon progress
through an LBV stage and eventually become supernovae at a rate of ≈ 1 per 20,000 years for
the next several Myr.

1. Introduction

Massive stars represent one of the last great frontiers of stellar evolution theory. Because of their incredible burning rate, their interiors evolve much more quickly than those
of lower mass stars. Due to their prodigious winds, their outward appearance changes
drastically over short timescales, i.e. changes in $T_{\rm eff}$ by factors of two or three in a time
span of hundreds of years. The combination of these factors makes it difficult to model
the evolution of massive stars, so their evolutionary progressions are largely uncertain.

Massive stars also present interesting test points for star formation theories, i.e. how
can stars which emit so much radiation accrete so much matter? Current theories suggest
that massive stars begin hydrogen burning while still accreting matter. Indeed, these
theories suggest that massive stars with $M_{\rm initial} \sim 85$–150 M$_\odot$ have accretion times which
are comparable to their hydrogen burning times, i.e. their photospheres are unobservable
throughout their main sequence lifetimes (Bernasconi & Maeder 1996). Presently, the
best estimate of the most massive star which can form (and be seen) is constrained by
observation.

The Galactic Center is a remarkable laboratory for studying massive star birth and
evolution. While occupying just 1% of the Galactic volume, the Galactic Center harbors
10% of the star formation in the Galaxy. Correspondingly, the population in the center

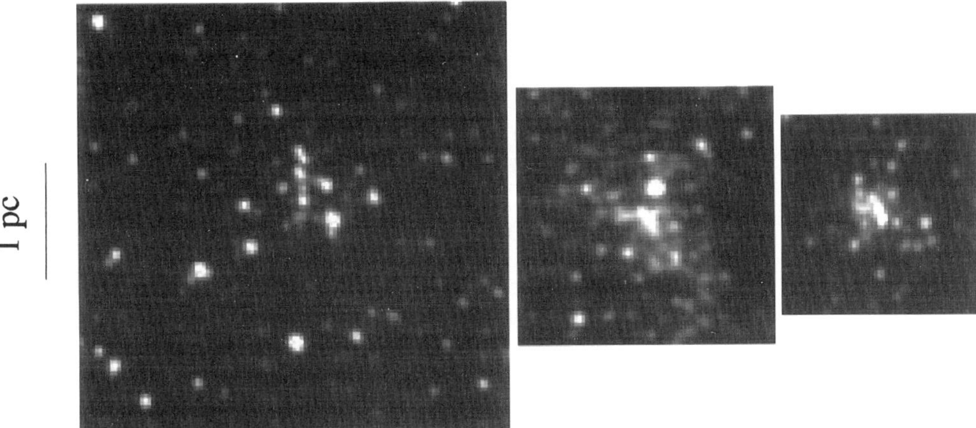

FIGURE 1. K'-band images of the Quintuplet, Central, and Arches clusters (left to right). See Figer 1995 for details.

contains at least 150 stars with $M_{\mathrm{initial}} > 20$ $M_\odot$. Figure 1 shows a collage of the three extraordinary clusters within the central 50 pc, each having $\gtrsim 10^4$ $M_\odot$ in stars and L $> 10^7$ $L_\odot$.

In addition to this large sample, the Galactic Center is an interesting region because the environmental factors which are thought to influence star formation and evolution are so extreme. The metallicity may be higher than anywhere else in the Galaxy, impacting both star formation and the evolution of massive stars with strong winds. The magnetic field strength is 300 times greater than the average in the Galactic disk, something which is expected to affect protostellar collapse. In addition to these environmental factors, the clusters in the Galactic Center provide an opportunity to investigate star birth in dense clusters.

We review the latest results concerning massive stars near the Galactic Center, using the Pistol Star as a case study. We conclude that the massive stars in the Galactic Center provide an extraordinary opportunity to test stellar evolution and star formation models.

2. The Central Cluster

The past decade has seen major advances in our understanding of the stars in the Central Cluster. After Forrest et al. (1987) and Allen, Hyland, & Hillier (1990) detected a blue supergiant in the center (the "AF star"), Krabbe et al. (1991) discovered about a dozen HeI emission-line stars. These, and subsequent, discoveries eventually solved the question of heating and ionization in the central parsec, and gave impetus to other studies of the massive star clusters in the central 50 pc (Figer 1995, Cotera 1995, Blum, Sellgren & Depoy 1995, Tamblyn et al. 1996).

The Central Cluster contains over 30 evolved massive stars having $M_{\mathrm{initial}} > 20$ $M_\odot$. A current estimate of the young population includes 9 WR stars, 20 stars with Ofpe/WN9-like K-band spectra, several red supergiants, and many luminous mid-infrared sources in a region of 1.6 pc in diameter centered on Sgr A* (Genzel et al. 1996). The spatial distribution of the early- and late-type stars has been the subject of several different studies (Becklin & Neugebauer 1968, Allen 1994, Rieke & Rieke 1994, Eckart et al. 1995, Genzel et al. 1996). The continuum surface distribution maps reveal that early-type stars

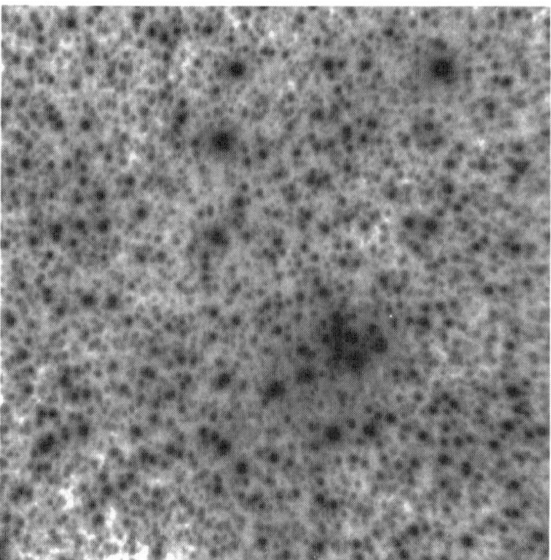

FIGURE 2. Alternate greyscale stretch of the K'-band image of Quintuplet Cluster in Figure 1.

concentrate towards the center, but red supergiants are mainly concentrated outside of a $4''$ void at the very center.

Najarro et al. (1994) investigated the AF star using a detailed spectroscopic analysis. They found that the star is a helium-rich blue supergigant/Wolf-Rayet star, characterized by a strong stellar wind and a moderate amount of Lyman ionizing photons. Subsequent spectroscopic studies of the brightest HeI emission-line stars by Najarro et al. (1997) confirmed the nature of these objects as "Ofpe/WN9" stars with extremely strong stellar winds ($\dot{M} \sim 5$ to $80 \times 10^{-5}\,M_\odot\,yr^{-1}$) and relatively small outflow velocities ($v_\infty \sim 300$ to $1,000\,km\,s^{-1}$). The effective temperatures of these objects were found to range from 17,000 K to 30,000 K with corresponding stellar luminosities of 1 to $30 \times 10^5\,L_\odot$. These results, together with the strongly enhanced helium abundances ($n_{He}/n_H > 0.5$), indicate that the HeI emission line stars power the central parsec and belong to a young stellar cluster of massive stars which formed a few million years ago.

3. The Quintuplet Cluster

The Quintuplet Cluster was first noted, and named for, its five bright infrared sources (Okuda et al. 1990, Nagata et al. 1990, Glass et al. 1990). Since the discovery of these luminous sources, many other luminous stars in the region have been identified. The cluster contains at least 30 stars having $M_{initial} > 20\,M_\odot$, including 4 WC types, 4 WN types, 2 LBVs, and many OBI stars; the Central Cluster is the only other Galactic cluster with so many bona fide WR stars. The two LBVs are added to the list of 6 LBVs in the Galaxy. They include the Pistol Star, one of the most luminous stars known (see below), and a newly identified LBV which is nearly as luminous as the Pistol Star (Figer, McLean, & Morris 1998b). Most of the luminous stars are thought to be 3–5 Myr old, but significant age differences remain, i.e. the Pistol Star is thought to be ≈ 2 Myrs old. See Figer, McLean, & Morris (1998b) for a review of the massive stars in the cluster.

Figure 2 shows a K-band image of the cluster. Most of the stars in the image do not belong to the cluster, but are simply part of the dense Galactic Center stellar population. Figure 3 shows a map of massive stars in the cluster, symbolized by subtype.

The Quintuplet proper members (QPMs) represent a rare phenomenon, having analogs only in the Central Cluster. The QPMs have very red intrinsic spectral energy distribu-

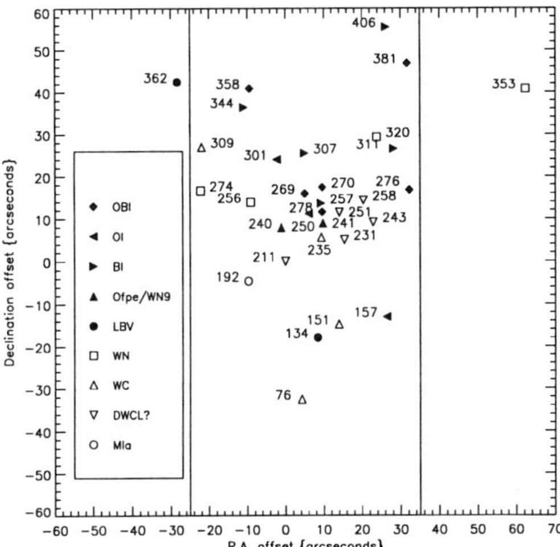

FIGURE 3. Massive stars in the Quintuplet Cluster, symbolized by subtype. The rectangular box indicates the region observed in a K-band slit scan. See Figer, McLean, & Morris (1998b) for details.

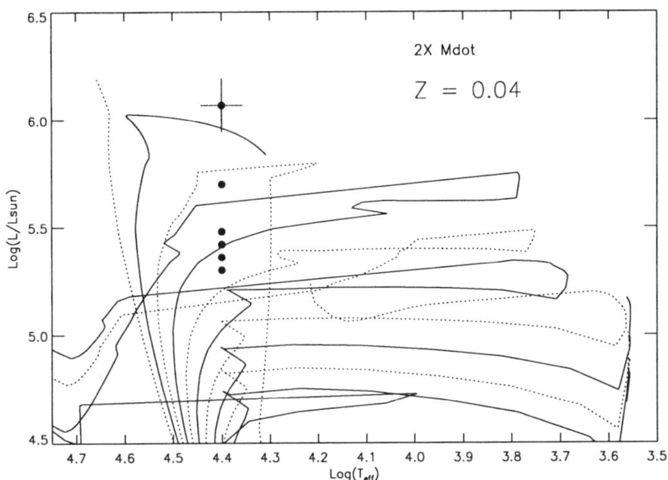

FIGURE 4. Model isochrones from the Geneva models for $Z = 2\,Z_\odot$ with "2×" $\dot{M}$. Isochrones begin at 1 Myr and are spaced by 1 Myr. Data for early B supergiants in the Quintuplet Cluster are overplotted.

tions which can be fit by cool blackbodies, $T_{\mathrm{eff}} \approx 800$ to 1,300 K. Being spectroscopically featureless at all observed wavelengths, their true nature has been elusive. Figer, McLean, & Morris (1998b) have summarized the hypotheses for the nature of these stars, and favor the "DWCL" hypothesis. DWCL stars (dusty Wolf-Rayet late-type carbon stars) are in a short-lived stage of evolution marked by massive circumstellar dust production (Williams, van der Hucht & The 1987). The classical WR emission-line spectrum has yet to be observed for these stars, so this hypothesis remains unproven.

Figure 4 shows isochrones from the Geneva models (Meynet et al. 1994) with data points for the B supergiants in the cluster overplotted. The stars span a large range in luminosity; however, the most luminous points could represent binary systems. If that is the case, then the plot suggests that the cluster is 4 Myrs old, in fair agreement with the expected age for a coeval cluster having equal numbers of WN and WC types. It is

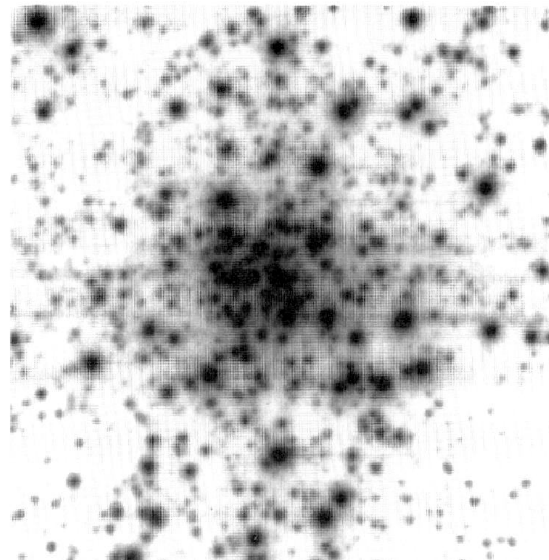

FIGURE 5. K'-band image of the Arches Cluster obtained with NIRC on Keck I. See Serabyn, Shupe, & Figer (1998) for details.

also consistent with the age given by the existence of the red supergiant in the cluster, ≈ 4 Myrs. The age is somewhat high compared to our determination of the age of the Pistol Star (see below).

4. The Arches Cluster

The Arches Cluster (Figure 5) is, perhaps, the most massive cluster in the Galaxy (Serabyn, Shupe, & Figer 1998). It certainly appears to be the most dense, $\rho_{stars,\,ave} \sim 3 \times 10^5$ $M_\odot$ pc^{-3}, substantially exceeding the core density in R136 and dwarfing any other massive cluster in the Galaxy, i.e. NGC 3603. Figure 6 shows a color-magnitude diagram of the brightest cluster members; the average apparent color ensures a location at a distance of the Galactic Center. Figure 7 shows the K-band luminosity function for the cluster with spectral types indicated. Cotera et al. (1996) presented H- and K-band spectroscopy of the brightest dozen or so members of the cluster, showing that they are similar to WN types. Unfortunately, it is difficult to distinguish between WNL stars and OI$^+$ stars solely on the basis of K-band spectra (Conti et al. 1995, Figer et al. 1997).

The Arches appears to be much more compact than the other GC clusters, something which is likely to be due to its relative youth. The absence of WC stars in the cluster argues for an age less than ≈ 3 Myr. The presence of stars with WN-like spectra would normally indicate an age > 2 Myr; however, the emission-line stars are much younger if they are O supergiants still on the main sequence. The cluster is probably 1–2 Myrs old, i.e. the Arches Cluster and R136 are probably closer in content and age than any other two clusters in the Galaxy or Magellanic Clouds. See Massey & Hunter (1998) for a discussion of the massive star content of R136.

5. The Pistol Star: A case study in stellar evolution

The Pistol Star poses an intriguing puzzle to the theory of stellar evolution (see Figure 8). Its position in the Hertzsprung-Russell diagram, i.e. very high luminosity but rather cool surface temperature, is seemingly in conflict with present ideas of massive star evolution. This can be seen in Figure 9, where the position of the Pistol Star is compared to two sets of stellar evolution tracks in the Hertzsprung-Russell diagram. The

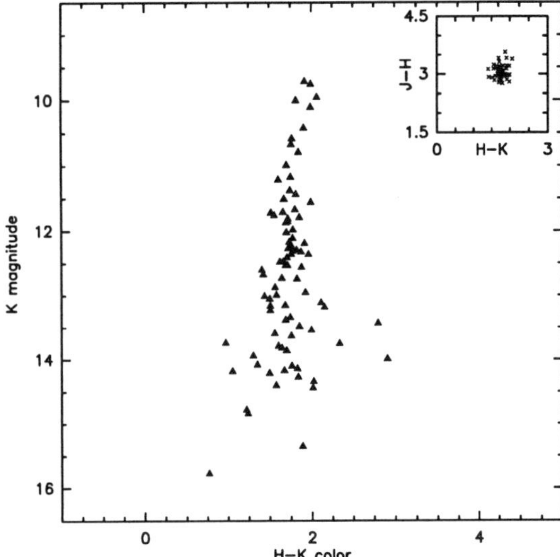

FIGURE 6. Color-magnitude diagram of the Arches Cluster. See Serabyn, Shupe, & Figer (1998) for details.

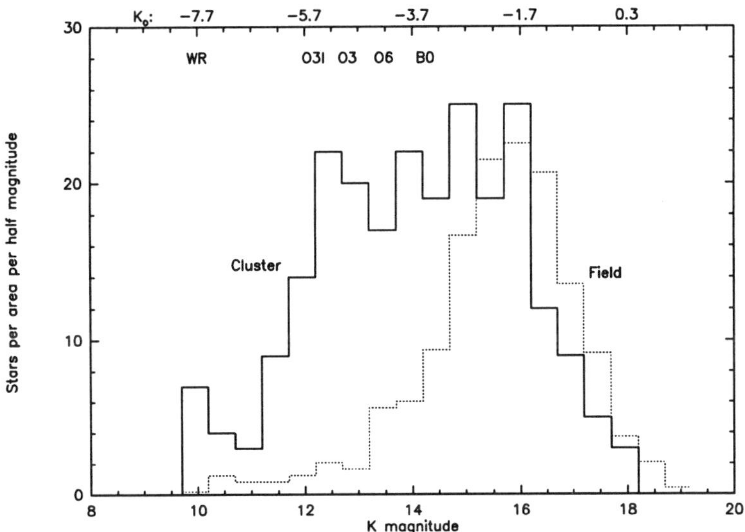

FIGURE 7. K-band luminosity function of the Arches Cluster. See Serabyn, Shupe, & Figer (1998) for details.

thin drawn tracks of Schaller et al. (1992) for 60, 85, and 120 $M_\odot$ avoid the upper right corner of the diagram, which is in agreement with the idea that "normal" massive stars do not cross the so called Humphreys-Davidson (1979) limit in the HR diagram or at least become unstable when they do so (cf. Langer 1997). The reason the tracks of very massive stars avoid the cool side of the HR diagram and quickly turn towards hotter surface temperatures is the enrichment of the envelope and surface with helium. This can, in principle, be achieved by two ways, either mass loss or internal mixing. The more massive a star is, the more efficient both of these processes become. E.g. a 120 $M_\odot$ zero age main sequence star contains more than 100 $M_\odot$ in its convective core, so only 20 $M_\odot$ need to be lost in a stellar wind for helium to be enriched at the surface of the star.

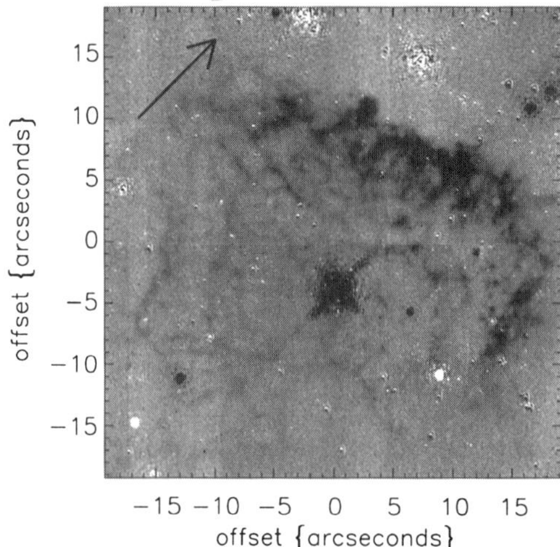

FIGURE 8. Negative greyscale Paschen-α minus continuum image of the Pistol Nebula obtained with NICMOS/HST. See Figer et al. (1998a) for details.

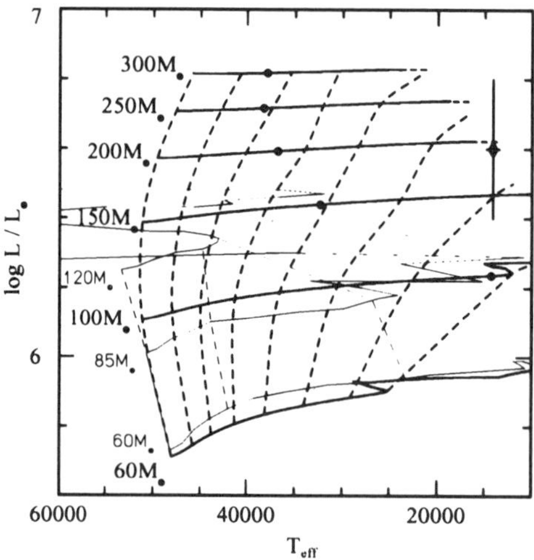

FIGURE 9. Stellar evolutionary tracks in the HR diagram for non-rotating stars in the initial mass range 60–300 $M_\odot$ and a metallicity Z of 2% (thick continuous lines). Thick dashed lines connect models with central helium mass fractions of 0.28 (leftmost track, ZAMS), 0.4, 0.5, 0.6, 0.7, 0.8, 0.9, and 0.98. Black dots mark the first appearance of hydrogen burning products at the stellar surface. The tracks of the 200, 250, and 300 $M_\odot$ models end due to the occurrence of surface instabilities. Thin continuous lines show evolutionary tracks for 60, 85, and 120 $M_\odot$ and Z = 0.02 obtained by Schaller et al. (1992)—without their effective temperature correction—with thin dashed lines connecting models with central helium concentrations of 0.30, 0.60, and 0.90. The position of the Pistol Star is marked by a diamond, together with the error bar.

Moreover, rotational mixing is thought to be important in massive stars (Langer et al. 1997, Maeder 1997), and more so with larger initial mass. Even though considerable uncertainties in the treatment of rotational mixing persist, it appears likely that this mixing will turn stars with $M_{\mathrm{initial}} \gtrsim 60\ M_\odot$ and average rotation rates into Wolf-Rayet stars while still in their core hydrogen burning evolution, so that they may not only avoid a red supergiant stage but perhaps even a Luminous Blue Variable stage (Maeder 1987, Langer 1992, Fliegner & Langer 1995, Meynet 1997).

How can a very luminous and cool star as the Pistol Star be understood? First, its high luminosity suggests a very high mass, but strong internal mixing can appreciably increase the luminosity of a star (cf. Fliegner et al. 1996). However, strong internal mixing appears to be excluded in the case of the Pistol Star from the arguments above. Second, the previous mass loss must not have been too large. E.g. the tracks of Schaller et al. (1992) shown in Figure 9, which include moderate convective core overshooting but no rotational mixing, and which use the empirical mass loss formula of de Jager et al. (1988), avoid the cool side of the HR diagram for $M_{\mathrm{initial}} \gtrsim 100\ \mathrm{M_\odot}$.

There are basically no empirical mass loss rates for stars with masses $\gtrsim 100\ \mathrm{M_\odot}$,— i.e. the use of empirical mass loss rates always involves an extrapolation in those cases. We have calculated evolutionary tracks with a metallicity of 2% and initial masses in the range 60–300 $\mathrm{M_\odot}$, applying the radiation driven wind theory of Kudritzki et al. (1989), with wind parameters $k = 0.085$, $\alpha = 0.657$, $\delta = 0.095$, and $\beta = 1$ according to Pauldrach et al. (1994). Our models lose less mass than those of Schaller et al. (1992); the 60, 100, and 150 $\mathrm{M_\odot}$ sequences have, at core hydrogen exhaustion, 54, 84, and 110 $\mathrm{M_\odot}$ respectively. The 200, 250, and 300 $\mathrm{M_\odot}$ sequences are terminated at central hydrogen mass fractions of $X_c = 0.16$, 0.22, and 0.29 where the remaining masses are 158, 197, and 244 $\mathrm{M_\odot}$, due to a hydrodynamic instability occurring at the stellar surface (see below).

We did not invoke any "convective core overshooting." This has been applied in many massive star calculations in the recent years in order to obtain a wider main sequence band (cf. Schaller et al. 1992). However, as rotationally induced mixing has a very similar effect (e.g. Langer 1992, Fliegner et al. 1996), we argue that any main sequence widening may be due to rotation and thus that the convective cores of non-rotating stars are not extended over their sizes predicted by the Schwarzschild criterion.

Our finding that the most massive stellar models computed here become unstable at around $T_{\mathrm{eff}} \simeq 20,000$ K and $\log(L/\mathrm{L_\odot}) \gtrsim 6.5$ (cf. Stothers & Chin 1997), together with the coincidence of the position of the Pistol Star with these figures, leads us to propose the following scenario. The Pistol Star is unusually massive and may therefore have an unusual formation history. Apparently, it has obtained very little angular momentum, which perhaps allowed its mass to grow so much. Consequently, it evolved with less-than-average mass loss towards the cool side of the HR diagram. During core hydrogen burning, it arrived at its Eddington-limit and strongly increased its mass loss rate. This is its present evolutionary stage. As it is still burning hydrogen in its core, the probability for the star to be found in this stage is not small. According to our models, its initial mass is $\sim 200\ \mathrm{M_\odot}$, and its present age is in the range 1.7–2.1 Myrs. The amount of mass lost prior to the occurrence of the surface instability is 42–53 $\mathrm{M_\odot}$ (see Figer et al. 1998a, for more details). Figure 10 places the Pistol Star in an HR diagram among its stellar "peer group."

6. Implications for stellar evolution and star formation models

Stellar evolution and formation models make definite predictions of the content of starburst clusters as a function of time. These predictions can be tested by comparing their outputs to observations of the massive clusters in the Galactic Center. It is crucial that state of the art models predict the content of these clusters before they are used to predict the contents of more distant clusters, i.e. super-star clusters in other galaxies.

Morris (1993) has argued that the lower mass cutoff might be elevated in the Galactic Center. A Salpeter initial mass function (Salpeter 1955) suggests a few $\times 10^5$ lower mass stars in the GC clusters, assuming a lower mass cutoff of 0.1 $\mathrm{M_\odot}$. This cutoff should be directly observable in these clusters. Present observations are unable to detect

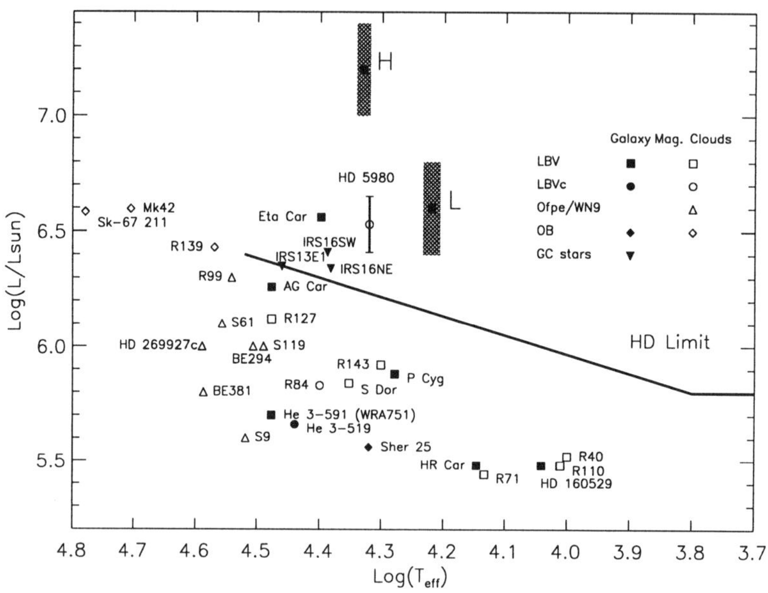

FIGURE 10. HR diagram reproduced from Figure 12 in Figer et al. (1998a). HR diagram of the Pistol Star ("L" and "H" models) and other hot, luminous stars in the Galaxy and Magellanic Clouds, in quiescence. The data were taken from Humphreys & Davidson (1994), unless otherwise noted. When possible, the LBV data are at quiescence. LBVs are shown as squares, LBVc's as circles, Ofpe/WN9 as triangles, and OB stars as diamonds. Filled symbols are for stars in the Galaxy, and open symbols are for stars in the Magellanic clouds. LBVs make redward excursions in this diagram during eruption. For example, η Car might have been as cool as 5,000 K during its last major eruption. See Humphreys & Davidson (1994) for more information concerning the locations of LBVs during eruption. Note that η Car may be a binary (Damineli, Conti, & Lopes 1997). Data for Sher 25 in NGC 3603 were taken from Moffat (1983). Data for HD 5980 are from Koenigsberger, Auer, & Guinan (1997); it is not certain which component of this binary system is the LBV, so the luminosity spans a large range. Data for He 3-591 (WRA 751) are from Hu et al. (1990). Data for He 3-519 are from Smith, Crowther, & Prinja (1994). Data for R71 and R84 were taken from Crowther, Hillier, & Smith (1995). Data for Sk −67?211 and Mk42 were taken from Kudritzki et al. (1996). Data for the Ofpe/WN9 stars are from Pasquali (1997). Data for R139 (O7 Iafp) are from Walborn & Blades (1997). Data for the "GC stars" are from Najarro et al. 1997. Note that IRS16NE has some LBV-like spectral characteristics (Tamblyn et al. 1996).

lower mass stars in the clusters, but HST/NICMOS and adaptive optics observations will be able to detect them. Once these observations are in hand, the Arches Cluster will provide the best testpoint for determining whether the initial mass function depends upon metallicity, galactocentric radius, or stellar density. Current evidence suggests that there is little, if any, correlation between the IMF and these environmental factors, at least to within measurement errors (Scalo 1998).

Stellar evolution models predict that massive stars should evolve into Wolf-Rayet stars, perhaps via an LBV stage, and onward to supernovae. The exact mixture of various WR subtypes and O-stars is sensitive to assumed mass-loss rates and metallicity. A key challenge for these models is to reproduce the content in the Quintuplet Cluster. Arguments against coevality are suspect because tidal shear and the strong winds from massive stars will tend to dissipate a natal cloud over short timescales. Note that the Quintuplet Cluster should have orbited the Galactic Center about 1.5 times since its birth. It will be more difficult to use the Central Cluster as a test point for stellar evolution models because it appears not to be coeval. In fact, star formation appears to be currently ongoing (Genzel et al. 1996).

The process of triggered star formation may be more prevalent in the Galactic Center than anywhere else in the Galaxy. In fact, the Jean's mass is so large, triggered star formation might be necessary to form stars in the Galactic Center (Morris 1993). Triggering could come from cloud-cloud collisions, or from supernova shocks. Meynet (1995) gives $\approx 5 \times 10^{-7} \times$ $N_{O-stars}$ supernovae per year during the peak supernovae period for a starburst cluster, where $N_{O-stars}$ is the number of O-stars in the cluster at $\tau_{age} = 0$. These periods are expected to last a few Myrs, starting at $\tau_{age} \approx 4$ Myrs. Using $N_{O-stars} \approx 100$, we find a supernovae rate of 1 per 20,000 years. Stars in the Quintuplet and Central clusters are closest to the onset of supernovae production, and their massive stars probably number around 100 total. Once the Arches Cluster becomes old enough to participate in supernovae production, the massive stars in the Quintuplet and Central clusters should be gone, so that the total supernovae rate in the Galactic Center should remain fairly constant over the next 5 Myrs or so ($N_{O-stars,\,Arches} \approx 100$).

Finally, we note that some very interesting single objects in the Galactic Center are still unexplained. The Quintuplet proper members are unique, although fainter mid-infrared sources are located in the Central Cluster. Why are there so many such objects in the Galactic Center? Why don't we see QPM-like objects in other massive clusters?

We thank the late Chris Skinner of ST ScI for providing assistance in performing the observations of the Pistol Star. We thank Christine Ritchie of ST ScI for assisting in the data reduction for the Pistol Star image.

REFERENCES

ALLEN, D. A., HYLAND, A. R. & HILLIER, D. J. 1990 *MNRAS* **244**, 706.

ALLEN, D. A. 1994. In *The Nuclei of Normal Galaxies* (eds. R. Genzel & A. I. Harris). p. 293. Kluwer.

BECKLIN, E. E. & NEUGEBAUER, G. 1968 *ApJ* **151**, 145.

BERNASCONI, P. A. & MAEDER, A. 1996 *A&A* **307**, 829.

BLUM, R. D., DePOY, D. L. & SELLGREN, K. 1995 *ApJ* **441**, 603.

CONTI, P. S., HANSON, M. M., MORRIS, P. W., WILLIS, A. J., & FOSSEY, S. J. 1995 *ApJL* **445**, 35.

COTERA, A. S. 1995. PhD Thesis. Stanford University.

COTERA, A. S., ERICKSON, E. F., COLGAN, S. W. J., SIMPSON, J. P., ALLEN, D. A., & BURTON, M. G. 1996 *ApJ* **461**, 750.

CROWTHER, P. A., HILLIER, D. J., & SMITH, L. J. 1995 *A&A* **293**, 172.

DAMINELI, A., CONTI, P. S., & LOPES, D. F. 1997 *New Astronomy* **2**, 107.

DE JAGER, C., NIHEUVENHUIJZEN, H., VAN DER HUCHT, K. A. 1988 *A&AS* **72**, 259.

ECKART, A., GENZEL, R., HOFMANN, R. SAMS, B. J., & TACCONI-GARMAN, L. E. 1995 *ApJL* **445**, 26.

FIGER, D. F. 1995. PhD Thesis. University of California, Los Angeles.

FIGER, D. F., McLEAN, I. S. & MORRIS, M. 1995 *ApJL* **447**, 29.

FIGER, D. F., McLEAN, I. S., & NAJARRO, F. 1997 *ApJ* **486**, 420.

FIGER, D. F., NAJARRO, F., MORRIS, M., McLEAN, I. S., GEBALLE, T. R., GHEZ, A. M., & LANGER, N. 1998a *ApJ*, in press.

FIGER, D. F., McLEAN, I. S., & MORRIS, M. 1998b *ApJ*, submitted.

FLIEGNER J. & LANGER N. 1995. In *Wolf-Rayet Stars: Binaries, Colliding Winds, Evolution* (eds. K. A. van der Hucht & P. M. Williams, IAU Symp. No. 163, p. 326. Kluwer.

FLIEGNER, J., LANGER, N., & VENN, K. 1996 *A&AL* **308**, 13.

FORREST, W. J., SHURE, M. A., PIPHER, J. L., WOODWARD, C. A. 1987. In *The Galactic Center*, (ed. D. Backer), AIP Conf. 155. p. 153. AIP.

GENZEL, R., THATTE, N., KRABBE, A., KROKER, H. & TACCONI-GARMAN 1996 *ApJ* **472**, 153.

GLASS, I. S., MONETI, A. & MOORWOOD, A. F. M. 1990 *MNRAS* **242**, 55P.

HUMPHREYS, R. M. & DAVIDSON, K. 1994 *PASP* **106**, 1025.

KOENIGSBERGER, G., AUER, L. H., & GUINAN, E. 1997 *ApJ* **496**, 934.

KRABBE, A., GENZEL, R., DRAPATZ, S. & ROTACIUC, V. 1991 *ApJL* **382**, 19.

KUDRITZKI, R.-P., LENNON, D. J., HASER, S. M., PULS, J., PAULDRACH, A. W. A., VENN, K., & VOELS, S. A. 1996. In *Science with the Hubble Space Telescope—II*, (eds. P. Benvenuti, F. D. Machetto, & E. J. Schreier). p. 135. ST ScI.

KUDRITZKI, R.-P., PAULDRACH, A., PULS, J., & ABBOTT, D. C. 1989 *A&A* **219**, 205.

LANGER N. 1992 *A&AL* **265**, 17.

LANGER N. 1997. In *Luminous Blue Variables: Massive Stars in Transition*, (eds. A. Nota & H. J. G. L. M. Lamers). p. 83. ASP.

LANGER, N., HEGER, A., FLIEGNER, J. 1997. In *Fundamental Stellar Properties: The Interaction between Observation and Theory*, (eds. T. R. Bedding, A. J. Booth, and J. Davis), IAU Symp. 189. p. 343. Kluwer.

MAEDER, A. 1987 *A&A* **178**, 159.

MAEDER, A. 1997. In *2nd Boulder-Munich Workshop*, (ed. I. Howarth). p. 85.

MASSEY, P. & HUNTER, D. A. 1998 *ApJ* **493**, 180.

MEYNET, G., MAEDER, A., SCHALLER, G., SCHAERER, D., & CHARBONNEL, C. 1994 *A&AS* **103**, 97.

MEYNET, G. 1995 *A&A* **298**, 767.

MEYNET, G. 1997. In *2nd Boulder-Munich Workshop*, (ed. I. Howarth). p. 96.

MOFFAT, A. F. J. 1983 *A&A* **124**, 273.

MORRIS, M. 1993 *ApJ* **408**, 496.

NAGATA, T., WOODWARD, C. E., SHURE, M., PIPHER, J. L. & OKUDA, H. 1990 *ApJ* **351**, 83.

NAJARRO, F., HILLIER, D. J., KUDRITZKI, R. P., KRABBE, A., GENZEL, R., LUTZ, D., DRAPATZ, S. & GEBALLE, T. R. 1994 *A&A* **285**, 573.

NAJARRO, F., KRABBE, A., GENZEL, R., LUTZ, D., KUDRITZKI, R. P., & HILLIER, D. J. 1997 *A&A* **325**, 700.

OKUDA, H., SHIBAI, H., NAKAGAWA, T., MATSUHARA, H., KOBAYASHI, Y., KAIFU, N., NAGATA, T., GATLEY, I. & GEBALLE, T. R. 1990 *ApJ* **351**, 89.

PASQUALI, A. 1997. In *Luminous Blue Variables: Massive Stars in Transition*, (eds. A. Nota & H. J. G. L. M. Lamers). p. 13. ASP.

PAULDRACH, A. W. A., KUDRITZKI, R. P., PULS, J., BUTLER, K., HUNSINGER, J. 1994 *A&A* **283**, 525.

RIEKE, G. H. & RIEKE, M. J. 1994. In *The Nuclei of Normal Galaxies*, (eds. R. Genzel & A. I. Harris). p. 283. Kluwer.

SALPETER, E. E. 1955 *ApJ* **121**, 161.

SCALO, J. 1998. In *The Stellar Initial Mass Function*, (eds. G. Gilmore, I. Parry, & S. Ryan), in press. ASP.

SCHALLER, G., SCHAERER, D., MEYNET, G., & MAEDER, A. 1992 *A&AS* **96**, 296.

SERABYN, E., SHUPE, D., & FIGER, D. F. 1998 *Nature* **394**, 448.

SMITH, L. J., CROWTHER, P. A., & PRINJA, R. K. 1994 *A&A* **281**, 833.

STOTHERS, R. B. & CHIN, C.-W. 1997 *ApJ* **489**, 319.

TAMBLYN, P., RIEKE, G. H., HANSON, M. M., CLOSE, L. M., McCARTHY, D. W., JR., & RIEKE, M. J. 1996 *ApJ* **456**, 206.

WALBORN, N. R. & BLADES, J. C. 1997 *ApJS* **112**, 457.

WILLIAMS, P. M., VAN DER HUCHT, K. A. & THE, P. S. 1987 *A&A* **182**, 91.

FG Sge—An update

By GUILLERMO GONZALEZ

University of Washington

FG Sge continues to exhibit dramatic changes. Starting in 1992, FG Sge has suffered several deep brightness declines very similar to those seen in R CrB stars. The composition apparently has not changed since 1992, but a number of spectroscopic changes have occurred: the strength of the C_2 molecular bands continue to vary, a broad blue-shifted component of the Na I D lines appeared in 1993 and has steadily increased in expansion velocity, and the usual absorption spectrum changed into an emission spectrum starting with the deep May 1994 decline. At this time FG Sge exhibits all the characteristics of the R CrB class, except that most R CrB stars do not show excesses of s-process elements.

1. Introduction

Since it first began brightening in the 1890s, FG Sge has been racing across the HR diagram towards cooler temperatures. The early photometric evolution and the more recent spectroscopic observations were reported by Herbig & Boyarchuk (1968). Langer et al. (1974) followed FG Sge as the abundances of the s-process elements increased dramatically in its atmosphere in the late 1960s. By the mid-1980s the temperature and pulsation period stabilized. Additional details on the history of FG Sge can be found in Kipper (1996) and Gonzalez et al. (1998; G98). I will summarize the findings of G98 below and report on the most recent behavior of FG Sge.

2. Photometric behavior

Shown in Figure 1 is the visual light curve of FG Sge since just before the deep decline on 26 August 1992. The data are entirely from amateur variable star observations. In May 1998 began the first deep decline since mid-1996. In early 1998 it almost reached its pre-1992-decline mean magnitude. Perhaps the expelled mass since 1992 has resulted in a few tenths of a magnitude of visual extinction near the star.

As first reported by Arkhipova (1996) and confirmed by G98, the timing of the first four deep brightness declines matches the ephemeris constructed from the pre-1992 observations. This shows that the deep declines are causally linked to the semi-regular pulsations in its atmosphere.

3. Spectroscopic changes

Since they were first detected in the spectrum of FG Sge in 1980, the C_2 bands have varied in strength. G98 suggested that the strength correlates with the pulsation cycle, but the data is too sparse to reach a definite conclusion. The bands have gone into emission on two occasions: May 1994 and June 1996. They were not present in emission during the 1992 decline. Our most recent spectra, 21 June 1997 and 15 May 1998, both show the C_2 bands in absorption.

The Na I D lines go strongly into emission during the deep declines. As can be seen in Figure 2, a broad blue-shifted absorption component of the D lines appeared shortly after the 1992 decline. The features have persisted since then, and they have moved farther to the blue and broadened. The center of the absorption feature is shifted by

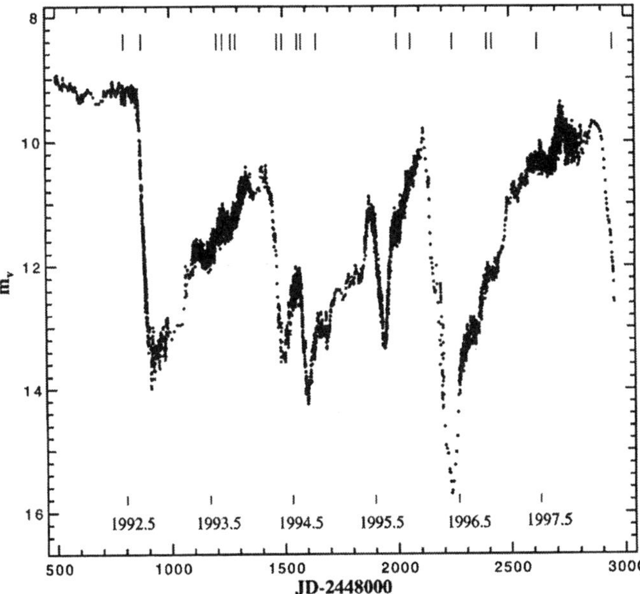

FIGURE 1. Visual magnitude estimates from the AAVSO, AFOEV, BAAVSS, VSOLJ. The original data has been smoothed with a five-point running average. The epochs of the spectroscopic observations are indicated along the top of the figure.

-160 km $^{-1}$ relative to the photospheric velocity in our June 1997 spectrum; this is down from -205 km $^{-1}$ from our December 1996 spectrum. This does not necessarily mean that the gas is decelerating, but rather that we are seeing changes in Na I D absorption resulting from episodic mass loss. The Na I D emission is strong in our May 1998 spectrum, obtained during the early part of the most recent decline (see Figure 1). This is in contrast to the appearance of the D lines during the early part of the 1992 decline, which did not display any emission components.

During the two deep declines of May 1994 and June 1996 the absorption spectrum was transformed into an emission spectrum (Figure 3). The early part of the 1994 decline was characterized by a near perfect reversal of the pre-decline spectrum. In the middle of the very deep 1996 decline, the continuum was weak and uniform, while the emission lines were sharp and well-separated. The CH G band was present very weakly in emission during the deep 1996 decline, in contrast to the strong C_2 bands (Figure 4).

4. Physical parameters

Montesinos et al. (1990) determined $T_{\rm eff}$ to be in the 6000–6500 K range in the late 1980s, based on analyses of the infrared colors and low resolution IUE UV spectra. The near constancy of the pulsation period since the 1980s argues for the near constancy in mean radius and hence in $T_{\rm eff}$. G98 showed that the variations in the C_2 bands and the range in the $T_{\rm eff}$ estimates in the 1980s and early 1990s can be accounted for by a variation in $T_{\rm eff}$ of about ±500 centered at about 6000 K. There is no evidence that the luminosity of FG Sge has changed significantly during this century; the secular change in its visual magnitude was attributed by Herbig & Boyarchuk and Langer et al. to a changing bolometric correction. Since the early 1970s it has not been possible to calculate an accurate bolometric correction due to the severe and unique line blanketing

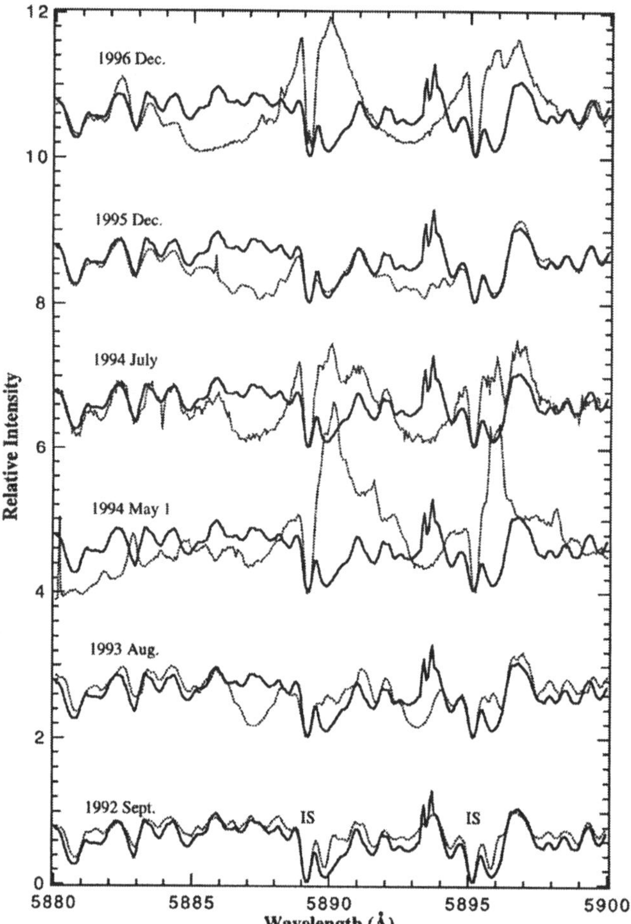

FIGURE 2. Evolution of the Na I D lines since 1992 June. The 1992 June spectrum (*solid curve*) is overplotted on the other spectra (*dotted curves*). The interstellar lines are indicated (IS).

in FG Sge's spectrum. G98 determined $\log g$ to be 2.0 by comparing the neutral and ionized Fe lines. This is larger than expected for the assumed value of the luminosity and T_{eff}. G98 suggested that a deficiency of H in FG Sge's atmosphere resulted in an overestimate of $\log g$.

5. Abundances

The abundances of the s-process elements have increased significantly since the early 1970s, when they were about 25 times higher than the early 1960s. Kipper & Kipper (1993) reported an s-process enhancement of about 1.8 dex, while G98 quoted an enhancement of about 3 dex. Both Kipper (1996) and G98 also reported an enhancement of carbon by about 0.7 dex based on analyses of the atomic carbon and molecular C_2 features; Langer et al. did not find any enhancement of carbon in their spectra. Compared to a normal F supergiant of solar metallicity, FG Sge's spectrum is very different (Figure 5). The lines of the light metals and of Fe are comparable in strength, but most

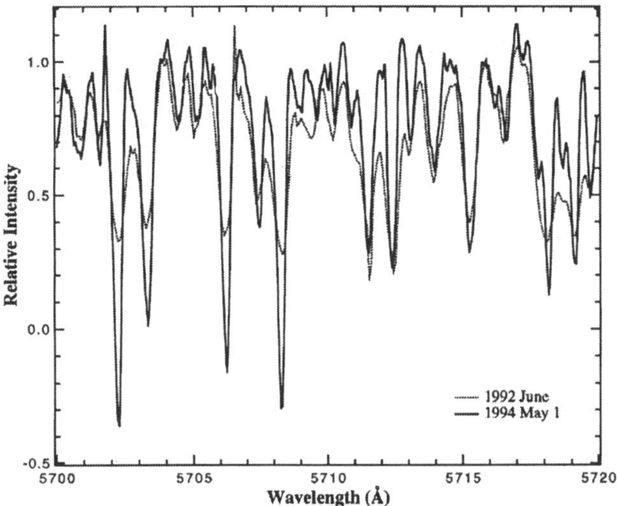

FIGURE 3. Emission-line spectrum from 1994 May 1 has been inverted and overplotted with the 1992 June absorption spectrum; the spectral resolutions have been matched.

of the heavy element lines in the spectrum of FG Sge are not detectable in the F5Ib star α Per.

The abundance pattern is a close match to the Solar System main s-process component (Figure 6). Even those elements in the Solar system that are mostly the product of the r-process, such as Eu, are greatly enhanced in FG Sge. Apparently, even Pb is enhanced, but the identification is not certain.

6. Discussion

6.1. *FG Sge as an R CrB star*

FG Sge now exhibits all three observational defining characteristics of the R CrB class: 1) sudden, deep brightness declines, 2) spectroscopic evidence of enhanced carbon either from atomic or molecular features, and 3) spectroscopic evidence of H-deficiency. The evidence for H-deficiency comes from the weakness of the CH G band and from the Hα line profile (Figure 7). Herbig & Boyarchuk found a normal H abundance for FG Sge in the 1960s. The similarity of the Na I D profiles in the spectra of FG Sge and R CrB during recovery from deep minima is additional evidence of FG Sge's close link to the R CrB class.

6.2. *Dust formation*

Woodward et al. detected an increase in the infrared flux at the time of the first deep decline in 1992. This was a unique event in FG Sge's history. FG Sge's circumstellar environment was altered by mass loss following the 1992 fading; no circumstellar emission lines were visible during this first fading. By the next deep decline, May 1994, the surrounding circumstellar material contributed emission lines when the continuum light was very weak. Our May 1998 spectrum, obtained in the early part the latest decline, already displays strong Na I D emission, which was lacking in our September 1992 spectrum.

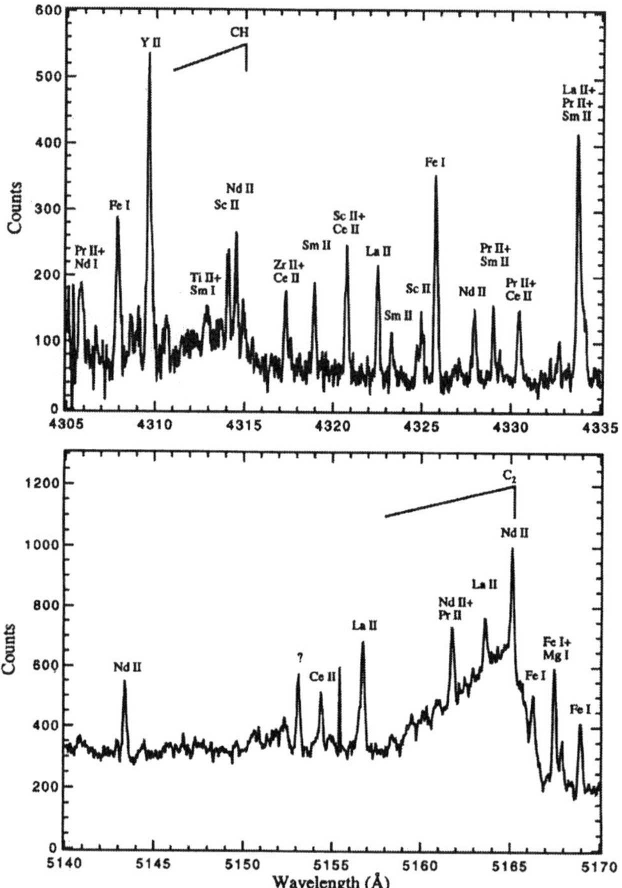

FIGURE 4. Two sections of the 1996 June Keck spectrum are shown. The CH (0-0) bandhead is weakly present in the upper diagram. The C_2 (0-0) bandhead is shown in the second diagram.

Hinkle et al. (1995) detected cool CO at the systemic velocity of FG Sge 3 months prior to its 1992 decline. This gas is thought to play an important role in cooling and carbon nucleation in a carbon-rich environment.

6.3. *Evolution of FG Sge*

The weight of the evidence argues for the occurrence of a deep mixing event in FG Sge bringing to the surface the products of s-processing similar to source of the main s-process component in the Solar System. This fits the very late He-shell flash scenario described by Iben et al. (1983) and Iben (1984). However, the details of the mixing are very complex, and current models have not been able to account for the surface composition changes in FG Sge. While the close link between FG Sge and the R CrB class implies a similar evolutionary history, the very different surface abundances of the s-process elements complicates the picture. The theorists need to tell us if s-process surface enhancement is expected to be a common or a rare event in a very late He-shell flash objects.

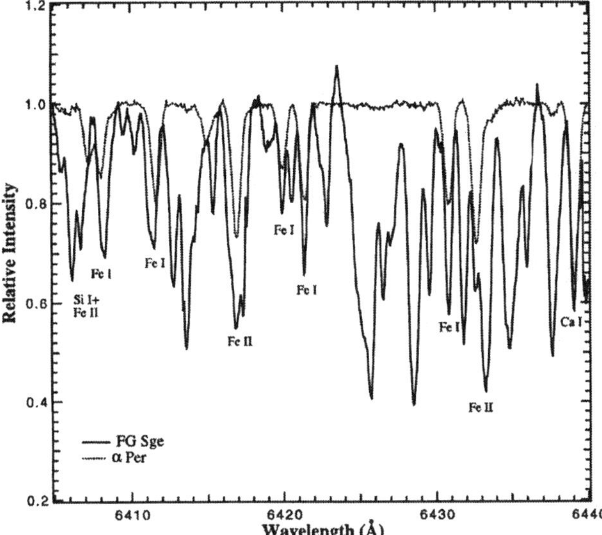

FIGURE 5. Comparison of the spectra of α Per and FG Sge. Both spectra were obtained during the same run in 1994 October. The strongest lines in the spectrum of α Per are identified.

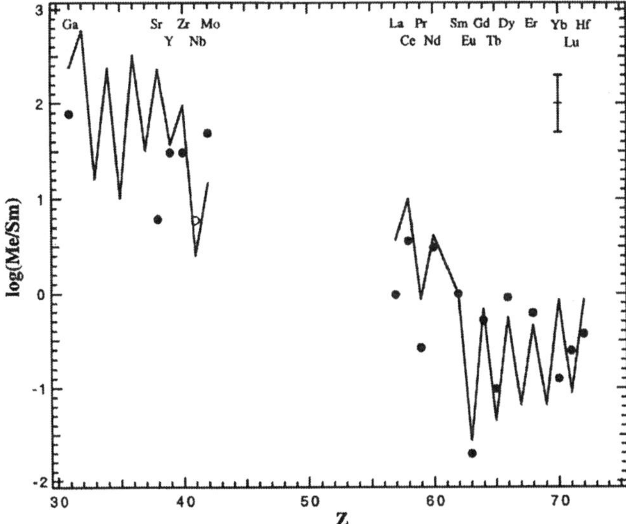

FIGURE 6. Logarithmic number density abundances of the s-process elements relative to Sm. The solid lines are the relative abundances of the Solar System main s-process component from John Cowan (private communication). The filled circles are the abundance estimates from G98. The open circle is an upper limit on the abundance of Nb.

6.4. *Next steps*

Analysis of FG Sge's spectrum revealed to us just how incomplete is the atomic data on the rare-earths. Only about half the visible absorption lines are identifiable. The inclusion of hfs in the syntheses would also improve the abundance estimates, but this is secondary in importance. However, inclusion of hfs could result in an independent confirmation of the source of the heavy elements in FG Sge due the difference in hfs for

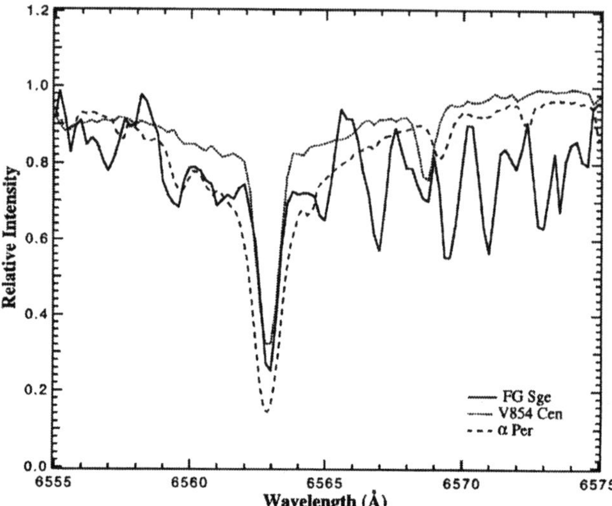

FIGURE 7. Hα region of FG Sge (1992 September), V854 Cen, and α Per. The spectral resolutions have been matched.

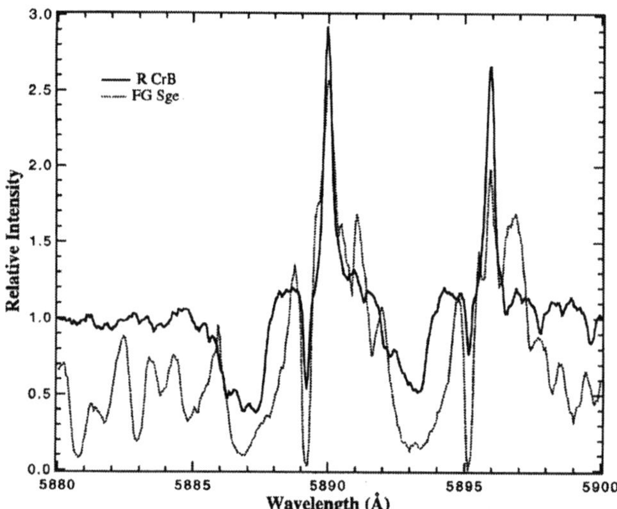

FIGURE 8. Na I D lines of FG Sge on 1994 August 16 compared to the same region in R CrB on 1996 May 9. At this time R CrB was recovering from the decline that began in 1995 October, and FG Sge was recovering from the minimum of 1994 May. Both spectra have similar spectral resolutions.

different isotopes. For example, in the Solar System 79% of the element Lu is from an r-process source (in the isotope 175Lu) according to calculations by John Cowan (private communication); in FG Sge the isotope 176Lu should be by far the most abundant one. The use of H-deficient model atmospheres would also be a major improvement.

REFERENCES

ARKHIPOVA, V. P. 1994 *Pis'ma Astron. Zh.* **22**, 828.

GONZALEZ ET AL. 1998 *ApJS* bf 114, 133 (G98).

HERBIG, G. H. & BOYARCHUK, A. A. 1968 *ApJ* **153**, 397.

HINKLE, K. H., JOYCE, R. R., & SMITH, V. V. 1995 *AJ* **109**, 808.

IBEN, I., JR. 1984 *ApJ* **277**, 333.

IBEN, I., JR., KALER, J. B., TRURAN, J. W., & RENZINI, A. 1983 *ApJ* **264**, 605.

KIPPER, T. 1996. In *Hydrogen-Deficient Stars* (eds. C. S. Jeffery & U. Heber). ASP Conf. Proc. 96. p. 329. ASP.

KIPPER, T. & KIPPER, M. 1993 *A&A* **276**, 389.

Langer, G., Kraft, R. P., & Anderson, K. S. 1974 *ApJ* **189**, 509.

MONTESINOS ET AL. 1990 *ApJ* **363**, 245.

In memory of Chris Skinner:
The nebula around the Carbon Star
IRC+10216

By MATTHEW BOBROWSKY,[1] CHRIS SKINNER,[2]
MARGARET MEIXNER,[3] DAVID AXON,[4]
AND DEAN C. HINES[5]

[1]Orbital Sciences Corporation, 7500 Greenway Center Drive, #700,
Greenbelt, MD 20770-3550, USA
mattb@oscsystems.com

[2]deceased

[3]Dept. of Astronomy, MC 221, Univ. of Illinois, 1002 W. Green Street,
Urbana, IL 61801, USA
meixner@astro.uiuc.edu

[4]Dept. of Astronomy, Univ. of Manchester, Manchester M13 9PL, England
dja@ast.man.ac.uk

[5]Steward Observatory, Univ. of Arizona, 933 N. Cherry Ave., Tucson, AZ 85721, USA
dhines@as.arizona.edu

Chris Skinner passed away on October 20, 1997, and will be remembered as an excellent astronomer and good friend. One of the last scientific programs in which Chris was involved was an investigation of IRC+10216. Here we present some results from that work.

The Carbon Star IRC+10216 appears to be in the final stages of red giant evolution. Almost 30 years ago, it was still highly obscured in the visible domain, although it is very bright in the infrared (IR). The near-IR shows a significant scattered (and polarized) component, while the mid- to far-IR comes from a dusty wind. The optical emission comes from a bipolar reflection nebula. Here we present IR images that show the bipolar structure. The spectral energy distribution can be modelled as an equatorially-enhanced dusty superwind.

1. Introduction

Red giant stars expel mass in the form of a stellar wind. After the hot stellar core is exposed, the circumstellar gas becomes ionized and becomes a planetary nebula (PN). The initial red giant winds are spherical, but most PNs are elliptical or bipolar (Zuckerman & Aller 1986, Olofsson 1993). The axial symmetry may result from the interaction of a fast stellar wind with a denser, equatorially-concentrated envelope left over from the earlier red giant mass loss (Kwok, Purton, & Fitzgerald 1978).

Soon after IRC+10216 was discovered in the IRC survey (Neugebauer & Leighton 1969), it was recognized as optically highly obscured, but bright in the IR (Becklin et al. 1969). It was then shown to be a carbon star with a dusty molecular envelope (Miller 1970, Herbig & Zappala 1970). Observations of millimeter-wavelength molecular spectral lines (Bieging & Rieu 1989, Groesbeck, Phillips, & Blake 1994, Dayal & Bieging 1995) showed that at large radii ($\geq 10^{16}$ cm), the envelope is roughly spherical and expanding at ~ 15 km s^{-1}. The mass-loss rate is 2–5×10^{-5} M$_\odot$/yr. Close to the star in IRC+10216, the envelope appears not spherical, but axially symmetric (Becklin et al. 1969, Kastner & Weintraub 1994).

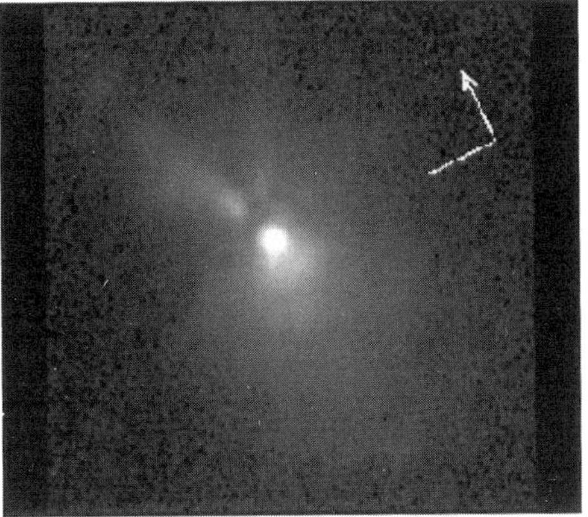

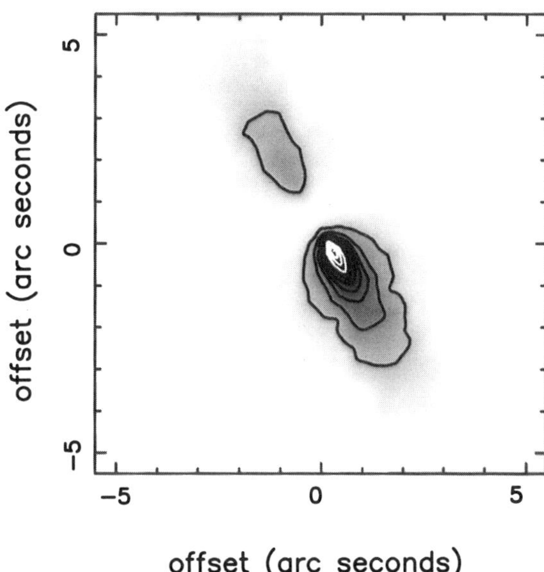

FIGURE 1. Upper panel: HST WFPC2 image of IRC+10216 using the F814W filter. The pixel scale is 0.046″, corresponding to 1.4×10^{14} cm at its estimated distance of 170 pc. The total field of view displayed here is 11.8″. A compass shows the directions of north (with an arrow) and east (no arrow) on the sky. Lower panel: simulated 0.8 μm image of IRC+10216's reflection nebulosity from our radiative transfer model (see text and Table 1). The direct light from the central star is not included in this simulated image. The field of the view of this image is identical to that shown in the upper panel.

2. Observations

Infrared images were obtained using two of Hubble Space Telescope's science instruments. The upper panel of Figure 1 shows a WFPC2 image acquired with filter F814W (0.827 μm). (This image has been previously presented by Skinner, Meixner, & Bobrowsky 1998.) A compact bipolar reflection nebula is apparent. The bright spot near the center is probably the central star. Figure 2 shows a WFPC2 image of the Cygnus Egg Nebula (CRL 2688), another bipolar reflection nebula in which the central star is hidden behind a dense, equatorial torus or disk. There are a number of morphological similarities between CRL 2688 and IRC+10216. Both have a pair of bipolar lobes. A set

FIGURE 2. HST WFPC2 image of the Cygnus Egg nebula, CRL 2688, via the F606W filter, for the purposes of comparison with IRC+10216. The pixel scale is $0.2''$, corresponding to 5.4×10^{15} cm at a distance of 1.8 kpc. The total field of view displayed is approximately $80''$. The image has been discussed by Sahai et al. 1998.

of faint outer concentric rings or arcs are seen around CRL 2688, which may be the result of periodic episodes of mass loss; there are hints of similar structure around IRC+10216. A pair of horns pointing north out of IRC+10216 resembles pairs of horns pointing out of both bipolar lobes of CRL 2688. Also, both have a dark band across the 'waist' of the nebula.

A 1.65 μm image obtained with NICMOS is shown in Figure 3. The diffraction spikes and saturated pixels are not part of IRC+10216 and should be ignored. What *is* part of IRC+10216 is the extension of gas between the diffraction spikes, and the concentric circular arcs. The central star is more clearly visible in this image than in the WFPC2 image. Figure 4 shows a polarization image at 2.1 μm. (The NICMOS polarimetry was processed using the algorithms described by Hines 1998.) The jet-like feature is strongly polarized and, about $\sim 30°$ away, there is a second polarized feature forming a pair of 'horns'. The polarization is very high ($\sim 50\%$) over much of the nebula. The polarization vectors show the symmetry that would be expected from bipolar outflows, and the vectors in the bipolar outflows are perpendicular to the direction toward the central star. The emission from the "horns" is primarily scattered light, and the high polarization of the "horns" implies that they are illuminated by the central star.

3. Axially-symmetric radiative transfer model

IRC+10216 was modelled using an axially-symmetric radiative transfer code that includes scattering, absorption, and re-emission by dust grains (Skinner, Meixner, & Bobrowsky 1998). This code had been used previously to model CRL 2688 with good success (Skinner et al. 1997). Good simulations of optical and IR images were also produced (Skinner et al. 1997).

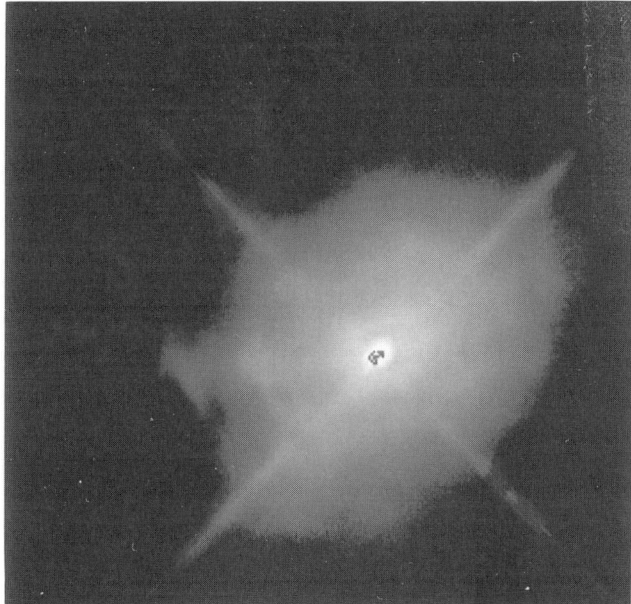

FIGURE 3. HST NICMOS image of IRC+10216 at 1.65 μm.

FIGURE 4. HST NICMOS image of IRC+10216 showing polarization vectors overlaid on an image of the total polarized intensity.

The model parameters used for IRC+10216 are shown in Table 1. These parameters are comparable to those of Groenewegen 1997. In the model, the number of grains with a given radius, r, is proportional to r^α between an inner radius, r_{min}, and an outer radius, r_{max}. The model dust shell is toroidal in the inner region, and spherical farther out.

Parameter	value	source
Stellar effective temperature	2010 K	WDS
Stellar radius	8.0×10^{13} cm	model
Distance	170 pc	KW,WDS
Luminosity	2.0×10^4 $L_\odot$	model
Dust shell inner radius	$4.0 R_*$	model
Outflow velocity	14.5 km/sec	GPB
Dust mass-loss rate	1.0×10^{-7} $M_\odot$/yr	model
amorphous carbon : SiC : $Mg_{0.9}Fe_{0.1}S$ by mass	92 : 3 : 5	model
Minimum grain radius, r_{min}	0.04 μm	model
Maximum grain radius, r_{max}	0.4 μm	model
Grain size distribution index, α	-1	model
Bipolar axis inclination to sky	$20\pm10°$	model
Equator/pole density ratio	10	model
Bicone opening angle	$15°$	model

TABLE 1. IRC+10216 Model Parameters

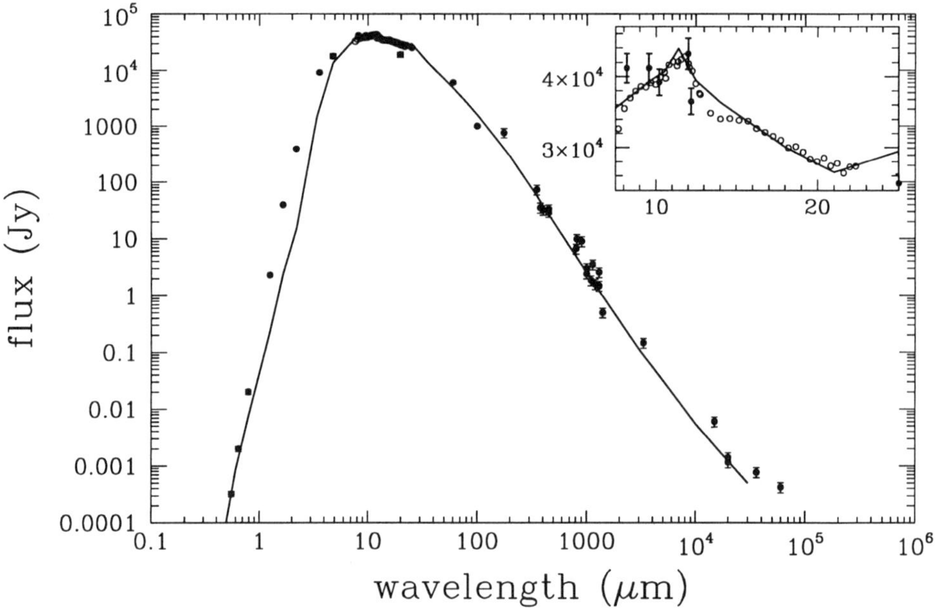

FIGURE 5. Spectral energy distribution of our axially symmetric radiative transfer model of IRC+10216 (solid line) compared with previously published observations (sources can be found in Bagnulo, Doyle, & Griffin 1995). The inset shows the 8–24 μm spectrum from the InfraRed Astronomical Satellite, IRAS, (plotted as open circles) as well as ground-based photometry (filled circles with errorbars) compared with the model (solid line): the upturn of the model at the long wavelength end is due to the 30 μm MgS dust emission feature.

The model image, for a wavelength of 0.8 μm, is shown in the lower panel of Figure 1. The bipolar axis is inclined $\sim 20°$ from the plane of the sky. The resulting model spectral energy distribution is shown in Figure 5. The model fits the data well in the optical and at wavelengths longward of 7.5 μm.

4. Results

The circumstellar envelope must be strongly toroidal close to the star. A sudden increase in density from the pole to the equator (by a factor of ten) is consistent with the presence of a bipolar outflow blowing polar holes through the envelope. Profiles of millimeter-wavelength lines hint that such an outflow may be present.

Molecular rotational lines show that, a few arcseconds from the star, the envelope is approximately spherical. Therefore, the mass-loss geometry has evolved from isotropic to equatorially-enhanced during the past 200 years (with little change in mass-loss rate). This may constitute the first known observation of the transition from a spherical to non-spherical outflow for a red giant without a known binary companion.

The evolution of IRC+10216 off of the AGB and into its PN phase is occurring with toroidal mass loss. The transition of mass-loss geometry from isotropic to equatorially enhanced need not be accompanied by a large and simultaneous increase in mass-loss rate. The morphological change in the outflow may be a regular part of red giant evolution, and the enormous mass-loss rate increases reported for some stars may be an unrelated phenomenon, or one that *follows* the onset of toroidal mass loss.

Some superwind models suggest that equatorially enhanced mass-loss could result from the engulfment of a binary companion star by the expanding red giant, with the transfer of angular momentum into the red giant wind (Morris 1987). With no significant change in mass-loss rate, IRC+10216 does not seem to be consistent with that model. For the same reason, our observations of IRC+10216 argue against a remnant protostellar disk, whose inertia would focus a spherical outflow into a bipolar one. The bipolar outflows may originate very close to the central star (Haniff & Buscher 1998). A change in the structure of the red giant envelope appears more likely to be responsible.

Few non-spherical mass flows from red giants are known. Two examples include OH231.8+4.2 and the carbon star V Hydrae, both close binaries. IRC+10216 is not a rapid rotator, and there is no evidence for a companion. The bipolar outflows do not appear to be faster than the general, isotropic outflow usually observed in molecular lines. It would be useful to observe other stars that are in the same stage of their evolution.

These observations were obtained at the Space Telescope Science Institute, which is operated by the Association of Universities for Research in Astronomy, Inc., under NASA contract NAS 5-26555. Bobrowsky was supported by NASA ST ScI Grant # GO-06364.01-95A. Meixner was supported by NASA ST ScI Grant # GO-06364.02-95A. Hines acknowledges support from the NASA grant NAG5-3042 to the NICMOS instrument definition team.

REFERENCES

BAGNULO, S., DOYLE, J. G., & GRIFFIN, I. P. 1995 *A&A*, **301**, 501.

BECKLIN, E. E., FROGEL, J. A., HYLAND, A. R., KRISTIAN, J., & NEUGEBAUER, G. 1969 *ApJ*, **158**, L133.

BIEGING, J. H. & RIEU, N-Q 1989 *ApJ*, **329**, L107.

DAYAL, A. & BIEGING, J. H. 1995 *ApJ* **439**, 996.

GROENEWEGEN, M. A. T. 1997 *A&A*, **317**, 503.

GROESBECK, T. D., PHILLIPS, T. G., & BLAKE, G. A. 1994 *ApJS*, **94**, 147.

HANIFF, C. A. & BUSCHER, D. F. 1998, *A&A*, **334**, L5.

HERBIG, G. H. & ZAPPALA, R. R. 1970 *ApJ*, **162**, L15.

HINES, D. C. 1998 in *NICMOS and the VLT A New Era of High Resolution Near Infrared Imaging and Spectroscopy* (eds. W. Freudling & R. Hook). ESO Conference and Workshop Proceedings No. 55.

KASTNER, J. H. & WEINTRAUB, D. A. 1994 *ApJ*, **434**, 719.

KWOK, S., PURTON, C. R., & FITZGERALD, P. M. 1978 *ApJ*, **219**, L125.

MILLER, J. S. 1970 *ApJ*, **161**, L95.

MORRIS, M. 1987 *PASP*, **99**, 1115.

NEUGEBAUER, G. & LEIGHTON, R. B. 1969 *Two micron sky survey: a preliminary catalog*, Calif. Inst. Technology.

OLOFSSON, H. 1993 in *Mass Loss on the AGB and Beyond*, Proc. 2nd ESO/CTIO Workshop, (ed. H. E. Schwartz). p. 330. ESO.

SAHAI, R. ET AL. 1998 *ApJ*, **493**, 301.

SKINNER, C. J., MEIXNER, M., BARLOW, M. J., COLLISON, A. J., JUSTTANONT, K., BLANCO, P., PIÑA, R., BALL, J. R., KETO, E., ARENS, J. F., & JERNIGAN, J. G. 1997 *A&A*, **328**, 290.

SKINNER, C. J., MEIXNER, M., & BOBROWSKY, M. 1998 *MNRAS*, in press.

ZUCKERMAN, B. & ALLER, L. H. 1986 *ApJ* **301**, 772.

Concluding remarks

By ICKO IBEN, JR.

Astronomy and Physics Departments, University of Illinois, 1002 W. Green St., Urbana, IL
61801; icko@astro.uiuc.edu

Random comments on several issues addressed at the conference are offered: the role of the
interstellar medium in stellar evolution, mass loss from AGB stars and the formation of planetary
nebulae, progenitors of blue stragglers, kicks during neutron-star formation, and why stars
become red giants.

1. Introduction

It is impossible to summarize all of the many insights offered by previous speakers at
this, one of the most entertaining conferences on stellar evolution held in recent years.
I will therefore confine my remarks to a few impressions formed while listening to the
talks.

2. The role of the interstellar medium in stellar evolution

For essentially all of my career in astronomy, I have thought of the dynamics of the
interstellar medium and the evolution of stars as two quite distinct subjects, with the
ISM playing the almost passive role of supplying matter out of which stars are formed
(in some unspecified way) and the stars contributing in an almost incidental way to the
enrichment of the ISM in elements heavier than hydrogen. In my universe, the evolution
of stars was the important thing, whereas the ISM was simply an interesting component
of galaxies, the physics of which people toyed with for amusement.

This conference has removed the scales from my eyes, and I now see the ISM as an all-
important phenomenon which makes stars in the mass spectrum essential for producing
heavy elements in the nearly universal distribution we call "solar system abundances."
That is, the ISM may be thought of as a giant womb which produces stars according
to a prescription that is (perhaps) a statistical characteristic of hierarchically structured
clouds and then happily accepts the enriched matter stars cough up at the end of their
nuclear burning lives (in a superwind on the AGB if single and initially less massive than
~ 11 M$_\odot$, in a common envelope event if in a close binary, or in a strong wind followed
by a supernova explosion if massive, whether single or in a binary). In a sense, stars are
but a phase in the life of the ISM!

3. Mass loss and planetary nebula formation

There are at least four mechanisms which are at work in producing the variety of
planetary nebulae which are born at a frequency of about one per year in our Galaxy.

(i) Single stars and stars in wide binaries become AGB stars if their initial mass is
less than some critical value (of order 11 M$_\odot$ for population I stars) and, at some point
in their lives, they become Miras and pulsate at large amplitude in a radial acoustical
mode which may be either the fundamental or first overtone. It is known empirically
that, when the period of pulsation reaches ~ 400 days, mass loss from the surface takes
place at roughly 10^{-4} M$_\odot$ yr^{-1}. This rate is from two to three orders of magnitude larger
than the rate at which matter is processed in the interior by nuclear burning. A popular

scenario for explaining this mass loss supposes that the acoustical pulsations create shock waves which heat up the atmosphere of the star and cause it to swell to many times the dimensions of a static atmosphere. During some portion of a pulsation cycle, a pocket of gas in the atmosphere passes through a part of the density-temperature plane favorable for dust-grain growth. Radiation pressure drives off the grains which drag the gas along. This scenario has been developed extensively by George Bowen and Lee Ann Wilson.

(ii) Alan Sweigert at this meeting shows in a quasistatic calculation that, when the electron-degenerate CO (or ONe) core of an AGB model star exceeds a critical mass, and the mass of hydrogen-rich matter outside of the core is smaller than another critical value, there is a region just outside the core in which, following the peak of a helium shell flash, gas pressure tends to zero (actually, $P_{gas}/P_{rad} \rightarrow 0$). This suggests that, in a dynamical calculation, the envelope of the model may actually detach from the core and float off into space, thus contributing another mechanism for the loss of the envelope.

(iii) In the course of a 3D hydrodynamic study of matter motions in the convective envelope of a red supergiant, David Porter, Steve Anderson, and Paul Woodward have found that oscillations are excited by a dipolar convective flow in a way which does not depend on the existence of the zones of partial ionization which are responsible for driving of the classical radial pulsations. The amplitude of the non-radial oscillations is quite large, and the surface of the star exhibits irregular deviations from sphericity which could have a profound effect on the nature of wind outflow and could in fact constitute a mechanism for mass loss which is completely different from that described in (i).

(iv) A fourth mode of mass loss takes place in close binaries when one of the components fills its Roche lobe and transfers matter to its companion more rapidly than it can be incorporated by the companion. The rejected material forms a hot blanket which eventually overflows the Roche lobe of the companion and a common envelope is formed. The "eggbeater" formed by the evolved core of the donor and the companion interacts with the matter in the common envelope and drives it off, imparting bipolar symmetry to the ejected matter. At this meeting, Howard Bond has suggested (presumably with tongue in cheek) that, since most ($\sim 80\%$) of all planetary nebulae are bipolar, perhaps most planetary nebulae are made by close binaries.

This is of course not true, as there are many AGB stars which are single or in wide binaries, losing mass as Miras at rates up to $\sim 10^{-4}$ $M_\odot$ yr^{-1}, evolving into OH/IR sources, and then becoming central stars of planetary nebulae. FG Sge and Sakurai's object are central stars of planetary nebulae formed in this way and are experiencing the consequence of a final helium shell flash to become born again AGB stars. During its century-long excursion to the red, FG Sge has not once bumped into a companion. The star ICR-10216 has no known companion, but it emits matter at $\sim 5.5 \times 10^{-5}$ $M_\odot$ yr^{-1} in a bipolar, equatorially enhanced flow which, as we have learned from Chris Skinner and Matt Bobrowsky at this meeting, magically becomes spherically symmetric at distances larger than ~ 100 stellar radii. It is further significant that OH/IR sources are basically round and yet some of them must evolve into bipolar nebulae, given the large frequency of such nebulae. Is the shaping due to interaction with a planetary system or an Oort-like cloud of cometary material? Or is the shaping due to the interaction of nebular material with the photon and particle radiation from the central star?

In summary, single stars as well as close binaries can produce bipolar planetary nebulae. A potential way of distinguishing between a single or a binary star progenitor is to determine the chemical abundances of the nebular material. Most planetary nebulae with known close binary central stars (found by Bond and others) have probably been made in a common envelope event which occurs when the primary fills its Roche lobe before it has begun the helium shell-flashing phase of evolution. This is known as an "early case C" common envelope event. The nebula made in such an event will not be carbon rich or s-process rich. In contrast, single stars have spent from 100,000 to 1,000,000 years in the helium shell-flashing phase, dredging up carbon and s-process elements into the envelope before this envelope is completely lost in the superwind.

4. Progenitors of blue stragglers

Circumstantial evidence for a mechanism which leads to the merger of components of a close, low mass binary is overwhelming. The mechanism is angular momentum loss by a magnetic stellar wind supported by one or both components of mass less than $\sim 1.5\ M_\odot$. For main-sequence binaries with masses larger than this, there is a continuum in the distribution of numbers versus orbital period which extends to the shortest possible period limited by the primary filling its Roche lobe. This continuum is reasonable if one supposes that there is a continuum in the distribution of specific angular momentum in protostellar clouds and that multiple stars form when the specific angular momentum is larger than that which a single star can possess.

In contrast, there is a distinct drop-off in the number of short period main-sequence binaries at a period large compared with the minimum period set by insisting that the Roche lobe of the primary be larger than the primary. Due to tidal coupling, components of close binaries with periods less than a few days typically spin with periods synchronous with the orbital period. From a study of stars in the Pleiades and Hyades clusters, it is known that single stars of low mass which possess deep convective envelopes spin down with time, and a theoretical analysis of the interaction between the Sun's wind and its magnetic field shows that it is the torquing of the wind by the field that is responsible for the spin down. Putting two and two together, one infers that low mass stars in close binaries try to spin down, but are prevented from doing so by tidal torquing, with the result that the loss of angular momentum from the system due to the magnetic stellar wind is made good by a reduction in the orbital angular momentum of the binary. The orbit shrinks and, eventually, the primary is forced to fill its Roche lobe. The system evolves into a contact binary (W UMa star) and ultimately merges.

This scenario is greatly strengthened by the simultaneous existence of W UMa stars and blue stragglers in many globular clusters and it is undoubtedly the dominant mode of blue straggler formation. Blue stragglers are, as you know, main-sequence stars brighter and therefore more massive than stars near cluster turnoff. Formation of a blue straggler by a merger achieved in a head-on collision, as detailed in another talk at this conference, is much, much less likely than the merger of a pre-existing binary mediated by a magnetic stellar wind.

It is worth noting that the scenario presented at this meeting for the formation of low mass single stars must be significantly modified if it is to account for the formation of close binary low mass stars. It is difficult to see how a stable disk could exist around either component in a close binary. Yet low mass close binaries are formed, and one wonders if the disk invoked in the single star formation scenario is really essential.

5. Neutron star kicks and pulsar velocities

In several talks dealing with supernovae and radio pulsars, an asymmetric "kick" has been treated as an established fact, necessary for understanding the high velocities of radio pulsars. In continues to be ignored that, actually, the space distributions and kinematics of most compact remnants of supernova explosions can be understood in the framework of a theory in which, to a good first approximation, type II and type Ib and Ic supernova explosions are spherically symmetric in the rest frame of the supernova precursor. For example, the space velocities of binary high mass X-ray sources and runaway OB stars and the space distribution of low mass X-ray binaries can be understood in terms of a momentum conserving recoil to the mass ejected by the neutron-star precursor with a mean velocity equal to the velocity of the precursor in its orbit.

The velocities of radio pulsars perpendicular to the line of sight are based on proper motions and estimates of distance inferred from a dispersion measure and a model of the free electron distribution in the Galactic plane. There is a strong bias toward high velocities at large distances due to the simple fact that, for a given perpendicular velocity, the accuracy of a proper motion measurement decreases with distance. Even when the bias is minimized by confining the sample of radio pulsars to a sphere of radius 500 pc about the Sun, however, there is a noticeable lack of pulsar velocities of order 0–10 km s^{-1} which would be predicted if neutron stars produced by single stars and stars in wide binaries were made in a spherically symmetric explosion. On the other hand, population synthesis models for the neutron stars produced by close binaries give a velocity distribution not too different from the observed one for radio pulsars. Furthermore, estimates of the pulsar and supernova birthrates suggest that the birthrate of radio pulsars is roughly equal to the birthrate of type Ib and Ic supernovae (originating in close binaries) and only a third of the birthrate of type II supernovae (made by single stars and stars in wide binaries). It is not a great stretch of the imagination to infer that perhaps single stars and stars in wide binaries do not make neutron stars which spin fast enough to be pulsars. The higher spin rate of neutron stars produced in close binaries could be ascribed to tidal forces which prevent angular momentum loss as great as that experienced by single stars and stars in wide binaries.

6. Why stars become red giants

For many decades, astrophysicists have attempted to describe why a star becomes a red giant. Some explanations are quite imaginative, such as the one which uses the properties of a polytrope of index > 5 ($P \propto \rho^{1+1/>5}$) to explain an object in which essentially all of the mass is contained in an electron-degenerate core and a hydrogen-rich convective envelope, each of which is described by a polytropic segment of index $3/2$ ($P \propto \rho^{1+2/3}$). Another imaginative explanation suggests that the atoms in the envelope of a star cooperate (absorbing photons to increase the number of occupied exited states and thereby increase the opacity) to force the macroscopic envelope to mimic the behavior of each of them (the atoms) in making a transition between a "ground (pre-red giant) state" to an "excited (red giant) state."

My own explanation is that evolution to the red giant state is the consequence of the interplay of different parts of a very complicated physical system such as that presented by, say, 10^{57} mass points each of different mass, connected by springs of different spring constants. The sensible way to see what happens when the latter system is struck with an arbitrary blow is to strike the blow and take a picture with a movie camera. The physical system is described by very simple equations, but there is no simple general solution. In the same way, to see what a star does when it exhausts hydrogen over a significant fraction of its interior, the sensible procedure is to solve the straightforward equations of stellar structure and take a picture. All of the "explanations" of "why" red giants form are basically convoluted descriptions of the picture which the detailed evolutionary calculations reveal.

To prepare a homily which I must deliver at a memorial service for my recently departed mother when I return home from Baltimore, I brought with me a number of poems my

mother has written over the years. I came across one which illustrates my point well:
some questions simply don't have a simple answer.

Did Anyone Ever Ask the Hen?

by *Kathryn T. Iben*

When a great tree falls
And people aren't near
Does it make a noise
If no one can hear?

And which came first,
The hen or the egg?
This impractical question
We ask and then beg.

Some wise men say
Its beyond their ken
Did anyone ever
Ask the Hen? — Anonymous

So I Asked The Hen

"What a silly question," answered the Hen,
"The kind that's thought up by idle men;
I invite you to come—just be my guest
And watch tomorrow as I sit on my nest.
I'll stare straight ahead and then give a twitch
And the miracle happens with nary a hitch.
You'll see me squirm and an egg drops out
And I'm its mother without a doubt!"

But a voice screams out from the top of the wall
And there sits Humpty before his great fall.
"If you don't believe that I came first
There's nothing to do but fall and burst!"
So, sobbing, he fell and gave up his life
To show he was first and so end the strife.

A two-footed creature emerged from the muck
And even as it toddled it started to cluck
"I ca-ca-ca-came first!"